62M   MATAS

London, Skoob Bks, 1 Oct 2016

£9
-10%
£8-10

9
25

Oxford Lecture Series in
Mathematics and its Applications

Series Editors
John Ball Dominic Welsh

Books in the series

The Factorization Method for Inverse Problems

(Inverse Scattering Problems)

Andreas Kirsch and Natalia Grinberg

Institute of Algebra and Geometry
University of Karlsruhe (TH)
Karlsruhe, Germany

OXFORD

UNIVERSITY PRESS

OXFORD
UNIVERSITY PRESS

Great Clarendon Street, Oxford OX2 6DP

Oxford University Press is a department of the University of Oxford.
It furthers the University's objective of excellence in research, scholarship,
and education by publishing worldwide in

Oxford New York

Auckland Cape Town Dar es Salaam Hong Kong Karachi
Kuala Lumpur Madrid Melbourne Mexico City Nairobi
New Delhi Shanghai Taipei Toronto

With offices in

Argentina Austria Brazil Chile Czech Republic France Greece
Guatemala Hungary Italy Japan Poland Portugal Singapore
South Korea Switzerland Thailand Turkey Ukraine Vietnam

Oxford is a registered trade mark of Oxford University Press
in the UK and in certain other countries

Published in the United States
by Oxford University Press Inc., New York

British Library Cataloguing in Publication Data

Data available

Library of Congress Cataloging in Publication Data

Data available

Typeset by Newgen Imaging Systems (P) Ltd., Chennai, India
Printed in Great Britain
on acid-free paper by
Biddles Ltd., King's Lynn, Norfolk

ISBN 978–0–19–921353–5

1 3 5 7 9 10 8 6 4 2

Preface

This book is devoted to the problem of *shape identification*. Problems of this type occur in a number of important fields which belong to the class of *inverse problems*. As the first of these fields, we mention *inverse scattering problems* where one wants to detect – and identify – unknown objects through the use of acoustic, electromagnetic, or elastic waves. Complex models in scattering theory involve boundary value problems for partial differential equations such as Maxwell's equations in electromagnetics, and one of the important problems in inverse scattering theory is to determine the shape of the obstacle from field measurements. Applications of inverse scattering problems occur in such diverse areas as medical imaging, material science, nondestructive testing, radar, remote sensing, or seismic exploration. A survey on the state of the art of the mathematical theory and numerical approaches for solving inverse time harmonic scattering problems until 1998 can be found in the standard monograph [43] by David Colton and Rainer Kress. We also refer to Chapter 6 of [106] and [155] for an introduction and survey on inverse scattering problems.

The second important area where the identification of unknown shapes plays an important role is *tomography*, in particular, electrical impedance tomography or optical tomography. *Electrical impedance tomography* is a technique to recover spatial properties of the interior of a conducting object from electrostatic measurements taken on its boundary. For example, a current through a homogeneous object will, in general, induce a different potential than the same current through the same object containing an enclosed cavity. The problem of impedance tomography, we are interested in, is to determine the shape of the cavity from measurements of the potential on the boundary of the object. For survey articles on this subject we refer to [16] by Liliana Borcea and [82] by Martin Hanke and Martin Brühl.

Shape identification problems are intrinsically *nonlinear*, i.e., the measured quantities do not depend linearly on the shape. Even the notion of linearity does not make sense since, in general, the set of admissible shapes does not carry a linear structure. Traditional (and still very successful) approaches describe the objects by appropriate parameterizations and compute the parameters by *iterative schemes* as, e.g., Newton-type methods. Besides the well-known advantages (fast convergence) and disadvantages (only local convergence properties) of iterative methods for nonlinear problems, these methods share the common drawback that important information on the unknown object such as the number of connectivity components or the type of the boundary condition has to be known in advance. Nevertheless, methods of this type are widely used – in particular because the first or second order derivatives can be characterized by using techniques from the shape optimization theory. We refer to [156, 173] for general references and [112, 134, 135, 84, 83, 86] for applications in inverse scattering theory.

In particular, classical iterative algorithms using explicit parameterizations of the objects are not able to change the number of connectivity components during the algorithm. This observation has led to the development of *level set methods* which are based on implicit representations of the unknown object involving an "evolution parameter" t. Since the pioneering work [154] by S. Osher and J. Sethian, this method has been further developed and applied in a huge number of papers. We refer to [22] for a recent survey. Since around 1995 iterative methods for solving problems in shape optimization have been developed which completely avoid the use of parameterizations and replace the classical Fréchet derivative by a geometrically motivated *topological derivative*, see, e.g., [171, 172]. These methods have also been applied to problems in inverse scattering theory in [15, 78]. We refer also to [67].

While very successful in many cases, iterative methods for shape identification problems – may they use classical tools as the Fréchet derivative or more recent techniques such as domain derivatives, level curves, or topological derivatives – are computationally very expensive since they require the solution of a direct problem in every step. Furthermore, for many important cases the convergence theory is still missing. This is due to the fact that these problems are not only nonlinear but also because their linearizations are *improperly posed*. Although there exist many results on the convergence of (regularized) iterative methods for solving nonlinear improperly posed problems (see, e.g., [62, 87]), the assumptions for convergence are not met in the applications to shape identification problems.[1]

These difficulties and disadvantages of iterative schemes gave rise to the development of different classes of *non-iterative* methods which avoid the solution of a sequence of direct problems. We briefly mention *decomposition methods* (according to the notion of [44]) which consist of an analytic continuation step (which is linear but highly improperly posed) and a nonlinear step of finding the boundary of the unknown domain by forcing the boundary condition to hold. We refer to [6, 3, 45, 46, 123, 50, 125, 132] for some versions of this approach. In Section 7.2 we will briefly recall the Dual Space Method of Colton and Monk (see below) which belongs to this class. There is also a close connection to the Point-Source Method of Roland Potthast in [157, 158, 160].

In this monograph, we will focus on a different class of non-iterative methods. The common idea is the construction of criteria on the known data to decide whether a given point z (or a curve or a set) is inside or outside the unknown domain D. By choosing a grid of "sampling" points z (or collection of curves or sets) in a region known to contain the unknown domain D one is therefore able to compute the (approximate) characteristic function of D. In the following, we will collect these approaches under the name *sampling methods*. They differ in the way of defining the criterion and in the type of test objects.

One of the first methods which falls into this class has been developed by David Colton and one of the authors of this book (A.K.) in 1996 ([39]), now known as the *Linear Sampling Method*. Its origin goes back to the *Dual Space Method* developed by

---

[1] Or, at least, it is unknown whether these assumptions are fulfilled or not.

David Colton and Peter Monk during 1985 and 1990 (see, e.g., [45, 46, 49] or [43]). We also mention the work [105] of Victor Isakov in which singular functions are used to prove uniqueness of an inverse scattering problem. The numerical implementation of the Linear Sampling Method is extremely simple and fast because sampling is done by points z only. For every sampling point z one has to compute the field of a point source in z with respect to the background medium[2] (if this is constant the response is even given analytically) and evaluate a series, i.e., a finite sum in practice.

In 1998 Masaru Ikehata published the paper [96] in which he presented a method now known as Ikehata's *Probe Method*. Instead of points, the region is probed by curves (called needles in [96]), and points on these curves are identified which belong to the boundary of the unknown domain D. In many subsequent papers, mainly by Ikehata and his collaborators, the probe method has been applied to several inverse scattering problems and inverse conductivity problems (see [95, 97, 98, 99, 100, 101, 102, 33, 103, 104]). We explicitly mention [33, 63] for numerical implementations of the probe method.

In 2000 Roland Potthast presented a sampling method in [159] which he later in [160], Chapter 6, called the *Singular Sources Method*. The idea, formulated here for the inverse scattering problem, is to approximate (from the given data) the scattered field $v^s(\cdot, z)$ which belongs to the field of a point source in z as incident field. This scattered field $v^s(z, z)$, evaluated at the same source point z, is unbounded if z approaches the boundary ∂D from the exterior. Therefore, the unknown region D is found as the set of points z where $v^s(z, z)$ becomes large. In this sense, we consider this method as a sampling method with respect to point sampling.

Sampling by sets is done in the *Range Test* and the *No Response Test*, developed by Roland Potthast, John Sylvester, and Steven Kusiak in [163] and Russell Luke and Roland Potthast in [141]. We refer also to [88].

A problem with all of the mentioned methods (except of Ikehata's probe method) from the mathematical point of view is that the computable criterion provides only sufficient conditions which are, in general, not necessary. The *Factorization Method*, developed by the authors in [114, 115, 76, 74] overcomes this drawback and provides a criterion for z which is both, necessary and sufficient. Therefore, this method provides a simple formula for the characteristic function of D which can easily be used for numerical computations. We emphasize, that these results hold in the resonance region, i.e., no asymptotic forms such as the Born approximation or the geometric or physical optics approximations are assumed. Compared to Ikehata's Probe Method the Factorization Method is much more direct, both from the theoretical point of view as well as with respect to the computational implementation. From the numerical point of view, the Linear Sampling Method and the Factorization Method are equally simple and fast. A typical feature of these two methods is that they make no explicit use of boundary conditions or topological properties of D. In other words, they determine the unknown domain without knowing in advance the type of boundary condition or the number of components.

---

[2] Essentially, one has to compute the Green's function.

Since their first presentations, the Linear Sampling Method and the Factorization Method have been developed for several problems in inverse scattering theory and tomography. We refer to [24, 23, 26, 27, 28, 35, 37, 38, 80, 65, 178] for some recent work on the Linear Sampling Method and [7, 116, 117, 129, 118, 77, 119, 120] for papers related to the Factorization Method in inverse scattering theory. The interesting papers [8] and [11] by Tilo Arens and Armin Lechleiter discovers a deeper relationship between these two methods. Stimulated by the first paper [114] Martin Brühl and Martin Hanke investigated the Factorization Method for problems in impedance tomography (see [18, 19, 20, 82, 21] and later [90, 92, 94]) by Nuutti Hyvönen. For a more general approach to elliptic equations we refer to [120] and [68]. There are several applications of the Factorization Method which are not covered in this monograph. We mention scattering problems for periodic surfaces or arcs (cf. [10, 9] and [129], respectively), for elastic media (see [7, 29]), for static problems (cf. [81, 133, 131]), or in optical tomography (cf. [14, 89, 91, 93]).

However, it should also be mentioned that the range of problems for which the Factorization Method has been justified from the mathematical point of view is considerably smaller than the one for the Linear Sampling Method or the other sampling methods.

While the main subject of this monograph is the Factorization Method, we will report on the Linear Sampling Method, Ikehata's Probe Method, and Potthast's Singular Sources Method in Chapter 7. We also refer to the survey articles [161, 162] by Roland Potthast on Sampling and Probe Methods and to [88] for an interesting relationship between sampling methods and iterative methods.

The monograph is organized as follows. Chapters 1–4 study the Factorization Method for scattering problems where the wave propagation is described by the three-dimensional scalar Helmholtz equation. In Chapters 1–3 impenetrable scatterers D are considered where boundary conditions on the boundary ∂D of D of Dirichlet, Neumann, impedance or mixed type are imposed. Chapter 4 is devoted to the penetrable case, and we show that the Factorization Method can be considered as an extension of the well-known MUSIC-algorithm from signal processing (cf. [60, 30]).

There exist several variants of the Factorization Method. The – in our opinion – most satisfactory version holds for scattering problems with non-absorbing media, such as the obstacle scattering case with Dirichlet or Neumann boundary conditions. The Factorization Method for this situation is investigated in detail in Chapter 1. The mathematically most important feature of these problems is the normality of the far field operator. This makes it possible to use the spectral theory for normal operators. The authors think that the Factorization Method is a particularly interesting and useful application of this theory. As an intermediate step we prove a characterization of the scatterer D by an inf-condition which is, although not as elegant as the final characterization by the solvability of an equation, the basis for characterizations of D for scattering problems for absorbing media.

We consider Chapter 1 also as an introduction into our method and emphasize that it can not be left out by the reader because it sets up the basis for all subsequent chapters.

A first example of a case where the far field operator fails to be normal is studied in Chapter 2. The impedance boundary condition serves as a simple model for an absorbing medium. Since the far field operator F is no longer normal the final characterization of Chapter 1 does not hold and has to be modified. This is done by considering a suitable combination $F_\#$ of the self-adjoint parts $(F+F^*)/2$ and $(F-F^*)/(2i)$ of F which finally leads to a characterization of D by the solvability of an equation involving $F_\#$ instead of F.

Chapter 3 is devoted to mixed boundary conditions. The obstacle is assumed to consist of several parts, and on some them we impose Dirichlet boundary conditions, on the others Neumann or, more general, impedance boundary conditions. Even for the Dirichlet–Neumann case, where the far field operator F is still normal, it is an open problem whether or not the Factorization Method (in any of its forms) can be justified. Numerical experiments indicate that this is indeed the case but a rigorous proof is not known. However, if we a priori know some domains which enclose the parts with the Dirichlet boundary condition and the impedance boundary condition we can modify the operator $F_\#$ appropriately to treat this case as well.

In Chapter 4 we study the penetrable case, i.e., scattering by an inhomogeneous medium where we allow the medium to be absorbing. The techniques developed in Chapters 1 and 2 allow us to prove a characterization of the shape of the contrast (which is the difference between the indices of refraction of the scattering medium and the background medium) by the same operators used in Chapters 1 and 2. We note already here that this implies in practice that one does not need to know the type of obstacle – penetrable or impenetrable – in advance.

While Chapters 1–4 treat scattering problems for the scalar Helmholtz equation we investigate the Factorization Method in Chapters 5 and 6 for the scattering of time harmonic electromagnetic waves and the problem of impedance tomography, respectively. One assumption for the validity of the Factorization Method is that the square k^2 of the wavenumber k is not an eigenvalue of a corresponding eigenvalue problem. In the case of an impenetrable obstacle with Dirichlet or Neumann boundary conditions this eigenvalue problem is just the classical eigenvalue problem for $-\Delta$ in the domain D with respect to the boundary conditions. For the scattering by an inhomogeneous medium, however, a new type of eigenvalue problem (the "interior transmission eigenvalue problem") occurs which fails to be self-adjoint (cf. [55]). In Sections 4.5 and 5.5 we show under certain assumptions on the index of refraction that the eigenvalues form at most a countable set. The question of existence of eigenvalues is only settled for the spherically stratified case.

In Chapter 6 we investigate the problem of impedance tomography. In contrast to the scattering problems this problem is set up as a boundary value problem in a bounded domain B. The inverse problem we are interested in is to determine the shape D of an inclusion with different electrical properties than the background medium. In this application, the Factorization Method is set up for the difference of the Neumann–Dirichlet operators for the cases with and without inclusion rather than for the far field operator.

As mentioned above, the Factorization Method is only one of a class of new approaches for solving "geometric" inverse problems. In Chapter 7 we introduce the reader to some related sampling methods. In contrast to the Factorization Method they all use heavily the fact that every solution of the Helmholtz equation in some domain G can be approximated arbitrarily well by solutions of the Helmholtz equation in larger domains. We summarize two of such approximation theorems in Section 7.1. The Linear Sampling Method can be considered as the precursor of the Factorization Method and is closely related to the latter one. We present this method in Section 7.2 and show its relationship to the Dual Space Method of Colton and Monk. In Section 7.3 we present the basic ideas of the Singular Sources Method of Roland Potthast. Finally, in Subsection 7.4.1 of Section 7.4 we explain the Probe Method of Masaru Ikehata for the impedance tomography problem and extend it in Subsection 7.4.2 to the inverse scattering problem with boundary conditions of mixed type. Here we follow the presentation of [74].

In this monograph we use several spaces of functions on domains G or their boundaries ∂G. We try to follow the standard notations for these spaces. The boundary value problems in bounded domains are set up in the Sobolev space $H^1(G)$ of (Lebesgue) measurable functions such that their derivatives (in the sense of distributions) are regular and belong to $L^2(G)$. This space is equipped with the inner product

$$(u, v)_{H^1(G)} \; = \; (u, v)_{L^2(G)} \; + \; \big(\nabla u, \nabla v\big)_{L^2(G)}$$

where we denote by $(u, v)_{L^2(G)} = \iint_G u(x)\,\overline{v(x)}\,dx$ the inner product in $L^2(G)$. By $\overline{v(x)}$ we denote the complex conjugate of $v(x)$. If u and v are vector fields then $u(x)\overline{v(x)}$ has to be understood as the scalar product $\sum_{j=1}^3 u_j(x)\overline{v_j(x)}$. The corresponding norms are denoted by $\| \cdot \|_{L^2(G)}$ and $\| \cdot \|_{H^1(G)}$

For $k \in \mathbb{N}$ we denote by $C^k(\overline{G})$ the space of functions for which all partial derivatives up to order k exist in G and are continuously extendable to the closure $\overline{G}$ of G. The norms in $C^k(\overline{G})$ are denoted by $\| \cdot \|_{C^k(\overline{G})}$. We set $C^\infty(\overline{G}) = \bigcup_{k \in \mathbb{N}} C^k(\overline{G})$. Then we can equivalently define $H^1(G)$ as the completion of $C^1(\overline{G})$ with respect to the inner product $(\cdot, \cdot)_{H^1(G)}$. Spaces of vector fields are denoted by $H^1(G, \mathbb{C}^3)$ or $C^k(\overline{G}, \mathbb{C}^3)$.

We assume that G is a Lipschitz domain. Sometimes we assume that even $\partial G \in C^{1,\alpha}$ or $\partial G \in C^2$. For definitions of these notations as well as of spaces $C^{j,\alpha}(G)$ or $C^{j,\alpha}(\partial G)$ of Hölder continuous functions we refer to [146] (Section 3.2), [144], Chapter 3, [43], Section 2.2, or [167], Section 6.4. Then the spaces $C(\partial G)$ and $L^2(\partial G)$ are defined in the usual way using local coordinates. It can be shown (see, e.g., [17], Section 3) that the trace operator $\gamma : C^1(\overline{G}) \to L^2(\partial G)$, $\gamma u = u|_{\partial G}$, has a bounded extension on $H^1(G)$. Its range space

$$\mathcal{R}(\gamma) \; = \; \big\{\psi \in L^2(\partial G) : \text{there exists } u \in H^1(G) \text{ with } \gamma u = \psi\big\}$$

is denoted by $H^{1/2}(\partial G)$ and equipped with the norm

$$\|\psi\|_{H^{1/2}(\partial G)} \; = \; \inf\big\{\|u\|_{H^1(G)} : u \in H^1(G) \text{ with } \gamma u = \psi\big\}.$$

We define $H^{-1/2}(\partial G)$ as the dual space of $H^{1/2}(\partial G)$. We denote by $\langle \cdot, \cdot \rangle$ the dual form in $\langle H^{-1/2}(\partial G), H^{1/2}(\partial G)\rangle$, which is the extension of the inner product $(\cdot, \cdot)_{L^2(\partial G)}$:

$L^2(\partial G) \times L^2(\partial G) \to \mathbb{C}$ to $\langle \cdot, \cdot \rangle : H^{-1/2}(\partial G) \times H^{1/2}(\partial G)) \to \mathbb{C}$. We note that both, the inner product $(\cdot, \cdot)_{L^2(\partial G)}$ and the dual form $\langle \cdot, \cdot \rangle$ are sesqui-linear forms. Furthermore, $H_0^1(G) = \{u \in H^1(G) : \gamma u = 0 \text{ on } \partial G\}$ can be constructed as the closure in $H^1(G)$ of the space $C_0^1(G)$ of C^1-functions with compact support in G. We denote by $\mathcal{R}(F)$ and $\mathcal{N}(F)$ the range space and the null space, respectively, of an operator F.

Finally, we come to the pleasant task of thanking those who have supported and encouraged us for starting – and finishing – this project. Here we want to mention three working groups who have influenced the research on sampling methods in a very essential way. First of all, we would like to thank the research group of Fioralba Cakoni, David Colton, Russell Luke, and Peter Monk from the University of Delaware who not only made important contributions to the field but also provided a warm and hospitable environment for one of the authors (A.K.) to spend several weeks in Newark during the past years. Second, the working group at the University of Göttingen of Rainer Kress, Thorsten Hohage, and Roland Potthast (now at the University of Reading) are one of the most active groups in Germany in the field of inverse scattering problems and encouraged us to present the ideas of the Factorization Method to a broader audience of interested mathematicians, physicists, and engineers. Third, the working group of Martin Hanke-Bourgeois at the University of Mainz developed the Factorization Method for problems of electrical impedance tomography which started with the PhD thesis of Martin Brühl [18]. During a common BMBF-project funded by the German Federal Ministry of Education and Research many discussions resulted in new ideas, new generalizations, and new applications of the Factorization Method (see, e.g., the joint paper [69] and also [122]).

Last but not least particular thanks are given to the members of our own research group at the Department of Mathematics, in particular to Tilo Arens, Frank Hettlich, Armin Lechleiter, and Sebastian Ritterbusch for many fruitful discussions and arguments – and also for their willingness to shoulder "daily" duties during the preparation of this monograph.

The colored versions of the plots are available on my homepage
www.mathematik.uni-karlsruhe.de/iag1/~kirsch/en

Karlsruhe, Germany
May 2007

Andreas Kirsch
Natalia Grinberg

Contents

1

The simplest cases: Dirichlet and Neumann boundary conditions

As pointed out in the introduction this chapter is devoted to the analysis of the factorization methods for the most simplest case in scattering theory. We consider the scattering of time-harmonic plane waves by an impenetrable obstacle D which we model by assuming Dirichlet boundary conditions on the boundary $\Gamma = \partial D$ of D. With respect to the factorization methods we will carry out all proofs in detail. We point out that the title of this chapter should not lead to the wrong conclusion that this "simplest case" could be left out by those readers interested in the factorization method for more complicated models. In this chapter we formulate the basic functional analytic results which form the basis of the method and will be referred to several times in later chapters.

After a short derivation of the Helmholtz equation from the basic equations in continuum mechanics we will repeat in Section 1.2 some well-known results on the direct scattering problem. We will omit the proofs but refer to the existing literature such as [43]. However, we will emphasize the important "ingredients": Rellich's Lemma and unique continuation for the problem of uniqueness of the scattering problem and Green's theorem for the derivation of important properties such as the reciprocity principles.

Section 1.3 will collect analytical results on the inverse scattering problem such as uniqueness of the inverse problem and properties of the far field operator. Here we will present the proofs of those results which are necessary for the factorization method.

We start Section 1.4 with the basic factorization of the far field operator. Then in Subsection 1.4.2 a quite general approach is discussed in which the domain D is characterized by those points z for which the infimum of a certain function (depending on z) is positive. For the special case where the scattering operator is unitary – which is the case for the scattering by an obstacle under Dirichlet boundary conditions – this representation can be transformed into the characterization of D by those points z for which a certain equation of the first kind (where the right-hand side depends on z) is solvable or not.

In Section 1.6 we will briefly treat the case of Neumann boundary conditions. The analysis is quite analogous to the Dirichlet case.

1.1   The Helmholtz equation in acoustics

In the first part of this monograph we consider acoustic waves that travel in a medium, such as a fluid. Let $v(x, t)$ be the velocity vector of a particle at $x \in \mathbb{R}^3$ and time t. Let $p(x, t)$, $\rho(x, t)$, and $S(x, t)$ denote the pressure, density, and specific entropy, respectively, of the fluid. We assume that no exterior forces act on the fluid. Then the movement of the particle is described by the following equations:

$$\frac{\partial v}{\partial t} + (v \cdot \nabla)v + \frac{1}{\rho} \nabla p = 0 \quad \text{(Euler's equation)}, \tag{1.1}$$

$$\frac{\partial \rho}{\partial t} + \operatorname{div}(\rho v) = 0 \quad \text{(continuity equation)}, \tag{1.2}$$

$$f(\rho, S) = p \quad \text{(equation of state)}, \tag{1.3}$$

$$\frac{\partial S}{\partial t} + v \cdot \nabla S = 0 \quad \text{(adiabatic hypothesis)}, \tag{1.4}$$

where the function f depends on the fluid. This system is nonlinear in the unknown functions v, ρ, p, and S. Let the *stationary case* be described by $v_0 = 0$, time-independent distributions $\rho = \rho_0(x)$ and $S = S_0(x)$, and constant p_0 such that $p_0 = f(\rho_0(x), S_0(x))$. The *linearization* of this nonlinear system is given by the (directional) derivative of this system at (v_0, p_0, ρ_0, S_0). For deriving the linearization, we set

$$v(x, t) = \varepsilon\, v_1(x, t) + \mathcal{O}(\varepsilon^2),$$

$$p(x, t) = p_0 + \varepsilon\, p_1(x, t) + \mathcal{O}(\varepsilon^2),$$

$$\rho(x, t) = \rho_0(x) + \varepsilon\, \rho_1(x, t) + \mathcal{O}(\varepsilon^2),$$

$$S(x, t) = S_0(x) + \varepsilon\, S_1(x, t) + \mathcal{O}(\varepsilon^2),$$

and we substitute this into (1.1), (1.2), (1.3), and (1.4). Ignoring terms with $\mathcal{O}(\varepsilon^2)$ leads to the linear system

$$\frac{\partial v_1}{\partial t} + \frac{1}{\rho_0} \nabla p_1 = 0, \tag{1.5}$$

$$\frac{\partial \rho_1}{\partial t} + \operatorname{div}(\rho_0\, v_1) = 0, \tag{1.6}$$

$$\frac{\partial f(\rho_0, S_0)}{\partial \rho} \rho_1 + \frac{\partial f(\rho_0, S_0)}{\partial S} S_1 = p_1, \tag{1.7}$$

$$\frac{\partial S_1}{\partial t} + v_1 \cdot \nabla S_0 = 0. \tag{1.8}$$

First, we eliminate S_1. Since

$$0 = \nabla f\big(\rho_0(x), S_0(x)\big) = \frac{\partial f(\rho_0, S_0)}{\partial \rho} \nabla \rho_0 + \frac{\partial f(\rho_0, S_0)}{\partial S} \nabla S_0,$$

we conclude by differentiating (1.7) with respect to t and using (1.8)

$$\frac{\partial p_1}{\partial t} = c(x)^2 \left[\frac{\partial \rho_1}{\partial t} + v_1 \cdot \nabla \rho_0 \right],$$

(1.9)

where the speed of sound c is defined by

$$c(x)^2 := \frac{\partial}{\partial \rho} f\big(\rho_0(x), S_0(x)\big).$$

Now we eliminate v_1 and ρ_1 from the system. This can be achieved by differentiating (1.9) with respect to time and using equations (1.5) and (1.6). This leads to the *wave equation* for p_1:

$$\frac{\partial^2 p_1(x,t)}{\partial t^2} = c(x)^2 \, \rho_0(x) \, \mathrm{div} \left[\frac{1}{\rho_0(x)} \nabla p_1(x,t) \right].$$

(1.10)

Now we assume that all quantities are time-periodic. In particular, p_1 is of the form

$$p_1(x,t) = \mathrm{Re} \left[u(x) \, e^{-i\omega t} \right]$$

with frequency $\omega > 0$ and some complex-valued function $u = u(x)$ depending only on the spatial variable. Substituting this into the wave equation (1.10) yields the three-dimensional *reduced equation* for u:

$$\rho_0(x) \, \mathrm{div} \left[\frac{1}{\rho_0(x)} \nabla u(x) \right] + \frac{\omega^2}{c(x)^2} \, u(x) = 0.$$

(1.11)

If $\nabla \rho_0$ is negligible then the reduced wave equation (1.11) reduces to the *Helmholtz equation*

$$\Delta u(x) + \frac{\omega^2}{c(x)^2} \, u(x) = 0,$$

i.e.,

$$\Delta u(x) + k^2 \, n(x)^2 \, u(x) = 0$$

(1.12)

where $k = \frac{\omega}{c_0}$ denotes the *wavenumber* and $n(x) = \frac{c_0}{c(x)}$ the *index of refraction* and c_0 the constant *speed of sound* of free space. In particular, in free space ρ_0 is constant and thus (1.12) holds with $n = 1$, i.e.,

$$\Delta u + k^2 \, u = 0.$$

We emphasize again, that in general the field u is complex valued and the physically relevant time-dependent field is given by

$$U(x, t) = \mathrm{Re}\left[u(x)\, e^{-i\omega t}\right] \tag{1.13}$$

where we write U instead of p_1. In scattering theory u is the sum of an incident field u^i and a scattered field u^s. The incident field is a solution of the Helmholtz equation in free space while the scattered part compensates for the inhomogeneous medium. We call this situation *scattering by an inhomogeneous medium*. Obstacle scattering occurs if the fields do not penetrate into an obstacle D. In the *sound-soft* case the pressure p vanishes on the boundary ∂D of D which leads to the *Dirichlet boundary condition* $u = 0$ on ∂D. Similarly, the scattering by a *sound-hard* obstacle leads to a Neumann boundary condition $\partial u/\partial \nu = 0$ on ∂D since here the normal component of the velocity v vanishes on ∂D. The vector $\nu = \nu(x)$ denotes the unit normal vector at $x \in \partial D$. More general boundary conditions can be formulated as *impedance boundary conditions* where the normal component of v is proportional to the pressure. This is formulated as $\partial u/\partial \nu + \lambda u = 0$ on ∂D with some (possibly space-dependent) impedance λ.

1.2   The direct scattering problem

Let $D \subset \mathbb{R}^3$ be an open and bounded domain with C^2-boundary Γ such that the exterior $\mathbb{R}^3 \setminus \overline{D}$ of $\overline{D}$ is connected. Here and throughout the monograph we denote by $\overline{D}$ the closure of the set D of points in $\mathbb{R}^3$. A confusion with the complex conjugate $\overline{z}$ of $z \in \mathbb{C}$ is not expected. Furthermore, let $k > 0$ be the (real-valued) wavenumber and

$$u^i(x, \theta) = \exp(ikx \cdot \theta), \quad x \in \mathbb{R}^3, \tag{1.14}$$

be the incident plane wave of direction $\theta \in S^2$. Here, $S^2 = \{x \in \mathbb{R}^3 : |x| = 1\}$ denotes the unit sphere in $\mathbb{R}^3$. The obstacle D gives rise to a scattered field $u^s \in C^2(\mathbb{R}^3 \setminus \overline{D}) \cap C(\mathbb{R}^3 \setminus D)$ which superposes u^i and results in the total field $u = u^i + u^s$ which satisfies the *Helmholtz equation*

$$\Delta u + k^2 u = 0 \quad \text{outside } D, \tag{1.15}$$

and the *Dirichlet boundary condition*

$$u = 0 \quad \text{on } \Gamma. \tag{1.16}$$

The scattered field u^s satisfies the *Sommerfeld radiation condition*

$$\frac{\partial u^s}{\partial r} - ik\, u^s = \mathcal{O}\!\left(r^{-2}\right) \quad \text{for } r = |x| \to \infty \tag{1.17}$$

uniformly with respect to $\hat{x} = x/|x|$.

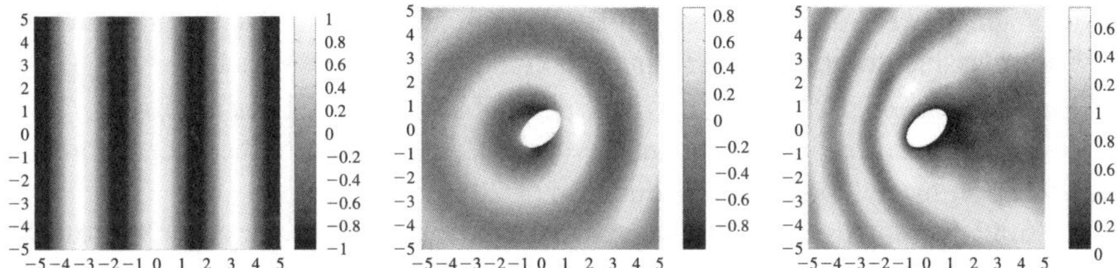

Figure 1.1 Incident, scattered, and total field

We illustrate this situation in Figure 1.1 which shows the real parts of the incident plane wave u^i (left picture), the scattered wave u^s (middle picture), and the total field u (right picture), respectively, in two dimensions[1].

Observing that the incident field u^i satisfies the Helmholtz equation (1.15) in all of $\mathbb{R}^3$ we note that the scattered field u^s solves the following exterior boundary value problem for $f = -u^i$:

Given $f \in H^{1/2}(\Gamma)$ find $v \in H^1_{loc}(\mathbb{R}^3 \setminus \overline{D})$ such that

$$\Delta v + k^2 v = 0 \quad \text{outside } D, \tag{1.18}$$

$$v = f \quad \text{on } \Gamma. \tag{1.19}$$

and

$$\frac{\partial v}{\partial r} - ik\,v = \mathcal{O}\!\left(r^{-2}\right) \quad \text{for } r = |x| \to \infty \tag{1.20}$$

uniformly with respect to $\hat{x} = x/|x|$.

We note that the solution of (1.18) is understood in the variational sense. Indeed, using the Helmholtz equation in Green's first formula for the region $D_R = \left\{ x \in \mathbb{R}^3 \setminus \overline{D} : |x| < R \right\}$

$$\iint\limits_{D_R} \left[\varphi\,\Delta v + \nabla\varphi \cdot \nabla v \right] dx = \int\limits_{\Gamma_R} \varphi\,\frac{\partial v}{\partial \nu}\,ds \tag{1.21}$$

yields

$$\iint\limits_{\mathbb{R}^3 \setminus \overline{D}} \left[\nabla\varphi \cdot \nabla v - k^2 \varphi v \right] dx = 0 \tag{1.22}$$

for any test function φ with compact support in $\mathbb{R}^3 \setminus \overline{D}$ (choose R such that that the ball of radius R contains the support of φ). Here and throughout this monograph, $\nu = \nu(x)$ denotes the unit normal vector at $x \in \partial D$ directed into the exterior of D. It follows from

---

[1] The setting of the two-dimensional scattering problem, i.e., where $x, \theta \in \mathbb{R}^2$ and $D \subset \mathbb{R}^2$, differs from the one in three dimensions only in the radiations condition which has now the form $\partial u^s/\partial r - ik\,u^s = \mathcal{O}\!\left(r^{-3/2}\right)$. We refer to [43] and Section 1.7.

interior regularity results for elliptic differential equations (cf. [70]) that any solution of (1.22) is a classical solution of the Helmholtz equation in $\mathbb{R}^3 \setminus \overline{D}$. We call $v \in H^1_{loc}(\mathbb{R}^3 \setminus \overline{D})$ a variational solution of (1.18), (1.19), and (1.20) if v solves (1.22) for all $\varphi \in H^1_0(\mathbb{R}^3 \setminus \overline{D})$ which vanish outside of some ball, and $v = f$ on Γ in the sense of the trace theorem and v satisfies Sommerfeld's radiation condition (1.20).

The fundamental results on uniqueness and existence are summarized in the following theorem.

Theorem 1.1 *For any $f \in H^{1/2}(\Gamma)$ there exists a unique (variational) solution $v \in H^1_{loc}(\mathbb{R}^3 \setminus \overline{D})$ of (1.18), (1.19), and (1.20).*

Furthermore, if the boundary data f is continuous on Γ then $v \in C^2(\mathbb{R}^3 \setminus \overline{D}) \cap C(\mathbb{R}^3 \setminus D)$. If f is even continuously differentiable on Γ then the normal derivative $\partial v/\partial v$ exists and is continuous on Γ.

For a *proof* we refer to [150] and [43] (see also Chapter 2, Section 2.1 where we show existence and uniqueness for the Robin boundary conditions). $\square$

Solutions of the Helmholtz equation (1.18) in some exterior domain which satisfy also Sommerfeld's radiation condition (1.20) will be referred to as *radiating solutions* of (1.18).

The uniqueness part in the proof of the previous theorem makes essential use of the following result which is due to Rellich [166] (cf. [43]).

Lemma 1.2 *Let v be a solution of the Helmholtz equation (1.18) in some region of the form $\{x \in \mathbb{R}^3 : |x| > R\}$ satisfying*

$$\lim_{r \to \infty} \int_{|x|=r} |v(x)|^2 \, ds = 0.$$

Then v vanishes for $|x| > R$.

For a *proof* we refer to [43]. $\square$

Green's formula is the essential tool also in the proof of the following representation formula for radiating solutions of the Helmholtz equation which sometimes is called "Green's representation formula".

Theorem 1.3 *Let $v \in C^2(\mathbb{R}^3 \setminus \overline{D}) \cap C(\mathbb{R}^3 \setminus D)$ be a radiating solution of (1.18) such that v posseses a normal derivative on the boundary Γ in the sense that the limit*

$$\frac{\partial v(x)}{\partial v} = \lim_{h \to +0} v(x) \cdot \nabla v\big(x + h v(x)\big), \quad x \in \Gamma,$$

exists uniformly with respect to $x \in \Gamma$. Then Green's formula holds in the form

$$v(x) = \int_\Gamma \left[v(y) \frac{\partial \Phi(x,y)}{\partial v(y)} - \frac{\partial v(y)}{\partial v} \Phi(x,y) \right] ds(y), \quad x \notin \overline{D}. \tag{1.23}$$

Again, $\nu(y)$ denotes the exterior unit normal vector at $y \in \Gamma$ and Φ the fundamental solution of the Helmholtz equation in $\mathbb{R}^3$ given by

$$\Phi(x,y) = \frac{\exp(ik|x-y|)}{4\pi|x-y|}, \quad x,y \in \mathbb{R}^3, \quad x \neq y. \tag{1.24}$$

For a *proof* we refer to [43, Theorem 2.4]. □

As a direct consequence of this theorem one has the following result. Its proof can again be found in [43], Theorems 2.5 and 2.6.

Theorem 1.4 *Let $v \in C^2(\mathbb{R}^3 \setminus \overline{D}) \cap C(\mathbb{R}^3 \setminus D)$ be a radiating solution of (1.18) such that v posses a normal derivative on the boundary Γ in the sense of Theorem 1.3.*

Then v is analytic in $\mathbb{R}^3 \setminus \overline{D}$ and has the asymptotic behavior

$$v(x) = \frac{\exp(ik|x|)}{4\pi|x|}\, v^\infty(\hat{x}) + \mathcal{O}(|x|^{-2}), \quad |x| \to \infty, \tag{1.25}$$

uniformly with respect to $\hat{x} = x/|x| \in S^2$. The function $v^\infty : S^2 \to \mathbb{C}$ is analytic and is called the far field pattern *of v. It has the form*

$$v^\infty(\hat{x}) = \int_\Gamma \left[v(y)\, \frac{\partial}{\partial\nu(y)} e^{-ik\hat{x}\cdot y} - \frac{\partial v(y)}{\partial\nu}\, e^{-ik\hat{x}\cdot y} \right] ds(y), \quad \hat{x} \in S^2. \tag{1.26}$$

As a consequence of the analyticity of v and Rellich's Lemma 1.2 we have

Corollary 1.5

(a) *If v vanishes on some open subset of $\mathbb{R}^3 \setminus \overline{D}$ then v vanishes everywhere in $\mathbb{R}^3 \setminus \overline{D}$. (Note that we always assume that the exterior of $\overline{D}$ is connected.)*

(b) *If v^∞ vanishes on an open part of S^2 (open relative to S^2) then v vanishes in the exterior of D.*

Application of these results to the scattering problem (1.15), (1.16), and (1.17) assures existence of a unique solution u for any incident field u^i. Its dependence on the incident direction is indicated by writing $u = u(\cdot, \theta)$. Analogously, the far field pattern u^∞ of u^s depends on the two angular values $\hat{x}$ and θ. We indicate this by writing $u^\infty = u^\infty(\hat{x}, \theta)$ and note that u^∞ depends analytically on both variables. This follows, e.g., from Theorem 1.6 below.

1.3   The far field patterns and the inverse problem

First, we will prove a reciprocity principle for u^∞. It states the (physically obvious) fact that it is the same if we illuminate an object from the direction θ and observe it in the direction $-\hat{x}$ or the other way around: illumination from $\hat{x}$ and observation in $-\theta$.

Theorem 1.6 *(First reciprocity principle) Let $u^\infty(\hat{x}, \theta)$ be the far field pattern corresponding to the direction $\hat{x}$ of observation and the direction θ of the incident plane wave. Then*

$$u^\infty(-\hat{x}, \theta) = u^\infty(-\theta, \hat{x}) \quad \textit{for all } \hat{x}, \theta \in S^2. \tag{1.27}$$

Proof: Application of Green's second theorem to u^i and u^s in the interior and exterior of D, respectively, yields

$$0 = \int_\Gamma \left[u^i(y,\theta) \frac{\partial}{\partial \nu} u^i(y,\hat{x}) - u^i(y,\hat{x}) \frac{\partial}{\partial \nu} u^i(y,\theta) \right] ds(y) \,,$$

$$0 = \int_\Gamma \left[u^s(y,\theta) \frac{\partial}{\partial \nu} u^s(y,\hat{x}) - u^s(y,\hat{x}) \frac{\partial}{\partial \nu} u^s(y,\theta) \right] ds(y) \,.$$

(More precisely, to prove the second equation, one applies Green's second theorem to u^s in the region $\{x \in \mathbb{R}^3 \setminus \overline{D} : |x| < R\}$ with R large enough and lets R tend to infinity.) Now we use the representation (1.26) for the far field patterns $u^\infty(-\hat{x},\theta)$ and $u^\infty(-\theta,\hat{x})$:

$$u^\infty(-\hat{x},\theta) = \int_\Gamma \left[u^s(y,\theta) \frac{\partial}{\partial \nu} u^i(y,\hat{x}) - u^i(y,\hat{x}) \frac{\partial}{\partial \nu} u^s(y,\theta) \right] ds(y) \,,$$

$$-u^\infty(-\theta,\hat{x}) = \int_\Gamma \left[u^i(y,\theta) \frac{\partial}{\partial \nu} u^s(y,\hat{x}) - u^s(y,\hat{x}) \frac{\partial}{\partial \nu} u^i(y,\theta) \right] ds(y) \,.$$

Adding these four equations yields

$$u^\infty(-\hat{x},\theta) - u^\infty(-\theta,\hat{x}) = \int_\Gamma \left[u(y,\theta) \frac{\partial}{\partial \nu} u(y,\hat{x}) - u(y,\hat{x}) \frac{\partial}{\partial \nu} u(y,\theta) \right] ds(y) \,. \quad (1.28)$$

So far, we have not used the boundary condition on Γ. With $u(y,\hat{x}) = 0$ and $u(y,\theta) = 0$ on Γ the assertion follows. $\square$

There exists an interesting second reciprocity principle which relates the scattered field $u^s = u^s(x,\theta)$ corresponding to the plane wave of direction θ as incident field to the far field pattern $v^\infty = v^\infty(\hat{x},z)$ which corresponds to the point source $\Phi(\cdot,z)$ as incident field. Indeed, by the same arguments as in Theorem 1.6 one can show (cf. [160], Section 2.1):

Theorem 1.7 *(Second or mixed reciprocity principle) Let $u^s(z,-\theta)$ be the scattered field at $z \in \mathbb{R}^3 \setminus \overline{D}$ which corresponds to the incident field $u^i(x,-\theta) = \exp(-ikx \cdot \theta)$, $x \in \mathbb{R}^3$. Furthermore, let $v^\infty(\theta,z)$ be the far field pattern of the scattered field v^s at θ which corresponds to the incident field $v^i(x,z) = \Phi(x,z)$, $x \in \mathbb{R}^3$. Then*

$$u^s(z,-\theta) = v^\infty(\theta,z) \quad \text{for all } \theta \in S^2 \text{ and } z \notin \overline{D}. \quad (1.29)$$

The far field patterns $u^\infty(\hat{x},\theta)$, $\hat{x},\theta \in S^2$, define the integral operator $F : L^2(S^2) \to L^2(S^2)$ by

$$(Fg)(\hat{x}) = \int_{S^2} u^\infty(\hat{x},\theta)\, g(\theta)\, ds(\theta) \quad \text{for } \hat{x} \in S^2 \,, \quad (1.30)$$

which we will call the *far field operator*. It is certainly compact in $L^2(S^2)$ since its kernel is analytic in both variables and is related to the *scattering operator* $S : L^2(S^2) \to L^2(S^2)$ by

$$S = I + \frac{ik}{8\pi^2} F \qquad (1.31)$$

where I denotes the identity. In the next theorem we collect some properties of these operators.

Theorem 1.8

(a) *The far field operator F satisfies*

$$F - F^* = \frac{ik}{8\pi^2} F^* F \qquad (1.32)$$

where F^* denotes the L^2 – adjoint of F.

(b) *The scattering operator $S = I + \frac{ik}{8\pi^2} F$ is unitary, i.e., $S^* S = S S^* = I$.*

(c) *The far field operator F is normal, i.e., $F^* F = F F^*$.*

(d) *Assume that there exists no non-trivial* Herglotz *wave function v_g, i.e., a function of the form*

$$v_g(x) = \int_{S^2} e^{ikx \cdot \theta} \, g(\theta) \, ds(\theta), \quad x \in \mathbb{R}^3 , \qquad (1.33)$$

with density $g \in L^2(S^2)$ which vanishes on Γ. In particular, such a function does not exists if k^2 is not a Dirichlet eigenvalue of $-\Delta$ in D.[2] Then F is one-to-one and its range $\mathcal{R}(F)$ is dense in $L^2(S^2)$.

Proof: (a) For $g, h \in L^2(S^2)$, define the Herglotz wave functions v^i and w^i by

$$v^i(x) = \int_{S^2} e^{ikx \cdot \theta} \, g(\theta) \, ds(\theta), \quad x \in \mathbb{R}^3 ,$$

$$w^i(x) = \int_{S^2} e^{ikx \cdot \theta} \, h(\theta) \, ds(\theta), \quad x \in \mathbb{R}^3 ,$$

respectively. Let v and w be the solutions of the scattering problem (1.15), (1.16), and (1.17) corresponding to incident fields v^i and w^i, respectively, with corresponding scattered fields $v^s = v - v^i$, $w^s = w - w^i$ and far field patterns v^∞, w^∞, respectively. Then, by linearity, $v^\infty = Fg$ and $w^\infty = Fh$. Green's formula in $D_R = \{x \in \mathbb{R}^3 \setminus \overline{D} : |x| < R\}$ and the boundary conditions yield

$$0 = \int_{D_R} \left[v \, \Delta \overline{w} - \overline{w} \, \Delta v \right] dx = \int_{|x|=R} \left[v \, \frac{\partial \overline{w}}{\partial \nu} - \overline{w} \, \frac{\partial v}{\partial \nu} \right] ds . \qquad (1.34)$$

---

[2] k^2 is called a Dirichlet eigenvalue of $-\Delta$ in D if there exists a non-trivial solution $u \in C^2(D) \cap C(\overline{D})$ of the Helmholtz equation in D such that u vanishes on Γ.

The integral on the right hand side is split into four parts by decomposing $v = v^i + v^s$ and $w = w^i + w^s$. The integral

$$\int_{|x|=R} \left[v^i \frac{\partial \overline{w}^i}{\partial \nu} - \overline{w}^i \frac{\partial v^i}{\partial \nu} \right] ds$$

vanishes by Green's second theorem in $\{x : |x| < R\}$ since v^i and $\overline{w}^i$ are solutions of the Helmholtz equation (1.18). We note that by our normalization of the far field pattern

$$v^s(x) \frac{\partial \overline{w}^s(x)}{\partial r} - \overline{w}^s(x) \frac{\partial v^s(x)}{\partial r} = -\frac{2ik}{(4\pi r)^2} v^\infty(\hat{x}) \overline{w^\infty(\hat{x})} + \mathcal{O}\left(\frac{1}{r^3}\right).$$

From this we conclude that

$$\int_{|x|=R} \left[v^s \frac{\partial \overline{w}^s}{\partial \nu} - \overline{w}^s \frac{\partial v^s}{\partial \nu} \right] ds \longrightarrow -\frac{ik}{8\pi^2} \int_{S^2} v^\infty \, \overline{w^\infty} \, ds = -\frac{ik}{8\pi^2} (Fg, Fh)_{L^2(S^2)}$$

as R tends to infinity. Finally, we use the definition of v^i and w^i and the representation (1.26) to compute

$$\int_{|x|=R} \left[v^i \frac{\partial \overline{w}^s}{\partial \nu} - \overline{w}^s \frac{\partial v^i}{\partial \nu} \right] ds$$

$$= \int_{S^2} g(\theta) \int_{|x|=R} \left[e^{ikx\cdot\theta} \frac{\partial \overline{w}^s(x)}{\partial \nu} - \overline{w}^s(x) \frac{\partial}{\partial \nu} e^{ikx\cdot\theta} \right] ds(x) \, ds(\theta)$$

$$= -\int_{S^2} g(\theta) \, \overline{w^\infty(\theta)} \, d(\theta) = -(g, Fh)_{L^2(S^2)}.$$

Analogously, we have that

$$\int_{|x|=R} \left[v^s \frac{\partial \overline{w}^i}{\partial \nu} - \overline{w}^i \frac{\partial v^s}{\partial \nu} \right] ds = (Fg, h)_{L^2(S^2)}.$$

Taking the limit $R \to \infty$ yields

$$0 = -\frac{ik}{8\pi^2} (Fg, Fh)_{L^2(S^2)} - (g, Fh)_{L^2(S^2)} + (Fg, h)_{L^2(S^2)}. \tag{1.35}$$

This holds for all $g, h \in L^2(S^2)$. From this the assertion (a) follows.

    (b) We compute

$$S^*S = \left(I - \frac{ik}{8\pi^2} F^* \right)\left(I + \frac{ik}{8\pi^2} F \right) = I + \frac{ik}{8\pi^2} (F - F^*) + \frac{k^2}{64\pi^4} F^*F$$

and thus $S^*S = I$ with part (a). This implies injectivity of S and thus also surjectivity since S is a compact perturbation of the identity. Therefore, $S^* = S^{-1}$ and thus also $SS^* = I$.

(c) This follows now by comparing the forms of $\mathcal{S}^*\mathcal{S}$ and $\mathcal{S}\mathcal{S}^*$.

(d) Let $g \in L^2(S^2)$ be such that $Fg = 0$ on S^2. From the definition of the far field operator we note that $Fg = v^\infty$ where v^∞ is the far field pattern which corresponds to the incident field $v^i(x) = \int_{S^2} \exp(ikx \cdot \theta)\, g(\theta)\, ds(\theta)$, $x \in \mathbb{R}^3$. Rellich's Lemma 1.2 and analytic continuation imply that the scattered field v^s vanishes outside of D.[3] From the boundary condition we conclude that the incident field v^i vanishes on Γ, i.e., v^i is a Dirichlet eigenfunction of $-\Delta$ in D. From the assumption on the wavenumber v^i has to vanish in D and thus everywhere by analytic continuation. Expansion into spherical wave functions by using the Jacobi–Anger expansion (see (1.82) below) yields that g vanishes on S^2. Finally, we show that the adjoint F^* of F is one-to-one as well which proves denseness of the range of F. Indeed, $F^*g = 0$ yields by using the reciprocity relation (1.27) that

$$0 = (F^*g)(\hat{x}) = \int_{S^2} \overline{u^\infty(\theta,\hat{x})}\, g(\theta)\, ds(\theta) \;=\; \int_{S^2} \overline{u^\infty(-\hat{x},-\theta)\, \overline{g(\theta)}}\, ds(\theta),$$

i.e., $F\tilde{g} = 0$ with $\tilde{g}(\theta) = \overline{g(-\theta)}$. Injectivity of F yields that $\tilde{g} = 0$ and thus also $g = 0$. $\square$

Now we turn to the formulation of the inverse scattering problem for which we will introduce the factorization methods in Section 1.4.

Inverse Scattering Problem: *Given the wavenumber $k > 0$ and the far field patterns $u^\infty(\hat{x},\theta)$ for all $\hat{x},\theta \in S^2$ determine the shape of the scattering obstacle D!*

The following uniqueness theorem, taken from [43], assures that in principle the data set $\left\{ u^\infty(\hat{x},\theta) : \hat{x},\theta \in S^2 \right\}$ is sufficient to determine D.

Theorem 1.9 *For fixed wavenumber $k > 0$ the far field patterns $u^\infty(\hat{x},\theta)$ for all $\hat{x},\theta \in S^2$ uniquely determine the shape of the scattering obstacle D, i.e., if there are two obstacles D_1 and D_2 with corresponding far field patterns $u_1^\infty(\hat{x},\theta)$ and $u_2^\infty(\hat{x},\theta)$, respectively, then $u_1^\infty(\hat{x},\theta) = u_2^\infty(\hat{x},\theta)$ for all $\hat{x},\theta \in S^2$ implies that $D_1 = D_2$.*

As in Figure 1.1 we want to illustrate also the inverse scattering problem with two examples, again in two dimensions. In this case, the far field patterns u^∞ depend on the two variables $\hat{x}$ and θ from the unit circle in $\mathbb{R}^2$ which we identify with the interval $[0,2\pi]$. Furthermore, we identify $\hat{x} = (\cos\phi, \sin\phi)^\top$ with $\phi \in [0,2\pi]$ and the unit vector θ with the angle $\theta \in [0,2\pi]$. Figures 1.2 and 1.3 show contour plots of the real and imaginary parts, respectively, of the far field patterns for two examples.

The inverse scattering problem is to identify the obstacle D from these plots.

In the first example (Figure 1.2) the contour lines are straight lines, i.e., u^∞ is constant along lines of the form $\phi - \theta = \text{const}$. In terms of the unit vectors $\hat{x}$ and θ this can be written in the form $\hat{x}\cdot\theta = \text{const}$. By the following result of Karp [108] (which is also true for the two-dimensional case) we conclude that this first example corresponds to D being a disk. The second example corresponds to a domain D which is certainly not a disk.[4]

---

[3] Here we make use of the assumption that the exterior of $\overline{D}$ is connected.

[4] This belongs to the domain D of Figure 1.7.

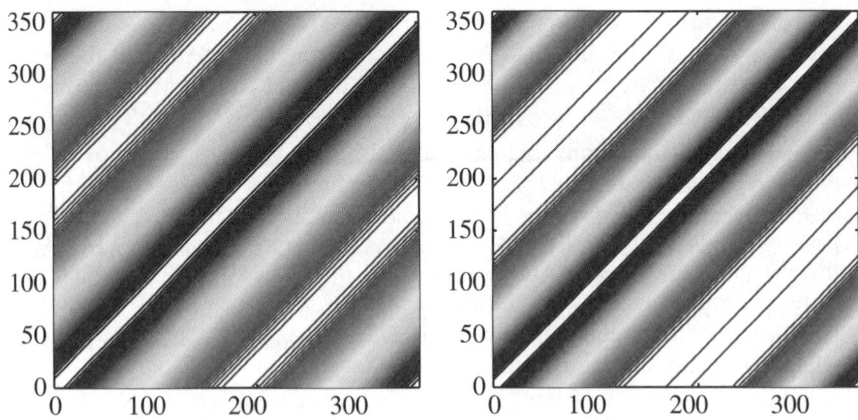

Figure 1.2 Real (left) and imaginary (right) parts of $u^\infty = u^\infty(\phi,\theta)$ for $\phi,\theta \in [0,2\pi]$

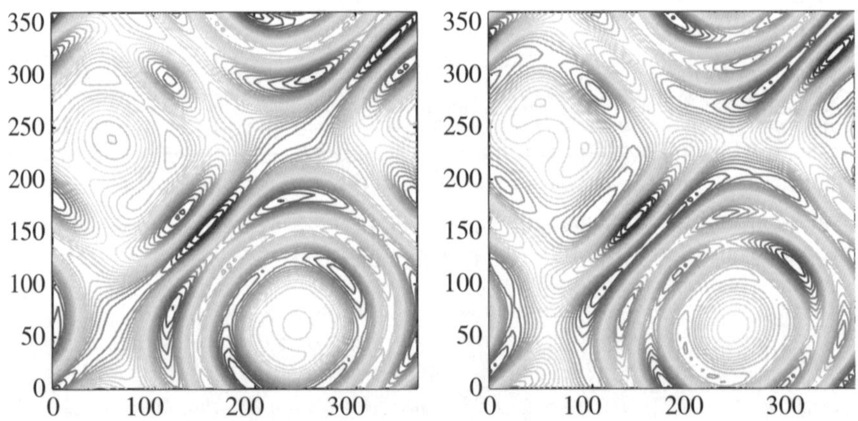

Figure 1.3 Real (left) and imaginary (right) parts of $u^\infty = u^\infty(\phi,\theta)$ for $\phi,\theta \in [0,2\pi]$

Theorem 1.10 *Let $k > 0$ and $u^\infty = u^\infty(\hat{x},\theta)$ for $\hat{x},\theta \in S^2$ be the far field patterns corresponding to some domain $D \subset \mathbb{R}^3$. If*

$$u^\infty(\hat{x},\theta) = u^\infty(Q\hat{x},Q\theta)$$

for all $\hat{x},\theta \in S^2$ and all rotations, i.e., all real orthogonal matrices $Q \in \mathbb{R}^{3\times3}$ with $\det Q = 1$, then D is a ball with center zero.

The *proof* is a simple consequence of the uniqueness result of Theorem 1.9, see Section 5.1 of [43].

Remarks:

(a) We note that by the analyticity of u^∞ with respect to both variables it is sufficient to require that $u_1^\infty(\hat{x}, \theta)$ and $u_2^\infty(\hat{x}, \theta)$ coincide for all $\hat{x}$ and θ from open subsets of S^2 or even from an infinite number of pairs $(\hat{x}_\ell, \theta_m)$. We refer to [130] for more details.

(b) The historically first uniqueness result for this inverse scattering problem is due to Schiffer (see remark in [138], Section V.5). The proof depends crucially on the compact imbedding property of $H_0^1(G)$ in $L^2(G)$ for any bounded open set G to ensure that the spectrum of $-\Delta$ in G with respect to Dirichlet boundary conditions is discrete. The analogous argument for other boundary conditions such as the Neumann boundary condition requires the compact imbedding property of $H^1(G)$ in $L^2(G)$ which holds only under smoothness assumptions on the boundary of G. Since the argument is applied to G being the (set-)difference of two obstacles D_1 and D_2 smoothness of this difference cannot be assured. Therefore, Schiffer's proof cannot be transfered to other boundary conditions. Based on results of Isakov [105] for penetrable obstacles Kirsch and Kress [124] obtained uniqueness results for several kinds of boundary conditions.

(c) To the authors knowledge it is an open problem whether the far field patterns $u^\infty(\hat{x}, \theta)$ for all $\hat{x} \in S^2$ but only *one* incident wave $u^i(x) = \exp(ikx \cdot \theta)$ uniquely determines D if no a priori information on D is available. Partial results are known if a priori information on the obstacle D is available. It has been shown by Colton and Sleeman in [57] (cf. [43], Theorem 5.2) that the scatterer is uniquely determined by the far field patterns of a finite number of incident plane waves provided a priori information on the size of the obstacle D is available. In recent papers by Elschner, Yamamoto, Liu, and others (see [61, 140]) uniqueness of the inverse scattering problem for one incident plane wave and polyhedral scatterers is shown.

(d) A similar theorem shows that the scatterer is uniquely determined by the far field patterns for an infinite number of incident plane waves with distinct wavenumbers from a bounded interval in $\mathbb{R}_{>0}$. We refer again to [130] for more details.

1.4 Factorization methods

We recall from (1.30) that the far field patterns u^∞ define the far field operator $F : L^2(S^2) \to L^2(S^2)$ by

$$(Fg)(\hat{x}) = \int_{S^2} u^\infty(\hat{x}, \theta)\, g(\theta)\, ds(\theta), \quad \hat{x} \in S^2. \tag{1.36}$$

With respect to the inverse problem, this operator contains the known data. It is the aim to give explicit characterizations of the unknown domain D by this "data operator" F. This section is organized as follows. In Subsection 1.4.1 we will derive a factorization of the operator F in the form

$$F = G\,T\,G^* \tag{1.37}$$

with some compact operator G and some isomorphism T between suitable spaces which depend on D, of course. This factorization is the basis of all versions of the Factorization Methods and is responsible for its name. From this factorization we observe already that the range of the operator F is contained in the range of G. There is a simple – and very explicit – relationship between the range of the operator G and the shape of D. Let us first define the operator G.

Definition 1.11 *Let the data-to-pattern operator $G : H^{1/2}(\Gamma) \to L^2(S^2)$ be defined by $Gf = v^\infty$ where $v^\infty \in L^2(S^2)$ is the far field pattern of the solution v of the exterior Dirichlet problem (1.18), (1.19), and (1.20) with boundary data $f \in H^{1/2}(\Gamma)$, i.e., $v \in H^1_{loc}(\mathbb{R}^3 \setminus \overline{D})$ solves*

$$\Delta v + k^2 v = 0 \quad \textit{outside } D, \tag{1.38}$$

$$v = f \quad \textit{on } \Gamma, \tag{1.39}$$

and

$$\frac{\partial v}{\partial r} - ik\,v = \mathcal{O}\!\left(r^{-2}\right) \quad \textit{for } r = |x| \to \infty \tag{1.40}$$

uniformly with respect to $\hat{x} = x/|x|$.

We note that we do not indicate the type of boundary condition by writing G_{Dir} for the Dirichlet boundary condition. Later in this chapter (in Section 1.6) we will introduce the analogous operator for the Neumann boundary condition and denote it also by G. In Chapter 2, however, we will indicate the type of boundary condition by writing G_{Dir} and G_{Neu}, respectively.

Then we have:

Theorem 1.12 *Let $G : H^{1/2}(\Gamma) \to L^2(S^2)$ be defined by Definition 1.11. For any $z \in \mathbb{R}^3$ define the function $\phi_z \in L^2(S^2)$ by*

$$\phi_z(\hat{x}) := e^{-ik\hat{x}\cdot z}, \quad \hat{x} \in S^2. \tag{1.41}$$

Then ϕ_z belongs to the range $\mathcal{R}(G)$ of G if, and only if, $z \in D$.

Proof: Let first $z \in D$ and define

$$v(x) := \Phi(x, z) = \frac{\exp(ik|x - z|)}{4\pi\,|x - z|}, \quad x \notin D,$$

and $f := v|_\Gamma$. Then $f \in H^{1/2}(\Gamma)$ and the far field pattern of v is given by

$$v^\infty(\hat{x}) = e^{-ik\hat{x}\cdot z}, \quad \hat{x} \in S^2,$$

which coincides with ϕ_z, i.e., $Gf = v^\infty = \phi_z$, i.e., $\phi_z \in \mathcal{R}(G)$.

Let now $z \notin D$ and assume on the contrary that there exists $f \in H^{1/2}(\Gamma)$ with $Gf = \phi_z$. Let v be the solution of the exterior Dirichlet problem with boundary data f and $v^\infty = Gf$ be its far field pattern. Since ϕ_z is the far field pattern of $\Phi(\cdot, z)$ we conclude by Rellich's Lemma 1.2 that $v(x) = \Phi(x, z)$ for all x outside of any sphere containing D and z. Finally, by analytic continuation[5] we conclude that v and $\Phi(\cdot, z)$ coincide on $\mathbb{R}^3 \setminus (\overline{D} \cup \{z\})$.

If $z \notin \overline{D}$ this contradicts the fact that v is analytic in $\mathbb{R}^3 \setminus \overline{D}$ and $\Phi(x, z)$ is singular at $x = z$.

If $z \in \Gamma$ we have that $\Phi(x, z) = f(x)$ for $x \in \Gamma, x \neq z$, i.e., the function $x \mapsto \Phi(x, z)$ is in $H^{1/2}(\Gamma)$. This contradicts the fact that this function is certainly not in $H^1(D)$ or $H^1_{loc}(\mathbb{R}^3 \setminus \overline{D})$ since $\nabla \Phi(x, z) = \mathcal{O}(1/|x - z|^2)$ as $x \to z$. $\quad\square$

The main work of the Factorization Methods is to relate the range of G to the known data operator F (or some operator which can be derived from F). In this chapter we will do this in two possible ways which will be presented in Subsections 1.4.2 and 1.4.3, respectively. Each of these subsections will begin with an abstract result from functional analysis, formulated in general Hilbert or Banach spaces. Application of this abstract result to the operators G and F will lead to a precise characterization of the range $\mathcal{R}(G)$ of G by the data operator F. Combination of this result with Theorem 1.12 will give fairly explicit formulas for the characteristic function χ_D of D which will solely be expressed by quantities computable from F.

1.4.1 Factorization of the far field operator

We recall Definition 1.11 of the crucial data-to-pattern operator $G : H^{1/2}(\Gamma) \to L^2(S^2)$. It is defined by $Gf = v^\infty$ where $v^\infty \in L^2(S^2)$ is the far field pattern of the solution v of the exterior Dirichlet problem (1.38), (1.39), and (1.40) with boundary data $f \in H^{1/2}(\Gamma)$.

Properties of G are collected in the following lemma.

Lemma 1.13 *The data-to-pattern operator $G : H^{1/2}(\Gamma) \to L^2(S^2)$ is compact, one-to-one with dense range in $L^2(S^2)$.*

Proof: First, injectivity is a direct consequence of Rellich's Lemma and analytic continuation, see Corollary 1.5.

To prove compactness we choose a ball $B = B(0, R)$ of radius R centered at 0 which contains $\overline{D}$ in its interior. Using the representation (1.26) we can decompose G as $G = G_2 G_1$ where $G_1 : H^{1/2}(\Gamma) \to C(\partial B) \times C(\partial B)$ and $G_2 : C(\partial B) \times C(\partial B) \to L^2(S^2)$ are defined by $G_1 f = (v|_{\partial B}, \partial v/\partial \nu|_{\partial B})$ and

$$G_2(g, h)(\hat{x}) = \int\limits_{\partial B} \left[g(y) \frac{\partial}{\partial \nu(y)} e^{-ik\hat{x}\cdot y} - h(y) e^{-ik\hat{x}\cdot y} \right] ds(y), \quad \hat{x} \in S^2,$$

respectively, and where again v denotes the solution of the exterior Dirichlet boundary value problem with boundary values f. Then G_1 is bounded by interior regularity results and G_2 compact which proves compactness of G.

---

[5] Note that it is here where we make use of the assumption that the exterior of D is connected.

To prove denseness of the range of G we recall the definition of the spherical harmonics Y_n^m defined by

$$Y_n^m(\hat{x}) = Y_n^m(\phi, \varphi) = \sqrt{\frac{2n+1}{4\pi} \frac{(n-|m|)!}{(n+|m|)!}}\, P_n^{|m|}(\cos\phi)\, e^{im\varphi}\,, \qquad (1.42)$$

for $-n \le m \le n$ and $n \in \mathbb{N}$. Here (ϕ, φ) are the spherical polar coordinates of $\hat{x} \in S^2$ and P_n^m the associated Legendre functions. The spherical harmonics Y_n^m are normalized such that they form a complete orthonormal system in $L^2(S^2)$. Furthermore, we denote by h_n the spherical Hankel functions of the first kind and order $n \in \mathbb{N}$. We refer to [43], Sections 2.3 and 2.4, for a brief introduction to these functions. Now we make use of the fact that every element f in $L^2(S^2)$ can be approximated by a finite linear combination of the spherical harmonics, i.e., for every $\varepsilon > 0$ there exists $N \in \mathbb{N}$ and constants $c_n^m \in \mathbb{C}$ with

$$\left\| f - \sum_{n=0}^{N} \sum_{|m|\le n} c_n^m\, Y_n^m \right\|_{L^2(S^2)} \le \varepsilon\,.$$

We choose the origin inside of D and define the function v by

$$v(x) = \frac{k}{4\pi} \sum_{n=0}^{N} \sum_{|m|\le n} c_n^m\, e^{i(n+1)\pi/2}\, h_n(k|x|) Y_n^m(\hat{x})\,, \qquad x \ne 0\,.$$

The far field pattern of $h_n(k|x|) Y_n^m(\hat{x})$ is determined by the asymptotic behavior of $h_n(t)$ as t tends to infinity and is given by $(4\pi/k)\exp\big(-i(n+1)\pi/2\big)\, Y_n^m(\hat{x})$ (cf. [43]). Therefore,

$$v^\infty = \sum_{n=0}^{N} \sum_{|m|\le n} c_n^m\, Y_n^m$$

and thus $\|f - v^\infty\|_{L^2(S^2)} \le \varepsilon$. Observing that $v^\infty = Gv|_\Gamma$ yields the assertion. $\square$

The operator T in (1.37) will be the adjoint of the single layer boundary operator S : $H^{-1/2}(\Gamma) \to H^{1/2}(\Gamma)$, defined by

$$(S\varphi)(x) := \int_\Gamma \Phi(x,y)\, \varphi(y)\, ds(y)\,, \qquad x \in \Gamma\,, \qquad (1.43)$$

where Φ denotes again the fundamental solution of the Helmholtz equation in three dimensions as defined in (1.24). Note that we write this as an ordinary integral although, strictly speaking, this has to be understood as the bounded extension of the classical operator S defined on $L^2(\Gamma)$.

In the following lemma we summarize some of the well-known properties of S which will imply, in particular, that T in (1.37) is an isomorphism. By $\langle H^{-1/2}(\Gamma), H^{1/2}(\Gamma)\rangle$ we denote the dual form which, in our setting, is the extension of the inner product of $L^2(\Gamma)$. In particular, this dual form is sesqui-linear, i.e., the mappings $\varphi \mapsto \langle \varphi, \psi \rangle$ and $\psi \mapsto \overline{\langle \varphi, \psi \rangle}$ are linear where the bar denotes the complex conjugation.

Lemma 1.14 *Assume that k^2 is not a Dirichlet eigenvalue of $-\Delta$ in D. Then the following holds.*

(a) *S is an isomorphism from the Sobolev space $H^{-1/2}(\Gamma)$ onto $H^{1/2}(\Gamma)$.*

(b) *$\operatorname{Im}\langle\varphi, S\varphi\rangle < 0$ for all $\varphi \in H^{-1/2}(\Gamma)$ with $\varphi \neq 0$. Again, $\langle\cdot,\cdot\rangle$ denotes the duality pairing in $\langle H^{-1/2}(\Gamma), H^{1/2}(\Gamma)\rangle$.*

(c) *Let S_i be the single layer boundary operator (1.43) corresponding to the wavenumber $k = i$. The operator S_i is self-adjoint and coercive as an operator from $H^{-1/2}(\Gamma)$ onto $H^{1/2}(\Gamma)$, i.e., there exists $c_0 > 0$ with*

$$\langle\varphi, S_i\varphi\rangle \geq c_0\|\varphi\|^2_{H^{-1/2}(\Gamma)} \quad \text{for all } \varphi \in H^{-1/2}(\Gamma). \tag{1.44}$$

(d) *The difference $S - S_i$ is compact from $H^{-1/2}(\Gamma)$ into $H^{1/2}(\Gamma)$.*

Proof: (a) The mapping properties of S in Sobolev spaces are intensively studied in [144].

(b) Define, for any $\varphi \in H^{-1/2}(\Gamma)$, the single layer potential v by

$$v(x) = \int_\Gamma \varphi(y)\,\Phi(x,y)\,ds(y), \quad x \in \mathbb{R}^3 \setminus \Gamma. \tag{1.45}$$

Then v is a solution of the Helmholtz equation in $\mathbb{R}^3 \setminus \Gamma$. From the theory of potentials with $H^{-1/2}(\Gamma)$–densities it is known (see [111, 144]) that $v \in H^1(D) \cap H^1_{loc}(\mathbb{R}^3 \setminus \overline{D})$, that the traces $v_\pm$ and $\partial v_\pm/\partial\nu$ exist in the variational sense with $v_\pm = S\varphi$ and $\varphi = \partial v_-/\partial\nu - \partial v_+/\partial\nu$. Here, $v_\pm$ denotes the limit from the exterior $(+)$ and interior $(-)$, respectively. Therefore, using Green's formula in D and in $D_R := \{x \in \mathbb{R}^3 \setminus \overline{D} : |x| < R\}$ we conclude that

$$\langle\varphi, S\varphi\rangle = \left\langle \frac{\partial v_-}{\partial\nu} - \frac{\partial v_+}{\partial\nu}, v \right\rangle \tag{1.46}$$

$$= \iint_{D\cup D_R} \left(|\nabla v|^2 - k^2|v|^2\right) dx - \int_{|x|=R} \overline{v}\,\frac{\partial v}{\partial r}\,ds \tag{1.47}$$

$$= \iint_{D\cup D_R} \left(|\nabla v|^2 - k^2|v|^2\right) dx - ik \int_{|x|=R} |v|^2 ds + \mathcal{O}\left(\frac{1}{R}\right) \tag{1.48}$$

as R tends to infinity. Taking the imaginary part yields

$$\operatorname{Im}\langle\varphi, S\varphi\rangle = -k \lim_{R\to\infty} \int_{|x|=R} |v|^2 ds = -\frac{k}{(4\pi)^2} \int_{S^2} |v^\infty|^2 ds \leq 0. \tag{1.49}$$

Let now $\operatorname{Im}\langle\varphi, S\varphi\rangle = 0$ for some $\varphi \in H^{-1/2}(\Gamma)$. Then $v^\infty = 0$. From (1.49), Rellich's Lemma 1.2, and unique continuation we conclude that v vanishes outside of D. Therefore, $S\varphi = 0$ on Γ by the trace theorem. Since S is an isomorphism φ has to vanish.

(c) For $k = i$ the same arguments as above yield

$$\langle \varphi, S_i \varphi \rangle = \iint\limits_{D \cup D_R} \left(|\nabla v|^2 + |v|^2 \right) dx + \int\limits_{|x|=R} |v|^2 ds + \mathcal{O}\left(\frac{1}{R}\right), \quad R \to \infty,$$

and thus as $R \to \infty$ (note that v decays exponentially):

$$\langle \varphi, S_i \varphi \rangle = \iint\limits_{\mathbb{R}^3} \left(|\nabla v|^2 + |v|^2 \right) dx = \|v\|^2_{H^1(\mathbb{R}^3)}.$$

The trace theorem and the boundedness of S_i^{-1} yields the existence of $c > 0$ and $c_0 > 0$ with

$$\langle \varphi, S_i \varphi \rangle \geq c \, \|v\|^2_{H^{1/2}(\Gamma)} = c \, \|S_i \varphi\|^2_{H^{1/2}(\Gamma)} \geq c_0 \, \|\varphi\|^2_{H^{-1/2}(\Gamma)}.$$

(d) This follows from the fact that the kernel of $S - S_i$ is of the form

$$\frac{\exp(ik|x-y|) - \exp(-|x-y|)}{4\pi \, |x-y|} = |x-y| \, A(|x-y|^2) + B(|x-y|^2)$$

with analytic functions A and B. $\square$

Now we are able to derive the fundamental factorization of F.

Theorem 1.15 *The following relation holds between F, G and S:*

$$F = -G \, S^* \, G^* \tag{1.50}$$

where $G^* : L^2(S^2) \to H^{-1/2}(\Gamma)$ and $S^* : H^{-1/2}(\Gamma) \to H^{1/2}(\Gamma)$ are the adjoints of G and S, respectively, with respect to $L^2(S^2)$ and the dual pairing[6] $\langle H^{-1/2}(\Gamma), H^{1/2}(\Gamma) \rangle$.

Proof: As an auxiliary operator we define $H : L^2(S^2) \to H^{1/2}(\Gamma)$ by,

$$(Hg)(x) := \int\limits_{S^2} g(\theta) \, e^{ikx \cdot \theta} \, ds(\theta), \quad x \in \Gamma. \tag{1.51}$$

Hg is the trace on Γ of the Herglotz wave function (1.33) with density g. Its adjoint $H^* : H^{-1/2}(\Gamma) \to L^2(S^2)$ is given by

$$(H^*\varphi)(\hat{x}) = \int\limits_{\Gamma} \varphi(y) \, e^{-ik\hat{x} \cdot y} \, ds(y), \quad \hat{x} \in S^2. \tag{1.52}$$

We note that by the asymptotic behavior of the fundamental solution $H^*\varphi$ is just the far field pattern of the single layer potential (1.45). The single layer potential (1.45) with continuous density φ is continuous in $\mathbb{R}^3 \setminus D$ and thus $H^*\varphi = GS\varphi$, i.e., by a density argument

$$H^* = GS \quad \text{and therefore} \quad H = S^* G^*. \tag{1.53}$$

---

[6] We recall again that in our setting the dual form is sesqui-linear rather than bi-linear, see the remark preceding Lemma 1.14.

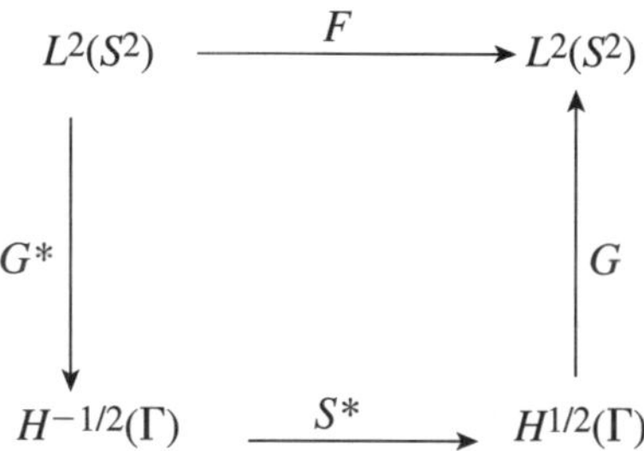

Figure 1.4 The factorization $F = -G\,S^*\,G^*$

Now we observe that Fg is the far field pattern of the solution of the exterior Dirichlet problem with boundary data

$$-\int_{S^2} g(\theta)\, e^{ikx\cdot\theta}\, ds(\theta) = -(Hg)(x)\,, \quad x \in \Gamma\,.$$

This shows that

$$Fg = -GHg\,. \tag{1.54}$$

Substituting H from (1.53) into (1.54) yields the assertion. $\square$

We sketched the factorization in Figure 1.4.

Remark: If k^2 is not a Dirichlet eigenvalue of $-\Delta$ in D we can solve (1.53) for G and arrive at the factorization

$$F = -H^* S^{-1} H \tag{1.55}$$

with the explicitly given operator H from (1.51).

1.4.2 The inf-criterion

Motivated by Theorem 1.12 we will now give a first expression of the range of G by the criterion which depends solely on F. Although not very helpful from the computational point of view it is nevertheless quite general and will lead to more elegant characterizations in the forthcoming subsections.

The method is based on the following result from functional analysis.

Theorem 1.16 *Let X, Y be (complex) reflexive Banach spaces with duals X^*, Y^*, respectively, and dual forms $\langle \cdot, \cdot \rangle$ in $\langle X^*, X \rangle$ and $\langle Y^*, Y \rangle$. Furthermore, let $F : Y^* \to Y$ and $B : X \to Y$ linear operators with*

$$F = BAB^* \quad \textit{for some linear and bounded operator} \quad A : X^* \to X \tag{1.56}$$

which satisfies a coercivity condition of the form: There exists $c > 0$ with

$$\big|\langle \varphi, A\varphi \rangle\big| \geq c\|\varphi\|_{X^*}^2 \quad \textit{for all } \varphi \in \mathcal{R}(B^*) \subset X^*\,. \tag{1.57}$$

Then, for any $\phi \in Y$, $\phi \neq 0$,

$$\phi \in \mathcal{R}(B) \quad \text{if and only if} \quad \inf\big\{|\langle \psi, F\psi \rangle| : \psi \in Y^*, \ \langle \psi, \phi \rangle = 1\big\} > 0. \qquad (1.58)$$

Here again, $\mathcal{R}(B)$ denotes the range of the operator $B : X \to Y$.

Furthermore, if $\phi = B\varphi_0 \in \mathcal{R}(B)$ for some $\varphi_0 \in X$ then

$$\inf\big\{|\langle \psi, F\psi \rangle| : \psi \in Y^*, \ \langle \psi, \phi \rangle = 1\big\} \geq \frac{c}{\|\varphi_0\|_X^2}. \qquad (1.59)$$

Proof: First, we observe that

$$\big|\langle \psi, F\psi \rangle\big| = \big|\langle B^*\psi, AB^*\psi \rangle\big| \geq c\|B^*\psi\|_{X^*}^2 \quad \text{for all } \psi \in Y^*. \qquad (1.60)$$

Let now $\phi = B\varphi_0$ for some $\varphi_0 \in X$. For $\psi \in Y^*$ with $\langle \psi, \phi \rangle = 1$ we have

$$\big|\langle \psi, F\psi \rangle\big| \geq c\|B^*\psi\|_{X^*}^2 = \frac{c}{\|\varphi_0\|_X^2}\,\|B^*\psi\|_{X^*}^2\,\|\varphi_0\|_X^2$$

$$\geq \frac{c}{\|\varphi_0\|_X^2}\,\big|\langle B^*\psi, \varphi_0 \rangle\big|^2 = \frac{c}{\|\varphi_0\|_X^2}\,\big|\langle \psi, \underbrace{B\varphi_0}_{=\phi} \rangle\big|^2 = \frac{c}{\|\varphi_0\|_X^2}.$$

This provides the lower bound of (1.59).

Second, assume that $\phi \notin \mathcal{R}(B)$. Define the closed subspace $V := \big\{\psi \in Y^* : \langle \psi, \phi \rangle = 0\big\}$. We show that $B^*(V)$ is dense in $\mathcal{R}(B^*) \subset X^*$. This is equivalent to the statement that the annihilators $\big[B^*(V)\big]^\perp$ and $\big[\mathcal{R}(B^*)\big]^\perp = \mathcal{N}(B)$ coincide. Therefore, let $\varphi \in \big[B^*(V)\big]^\perp$, i.e., $\langle B^*\psi, \varphi \rangle = 0$ for all $\psi \in V$, i.e., $\langle \psi, B\varphi \rangle = 0$ for all $\psi \in V$, i.e., $B\varphi \in V^\perp = \text{span}\,\{\phi\}$. Since $\phi \notin \mathcal{R}(B)$ this implies $B\varphi = 0$, i.e., $\varphi \in \mathcal{N}(B)$.

By a consequence of the Hahn–Banach Theorem one can find $\hat{\phi} \in Y^*$ with $\langle \hat{\phi}, \phi \rangle = 1$. Choose a sequence $\{\hat{\psi}_n\}$ in V such that

$$B^*\hat{\psi}_n \longrightarrow -B^*\hat{\phi} \quad \text{as } n \to \infty.$$

We set $\psi_n = \hat{\psi}_n + \hat{\phi}$. Then $\langle \psi_n, \phi \rangle = 1$ and $B^*\psi_n \to 0$. From the first equation of (1.60) we conclude that

$$\big|\langle \psi_n, F\psi_n \rangle\big| \leq \|A\|\,\|B^*\psi_n\|_{X^*}^2$$

and thus $\langle \psi_n, F\psi_n \rangle \longrightarrow 0$, $n \to \infty$, which proves that

$$\inf\big\{|\langle \psi, F\psi \rangle| : \psi \in Y^*, \ \langle \psi, \phi \rangle = 1\big\} = 0. \qquad \square$$

In our applications we will often prove the coercivity condition (1.57) with the help of the following lemma.

Lemma 1.17 *Let X be a reflexive Banach space and $A, A_0 : X^* \to X$ be linear and bounded operators such that*

(i) $\langle \varphi, A\varphi \rangle \in \mathbb{C} \setminus (-\infty, 0]$ for all $\varphi \in \text{closure } \mathcal{R}(B^*)$ with $\varphi \neq 0$,

(ii) $\langle \varphi, A_0\varphi \rangle$ is real-valued, and there exists $c_0 > 0$ with

$$\langle \varphi, A_0\varphi \rangle \geq c_0\|\varphi\|_{X^*}^2 \quad \text{for all } \varphi \in \mathcal{R}(B^*), \qquad (1.61)$$

(iii) $A - A_0$ is compact.

Then (1.57) holds, i.e., there exists $c > 0$ with

$$\left|\langle\varphi, A\varphi\rangle\right| \geq c\,\|\varphi\|_{X^*}^2 \quad \textit{for all } \varphi \in \mathcal{R}(B^*)\,. \tag{1.62}$$

Proof: If there exists no constant c with (1.62) then there exists a sequence $\{\varphi_n\}$ in $\mathcal{R}(B^*)$ with $\|\varphi_n\|_{X^*} = 1$ and $\langle\varphi_n, A\varphi_n\rangle \longrightarrow 0$ as n tends to infinity. Since the unit ball in X^* is weakly compact there exists a subsequence which converges weakly to some $\varphi \in$ closure $\mathcal{R}(B^*)$. We denote this subsequence again by $\{\varphi_n\}$. The compactness of $A - A_0$ yields that $(A - A_0)\varphi_n \to (A - A_0)\varphi$ in the norm of X. We conclude that $\langle\varphi_n, (A - A_0)(\varphi - \varphi_n)\rangle \longrightarrow 0$. By linearity,

$$\langle\varphi - \varphi_n, A_0(\varphi - \varphi_n)\rangle = \langle\varphi, A_0(\varphi - \varphi_n)\rangle - \langle\varphi_n, (A_0 - A)(\varphi - \varphi_n)\rangle$$
$$+ \langle\varphi_n, A\varphi_n\rangle - \langle\varphi_n, A\varphi\rangle\,.$$

The first three terms on the right hand side converge to zero, the forth term to $\langle\varphi, A\varphi\rangle$. Assumption (i) implies that φ vanishes. Therefore,

$$c_0\|\varphi_n\|_{X^*}^2 \leq \langle\varphi_n, A_0\varphi_n\rangle \leq \left|\langle\varphi_n, (A_0 - A)\varphi_n\rangle\right| + \left|\langle\varphi_n, A\varphi_n\rangle\right|$$

which tends to zero as $n \to \infty$. Therefore, also $\varphi_n \to 0$ which contradicts the assumption $\|\varphi_n\|_{X^*} = 1$. $\square$

We wish to apply Theorem 1.16 to the factorization (1.50) of the far field operator F. We choose $Y = L^2(S^2)$ and $X = H^{1/2}(\Gamma)$, identify Y^* with $L^2(S^2)$, and set $B = G$. We have to show the coercivity condition (1.62) for $A = -S^*$. This property follows immediately from the previous lemma in combination with Lemma 1.14 where we proved the properties of the operator S. We formulate the result as a corollary.

Corollary 1.18 *Assume that k^2 is not a Dirichlet eigenvalue of $-\Delta$ in D. Then there exists $c > 0$ with*

$$\left|\langle\varphi, S\varphi\rangle\right| \geq c\,\|\varphi\|_{H^{-1/2}(\Gamma)}^2 \quad \textit{for all } \varphi \in H^{-1/2}(\Gamma)\,. \tag{1.63}$$

Here again, $\langle\cdot,\cdot\rangle$ denotes the duality pairing in $\langle H^{-1/2}(\Gamma), H^{1/2}(\Gamma)\rangle$.

Application of Theorem 1.16 yields the following result.

Theorem 1.19 *Assume that k^2 is not a Dirichlet eigenvalue of $-\Delta$ in D. Then for any $\phi \in L^2(S^2)$ with $\phi \neq 0$ the following holds:*

$$\phi \in \mathcal{R}(G) \text{ if, and only if, } \inf\left\{\left|(\psi, F\psi)_{L^2(S^2)}\right| : \psi \in L^2(S^2),\, (\psi, \phi)_{L^2(S^2)} = 1\right\} > 0\,, \tag{1.64}$$

where again $G : H^{1/2}(\Gamma) \to L^2(S^2)$ is defined by $Gf = v^\infty$ and v solves the boundary value problem (1.38), (1.39), and (1.40).

Furthermore, if $\phi = Gf$ for some $f \in H^{1/2}(\Gamma)$ then

$$\inf\left\{\left|(\psi, F\psi)_{L^2(S^2)}\right| : \psi \in L^2(S^2),\ (\psi, \phi)_{L^2(S^2)} = 1\right\} \geq \frac{c}{\|f\|^2_{H^{1/2}(\Gamma)}} \tag{1.65}$$

for some $c > 0$ independent of ϕ.

Combining this with the characterization of Theorem 1.12 of D by the range of G yields immediately the main result of this subsection.

Theorem 1.20 *Assume that k^2 is not a Dirichlet eigenvalue of $-\Delta$ in D. For any $z \in \mathbb{R}^3$ define again $\phi_z \in L^2(S^2)$ by (1.41), i.e.,*

$$\phi_z(\hat{x}) := e^{-ik\hat{x}\cdot z}, \quad \hat{x} \in S^2.$$

Then $z \in D$ if, and only if,

$$\inf\left\{\left|(\psi, F\psi)_{L^2(S^2)}\right| : \psi \in L^2(S^2),\ (\psi, \phi_z)_{L^2(S^2)} = 1\right\} > 0. \tag{1.66}$$

Therefore, the characteristic function of D is given by

$$\chi_D(z) = \operatorname{sign} \inf\left\{\left|(\psi, F\psi)_{L^2(S^2)}\right| : \psi \in L^2(S^2),\ (\psi, \phi_z)_{L^2(S^2)} = 1\right\}, \quad z \in \mathbb{R}^3.$$

Furthermore, for $z \in D$ we have the estimate:

$$\inf\left\{\left|(\psi, F\psi)_{L^2(S^2)}\right| : \psi \in L^2(S^2),\ (\psi, \phi_z)_{L^2(S^2)} = 1\right\} \geq \frac{c}{\|\Phi(\cdot, z)\|^2_{H^{1/2}(\Gamma)}} \tag{1.67}$$

for some constant $c > 0$ which is independent of z.

Proof: It remains to prove the estimate (1.67). It follows directly from (1.65) and the observation that $\phi_z = G\Phi(\cdot, z)|_\Gamma$ for $z \in D$ (see proof of Theorem 1.12). $\square$

We note again that the evaluation of the form of $\chi_D(z)$ uses only known information on the far field operator F. Although satisfactory from the theoretical point of view there is a major drawback with respect to the computationally point of view since it is very time consuming to solve a minimization problem for every sampling point z.

We conclude this subsection with the remark that Theorem 1.20 provides an alternative – and very explicit – proof of uniqueness of the inverse scattering problem under the assumption that k^2 is not a Dirichlet eigenvalue of $-\Delta$ in D.

1.4.3   The $(F^*F)^{1/4}$-method

As an important observation from Theorem 1.16 we note that the inf-criterion in the characterization (1.58) depends only on F and not on the factorization itself. This observation leads directly to the first part of the following result (see [71]).

Theorem 1.21 *Let H be a Hilbert space and let $F : H \to H$ have two factorizations of the form*

$$F = B_1 A_1 B_1^* = B_2 A_2 B_2^* \tag{1.68}$$

with linear operators $B_j : X_j \to H$, $j = 1, 2$, from reflexive Banach spaces X_j into H and linear operators $A_j : X_j^* \to X_j$, $j = 1, 2$, which both satisfy the coercivity condition (1.57), i.e.,

$$\left| \langle \varphi, A_j \varphi \rangle \right| \geq c \|\varphi\|_{X_j^*}^2 \quad \text{for all } \varphi \in \mathcal{R}(B_j^*) \text{ and } j = 1, 2. \tag{1.69}$$

Then the ranges of B_1 and B_2 coincide. If in addition B_1 and B_2 are one-to-one then $B_2^{-1} B_1$ and $B_1^{-1} B_2$ are (topological) isomorphisms from X_1 onto X_2 and from X_2 onto X_1, respectively.

Proof: It remains to prove the second part. From $\left(B_2^{-1} B_1\right)\left(B_1^{-1} B_2\right) = I_{X_2}$ and $\left(B_1^{-1} B_2\right)\left(B_2^{-1} B_1\right) = I_{X_1}$ we observe that $B_2^{-1} B_1$ and $B_1^{-1} B_2$ are algebraical isomorphisms. It remains to show that they are bounded. Let $\varphi_1 \in X_1$ and set $\varphi_2 = B_2^{-1} B_1 \varphi_1$. Then $B_2 \varphi_2 = B_1 \varphi_1$ and thus $B_2 \varphi_2 \in \mathcal{R}(B_1)$. In particular, from Theorem 1.16 we conclude that

$$\frac{c}{\|\varphi_1\|_{X_1}^2} \leq \inf\left\{ \left| (\psi, F\psi)_H \right| : \psi \in H, \ (\psi, B_2 \varphi_2)_H = 1 \right\}$$

$$= \inf\left\{ \left| \langle B_2^* \psi, A_2 B_2^* \psi \rangle \right| : \psi \in H, \ \langle B_2^* \psi, \varphi_2 \rangle = 1 \right\}$$

$$= \inf\left\{ \left| \langle \phi, A_2 \phi \rangle \right| : \phi \in X_2^*, \ \langle \phi, \varphi_2 \rangle = 1 \right\}$$

since the range of B_2^* is dense in X_2^*. By a well-known result from functional analysis (application of the theorem of Hahn–Banach) there exists $\phi_2 \in X_2^*$ with $\|\phi_2\|_{X_2^*} = 1$ and $\langle \phi_2, \varphi_2 \rangle = \|\varphi_2\|_{X_2}$. Choosing $\phi = \phi_2 / \|\varphi_2\|_{X_2}$ in the last estimate yields (since $\langle \phi, \varphi_2 \rangle = 1$)

$$\frac{c}{\|\varphi_1\|_{X_1}^2} \leq \frac{1}{\|\varphi_2\|_{X_2}^2} \left| \langle \phi_2, A_2 \phi_2 \rangle \right| \leq \frac{\|A_2\|}{\|\varphi_2\|_{X_2}^2} \|\phi_2\|_{X_2^*}^2 = \frac{\|A_2\|}{\|\varphi_2\|_{X_2}^2}.$$

This yields

$$\|\varphi_2\|_{X_2} \leq \sqrt{\frac{\|A_2\|}{c}} \, \|\varphi_1\|_{X_1}$$

which proves continuity of $B_2^{-1} B_1$. The proof for the continuity of $B_1^{-1} B_2$ follows by interchanging the roles of B_1 and B_2. $\square$

As a first application of this theorem we formulate a result for self-adjoint and non-negative operators F which will be applied for problems in impedance tomography (see Chapter 6).

Corollary 1.22 *Let $F : H \to H$ be a compact and self-adjoint operator from the Hilbert space H into itself which has a factorization of the form*

$$F = B A B^*$$

with some operator $B : X \to H$ (where again X is a reflexive Banach space) and some self-adjoint operator $A : X^* \to X$ which is coercive on $\mathcal{R}(B^*)$, i.e., there exists $c > 0$ with $\langle \varphi, A\varphi \rangle \geq c \|\varphi\|_{X^*}^2$ for all $\varphi \in \mathcal{R}(B^*)$. Then the ranges of B and $F^{1/2}$ coincide.

Remark: The operator $F^{1/2}$ can be defined by using an eigensystem of F. Indeed, if $\lambda_j \geq 0$ are the eigenvalues of the non-negative (!) operator F with corresponding normalized eigenfunctions $\psi_j \in H$ then F has the form

$$F\psi = \sum_j \lambda_j\,(\psi,\psi_j)_H\,\psi_j\,, \quad \psi \in H\,, \tag{1.70}$$

and thus

$$F^{1/2}\psi = \sum_j \sqrt{\lambda_j}\,(\psi,\psi_j)_H\,\psi_j\,, \quad \psi \in H\,.$$

Proof of the corollary: The operator F admits a second factorization in the form $F = F^{1/2}F^{1/2}$. The assertion follows directly from Theorem 1.21 because the operators A and the identity satisfy both the coercivity condition (1.69). $\square$

The far field operator $F : L^2(S^2) \to L^2(S^2)$ for the Dirichlet boundary condition fails to be self-adjoint and this corollary is not applicable. However, it is normal and – even more – the operator $I + \frac{ik}{8\pi^2}F$ is unitary. For this situation there exists a corresponding result which we first formulate and prove in the general setting before we apply it to the factorization of the far field operator.

Theorem 1.23 *Let H be a Hilbert space, X a reflexive Banach space and let the compact operator $F : H \to H$ have a factorization of the form*

$$F = BAB^*$$

with operators $B : X \to H$ and $A : X^* \to X$ such that $\mathrm{Im}\langle\varphi, A\varphi\rangle \neq 0$ for all $\varphi \in \text{closure } \mathcal{R}(B^*)$ with $\varphi \neq 0$. Let furthermore A be of the form $A = A_0 + C$ for some compact operator C and some self-adjoint operator A_0 which is coercive on $\mathcal{R}(B^*)$ in the sense of (1.61). Finally, assume that F is one-to-one and $I + irF$ is unitary for some $r > 0$. Then the ranges of B and $(F^*F)^{1/4}$ coincide. Furthermore, the operators $(F^*F)^{-1/4}B$ and $B^{-1}(F^*F)^{1/4}$ are isomorphisms from X onto H and from H onto X, respectively.

Proof: First we note that by Lemma 1.17 the operator A satisfies the coercivity condition (1.69), i.e., there exists $c > 0$ with

$$\left|\langle\varphi, A\varphi\rangle\right| \geq c\|\varphi\|_{X^*}^2 \quad \text{for all } \varphi \in \mathcal{R}(B^*)\,. \tag{1.71}$$

The unitarity of $I + irF$ implies that F is normal. Therefore, there exists a complete set of orthonormal eigenfunctions $\psi_j \in H$ with corresponding eigenvalues $\lambda_j \in \mathbb{C}$, $j = 1, 2, 3, \ldots$ (see, e.g., [168]). Furthermore, since the operator $I + irF$ is unitary the eigenvalues λ_j of F lie on the circle of radius $1/r$ and center i/r. The spectral theorem for normal operators yields that F has the form (1.70), i.e.,

$$F\psi = \sum_{j=1}^{\infty} \lambda_j(\psi,\psi_j)_H\,\psi_j\,, \quad \psi \in H\,. \tag{1.72}$$

From this we conclude that F has a second factorization in the form

$$F = (F^*F)^{1/4} A_2 (F^*F)^{1/4} \, , \tag{1.73}$$

where the operator $(F^*F)^{1/4} : H \to H$ is given by

$$(F^*F)^{1/4} \psi = \sum_{j=1}^{\infty} \sqrt{|\lambda_j|} \, (\psi, \psi_j)_H \, \psi_j \, , \quad \psi \in H \, , \tag{1.74}$$

and the signum $A_2 : H \to H$ of F is given by

$$A_2 \psi = \sum_{j=1}^{\infty} \frac{\lambda_j}{|\lambda_j|} \, (\psi, \psi_j)_H \, \psi_j \, , \quad \psi \in H \, . \tag{1.75}$$

In order to apply Theorem 1.21 with $X_2 = H$ and $B_2 = (F^*F)^{1/4}$ we have to show that also A_2 satisfies the coercivity condition (1.69) on H.

We set $s_j = \lambda_j / |\lambda_j|$ for abbreviation. From the facts that $\left| \lambda_j - \frac{i}{r} \right| = \frac{1}{r}$ and that λ_j tends to zero as j tends to infinity we conclude that the only accumulation points of the sequence $\{s_j\}$ can be $+1$ or -1. This situation is illustrated in Figure 1.5. The main part of the proof consists of showing that the only accumulation point is $+1$. Before we show this we define the functions $\varphi_j \in X^*$ by

$$\varphi_j = \frac{1}{\sqrt{\lambda_j}} B^* \psi_j \, , \quad j \in \mathbb{N} \, ,$$

where the branch of the square root is chosen such that $\mathrm{Im} \sqrt{\lambda_j} > 0$. The following argument proves a kind of orthogonality relation of φ_j:

$$\langle \varphi_j, A \varphi_\ell \rangle = \frac{1}{\sqrt{\lambda_j} \, \overline{\sqrt{\lambda_\ell}}} \, \langle B^* \psi_j \, , A B^* \psi_\ell \rangle$$

$$= \frac{1}{\sqrt{\lambda_j} \, \overline{\sqrt{\lambda_\ell}}} \, (\psi_j \, , B A B^* \psi_\ell)_H = \frac{\overline{\lambda_\ell}}{\sqrt{\lambda_j} \, \overline{\sqrt{\lambda_\ell}}} \, (\psi_j, \psi_\ell)_H \, ,$$

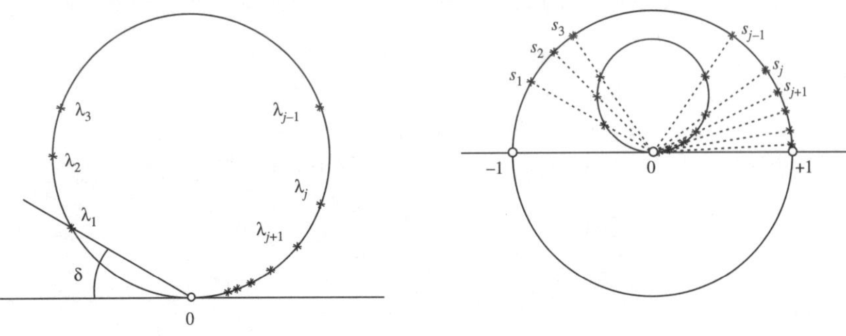

Figure 1.5 Eigenvalues $\{\lambda_j\}$ of F (left) and $s_j = \lambda_j / |\lambda_j|$ (right)

i.e.,

$$\langle \varphi_j, A\varphi_\ell \rangle = \overline{s_j}\, \delta_{j\ell} \quad \text{for } j, \ell \in \mathbb{N}. \tag{1.76}$$

From this condition and (1.71) we conclude that the sequence $\{\varphi_j\}$ is bounded:

$$c\, \|\varphi_j\|_{X^*}^2 \le \left| \langle \varphi_j, A\varphi_j \rangle \right| = |s_j| = 1 \quad \text{for all } j.$$

Now we assume that -1 is an accumulation point of $\{s_j\}$. Then there exists a subsequence of $\{s_j\}$ which converges to -1. We denote this by writing $s_j \to -1$ again. Since the sequence $\{\varphi_j\}$ is bounded there exists a further subsequence such that φ_j converges weakly in X^* to some $\varphi \in$ closure $\mathcal{R}(B^*)$. From (1.76) we have that

$$\langle \varphi_j, A\varphi_j \rangle = \langle \varphi_j, A_0\varphi_j \rangle + \langle \varphi_j, C\varphi_j \rangle \longrightarrow -1, \quad j \to \infty. \tag{1.77}$$

Since C is compact from X into X^* we conclude that $C\varphi_j$ converges to $C\varphi$ and thus

$$\langle \varphi_j, C\varphi_j \rangle = \langle \varphi_j, C\varphi \rangle + \langle \varphi_j, C(\varphi_j - \varphi) \rangle.$$

The second term on the right hand side converges to zero by the Cauchy-Schwarz inequality and the first term to $\langle \varphi, C\varphi \rangle$ by the weak convergence of φ_j to φ, i.e.,

$$\langle \varphi_j, C\varphi_j \rangle \longrightarrow \langle \varphi, C\varphi \rangle, \quad j \to \infty.$$

Comparing the imaginary parts of this and of (1.77) implies that $\operatorname{Im}\langle \varphi, A\varphi \rangle$ vanishes. From our assumption we conclude that φ has to vanish. Then (1.77) yields that

$$\langle \varphi_j, A_0\varphi_j \rangle \longrightarrow -1$$

which is impossible since the left-hand side is bounded below by zero. This proves that the sequence $\{s_j\}$ converges to $+1$. Now we proceed with the proof of the estimate (1.69) for A_2.

Let $\psi = \sum_{j=1}^{\infty} c_j \psi_j$ with $\|\psi\|_H^2 = \sum_{j=1}^{\infty} |c_j|^2 = 1$. We compute

$$\left| (A_2\psi, \psi)_H \right| = \left| \left(\sum_{j=1}^{\infty} s_j c_j \psi_j, \sum_{j=1}^{\infty} c_j \psi_j \right) \right| = \left| \sum_{j=1}^{\infty} s_j |c_j|^2 \right|.$$

The complex number $\sum_{j=1}^{\infty} s_j |c_j|^2$ belongs to the convex hull $M = \operatorname{conv}\{s_j : j \in \mathbb{N}\} \subset \mathbb{C}$ of all numbers s_j. We conclude that

$$\left| (A_2\psi, \psi)_H \right| \ge \inf\{|z| : z \in M\}.$$

The set M is contained in the part of the upper half-disk which is above the line $\ell = \{ts_1 + (1-t)1 : t \in \mathbb{R}\}$ passing through s_1 and 1.

The distance of the origin to this convex hull M is given by

$$\inf\{|z| : z \in M\} = \inf\{|z| : z \in \ell\} = \sin\frac{\delta}{2}, \tag{1.78}$$

where $\pi - \delta \in (0, \pi)$ is the argument of s_1, i.e., $s_1 = -\cos\delta + i\sin\delta$ (see Figure 1.6). Therefore, we arrive at the estimate

$$\left| (A_2\psi, \psi)_H \right| \ge \sin\frac{\delta}{2} \|\psi\|_H^2.$$

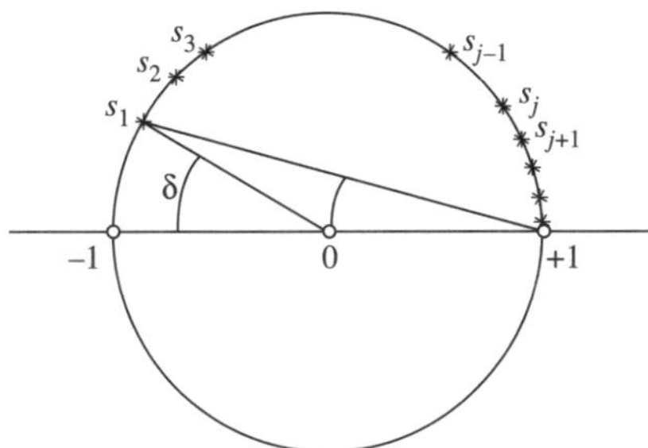

Figure 1.6 The distance from $\mathrm{conv}\{s_j\}$ to the origin is positive

Therefore, all the assumptions of Theorem 1.21 are satisfied. It's application yields the assertion. $\square$

We apply this abstract result to the factorization (1.50) with $H = L^2(S^2)$, $X = H^{1/2}(\Gamma)$, $B = G$, and $A = -S^*$. The assumptions on $A = -S^*$ are satisfied by Lemma 1.14. Furthermore, we note that $B = G$ and $(F^*F)^{1/4}$ are one-to-one by Lemma 1.13 and Theorem 1.8, respectively, if k^2 is not a Dirichlet eigenvalue. Therefore we have:

Theorem 1.24 *Assume that k^2 is not a Dirichlet eigenvalue of $-\Delta$ in D. Then the ranges of G and $(F^*F)^{1/4}$ coincide. Furthermore, the operators $(F^*F)^{-1/4}G$ and $G^{-1}(F^*F)^{1/4}$ are isomorphisms from $H^{1/2}(\Gamma)$ onto $L^2(S^2)$ and from $L^2(S^2)$ onto $H^{1/2}(\Gamma)$, respectively.*

We note that the range of G is expressed by a characterization which depends solely on the data operator F – just as in the case of Theorem 1.19. The combination of this result with Theorem 1.12 yields the main result of this subsection.

Theorem 1.25 *Assume that k^2 is not a Dirichlet eigenvalue of $-\Delta$ in D. For any $z \in \mathbb{R}^3$ define again $\phi_z \in L^2(S^2)$ by (1.41), i.e.,*

$$\phi_z(\hat{x}) := e^{-ik\hat{x}\cdot z}, \quad \hat{x} \in S^2.$$

Then

$$z \in D \Longleftrightarrow \phi_z \in \mathcal{R}\big((F^*F)^{1/4}\big) \tag{1.79}$$

$$\Longleftrightarrow W(z) := \left[\sum_j \frac{\left|(\phi_z, \psi_j)_{L^2(S^2)}\right|^2}{|\lambda_j|}\right]^{-1} > 0. \tag{1.80}$$

Here, $\lambda_j \in \mathbb{C}$ are the eigenvalues of the normal operator F with corresponding normalized eigenfunctions $\psi_j \in L^2(S^2)$.

Therefore, $\chi_D(z) = \mathrm{sign}\, W(z)$ is the characteristic function of D.

Proof: It remains to prove the characterization (1.80). We note from the characterization (1.79) that a point $z \in \mathbb{R}^3$ belongs to D if, and only if, the equation

$$(F^*F)^{1/4}g = \phi_z \tag{1.81}$$

is solvable in $L^2(S^2)$. We write ϕ_z in spectral form as

$$\phi_z = \sum_j (\phi_z, \psi_j)_{L^2(S^2)} \, \psi_j.$$

From (1.74) we observe that (1.81) is solvable if, and only if, the series

$$\sum_j \frac{\left|(\phi_z, \psi_j)_{L^2(S^2)}\right|^2}{|\lambda_j|}$$

converges,[7] and in this case

$$g = \sum_j \frac{(\phi_z, \psi_j)_{L^2(S^2)}}{\sqrt{|\lambda_j|}} \, \psi_j$$

is the solution of (1.81). Therefore, a point $z \in \mathbb{R}^3$ belongs to D if, and only if, the series

$$\sum_j \frac{\left|(\phi_z, \psi_j)_{L^2(S^2)}\right|^2}{|\lambda_j|}$$

converges which proves the characterization (1.80). $\square$

The essential assumption under which the characterization (1.80) had been derived was the normality of the far field operator F and the unitarity of the scattering operator S. In many cases the operator F fails to be normal. Examples of theses cases are "absorbing" media D or limited angle data $u^\infty(\hat{x}, \theta)$. If the far field operator fails to be normal not very much is known about eigenvalues. We refer to [43] for some results. In particular, a complete set of eigenfunctions usually does not exist. Therefore, the technique of this subsection does not work. As we will see later, the minimization approach is still applicable but, as we mentioned already, it is very time consuming from the computational point of view.

Before we modify the approach of this subsection in the next chapter appropriately we observe that the convergence of the series in (1.80) depends only on the rates of decay of the eigenvalues $|\lambda_j|$ and the expansion coefficients. From the obvious estimate

$$\frac{1}{\sqrt{2}}\big[|\operatorname{Re}\lambda_j| + |\operatorname{Im}\lambda_j|\big] \leq |\lambda_j| \leq |\operatorname{Re}\lambda_j| + |\operatorname{Im}\lambda_j|$$

and the observation that $\operatorname{Im}\lambda_j > 0$ we note that we can replace $|\lambda_j|$ in (1.80) by $|\operatorname{Re}\lambda_j| + \operatorname{Im}\lambda_j$. Furthermore, we observe that $|\operatorname{Re}\lambda_j| + \operatorname{Im}\lambda_j$ are the eigenvalues of the self-adjoint and positive operator

$$F_\# = |\operatorname{Re}F| + \operatorname{Im}F$$

where the self-adjoint parts $\operatorname{Re}F$ and $\operatorname{Im}F$ are defined by

$$\operatorname{Re}F = \frac{1}{2}\big[F + F^*\big] \quad \text{and} \quad \operatorname{Im}F = \frac{1}{2i}\big[F - F^*\big].$$

---

[7] This is just Picard's criterion, see [113].

Therefore, we can replace the operator $(F^*F)^{1/4}$ in (1.81) by $F_\#^{1/2}$. We will see in the next chapters that the characterization (1.81) by $F_\#^{1/2}$ has a much wider applicability.

We continue with the example of D being the unit ball in $\mathbb{R}^3$.

1.5 An explicit example

Let D be the unit ball in $\mathbb{R}^3$ centered at the origin. We will compute the quantities which appear in the series (1.80).

First, we expand the incident and scattered fields into spherical wave functions. Let j_n and h_n be the, respectively, spherical Bessel functions and spherical Hankel functions of the first kind and order $n \in \mathbb{N}$ and let $Y_n^m(\hat{x})$ be the spherical harmonics of order n normalized such that they form a complete orthonormal system in $L^2(S^2)$ (see (1.42)). The Jacobi–Anger expansion (cf. [43]) has the form

$$u^i(x) = e^{ikx\cdot\theta} = 4\pi \sum_{n=0}^{\infty} \sum_{|m|\le n} i^n j_n(k|x|)\, Y_n^m(\hat{x})\, \overline{Y_n^m(\theta)}, \quad x \in \mathbb{R}^3. \tag{1.82}$$

Again, the unit vector $\theta \in S^2$ denotes the direction of incidence and $\hat{x} = x/|x|$. It is immediately seen (at least formally) that the scattered field is given by

$$u^s(x) = -4\pi \sum_{n=0}^{\infty} \sum_{|m|\le n} i^n \frac{j_n(k)}{h_n(k)}\, h_n(k|x|)\, Y_n^m(\hat{x})\, \overline{Y_n^m(\theta)}, \quad |x| \ge 1. \tag{1.83}$$

The far field pattern of $h_n(k|x|)\, Y_n^m(\hat{x})$ is again given by $(4\pi/k)\exp\big(-i(n+1)\pi/2\big) Y_n^m(\hat{x})$ (compare proof of Lemma 1.13). As shown rigorously in [43] the far field pattern of u^s is derived by the term-by-term asymptotics of $h_n(k|x|)\, Y_n^m(\hat{x})$, i.e.,

$$u^\infty(\hat{x},\theta) = -\frac{(4\pi)^2}{k} \sum_{n=0}^{\infty} \sum_{|m|\le n} i^n \frac{j_n(k)}{h_n(k)}\, e^{-i(n+1)\pi/2}\, Y_n^m(\hat{x})\, \overline{Y_n^m(\theta)}$$

$$= \frac{(4\pi)^2 i}{k} \sum_{n=0}^{\infty} \sum_{|m|\le n} \frac{j_n(k)}{h_n(k)}\, Y_n^m(\hat{x})\, \overline{Y_n^m(\theta)}, \quad \hat{x},\theta \in S^2. \tag{1.84}$$

From this we observe that the far field operator $F : L^2(S^2) \to L^2(S^2)$ from (1.36) is given by

$$(Fg)(\hat{x}) = \int_{S^2} u^\infty(\hat{x},\theta)\, g(\theta)\, ds(\theta) = \frac{(4\pi)^2 i}{k} \sum_{n=0}^{\infty} \sum_{|m|\le n} \frac{j_n(k)}{h_n(k)}\, g_n^m\, Y_n^m(\hat{x}), \quad \hat{x} \in S^2,$$

$$\tag{1.85}$$

where

$$g_n^m = \int_{S^2} g(\theta)\, \overline{Y_n^m(\theta)}\, ds(\theta), \quad |m| \le n, \ n \in \mathbb{N},$$

are the expansion coefficients of $g \in L^2(S^2)$. From (1.85) we observe that

$$\lambda_n = \frac{(4\pi)^2 \, i}{k} \frac{j_n(k)}{h_n(k)}, \quad n \in \mathbb{N}, \tag{1.86}$$

are the eigenvalues of F of multiplicity $2n + 1$. The asymptotic behavior

$$j_n(k) = \frac{k^n}{1 \cdot 3 \cdots (2n+1)} \left(1 + \mathcal{O}\left(\frac{1}{n}\right)\right), \tag{1.87}$$

$$h_n(k) = \frac{1 \cdot 3 \cdots (2n-1)}{i \, k^{n+1}} \left(1 + \mathcal{O}\left(\frac{1}{n}\right)\right)$$

yields that

$$\lambda_n = -\frac{(4\pi)^2 \, k^{2n}}{(2n-1)!! \, (2n+1)!!} \left(1 + \mathcal{O}\left(\frac{1}{n}\right)\right), \tag{1.88}$$

where we used the convenient notation $p!! = 1 \cdot 3 \cdot 5 \cdots p$ for any odd number p. Next, we compute the expansion coefficients of the functions (1.41) by the Jacobi–Anger expansion (1.82), i.e.,

$$\phi_z(\hat{x}) = e^{-ikz \cdot \hat{x}} = 4\pi \sum_{n=0}^{\infty} \sum_{|m| \le n} (-i)^n j_n(k|z|) \, Y_n^m(\hat{x}) \, \overline{Y_n^m(\hat{z})}, \quad \hat{x} \in S^2,$$

where $\hat{z} = z/|z|$. From this we conclude that

$$(\phi_z, Y_n^m)_{L^2(S^2)} = 4\pi \, (-i)^n j_n(k|z|) \, \overline{Y_n^m(\hat{z})}.$$

Using the formula

$$\sum_{|m| \le n} |Y_n^m(\hat{z})|^2 = \frac{2n+1}{4\pi}$$

which is a special form of the addition theorem (see [43]) and the asymptotic form (1.87) of $j_n(k|z|)$ we conclude that

$$\sum_{|m| \le n} |(\phi_z, Y_n^m)_{L^2(S^2)}|^2 = 4\pi(2n+1) \frac{(k|z|)^{2n}}{[(2n+1)!!]^2} \left(1 + \mathcal{O}(1/n)\right).$$

Combining this with (1.88) yields

$$\sum_{|m| \le n} \frac{|(\phi_z, Y_n^m)_{L^2(S^2)}|^2}{|\lambda_n|} = \frac{2n+1}{4\pi \, k^{2n}} \frac{(2n-1)!!}{(2n+1)!!} (k|z|)^{2n} \left(1 + \mathcal{O}\left(\frac{1}{n}\right)\right)$$

$$= \frac{|z|^{2n}}{4\pi} \left(1 + \mathcal{O}\left(\frac{1}{n}\right)\right).$$

Here we observe directly that the series

$$\sum_{n=0}^{\infty} \sum_{|m|\leq n} \frac{\left|(\phi_z, Y_n^m)_{L^2(S^2)}\right|^2}{|\lambda_n|}$$

converges if, and only if, $|z| < 1$, i.e., z is inside D.

We finally remark that if $|z| < 1$ then the series behaves as $\sum_{n=0}^{\infty} \frac{|z|^{2n}}{4\pi} = \frac{1}{4\pi\,(1-|z|^2)}$, i.e., $W(z)$ behaves as $4\pi\,(1 - |z|^2)$ as z approaches the boundary of D.

1.6   The Neumann boundary condition

In this section we discuss briefly the obstacle scattering with respect to Neumann boundary conditions. In the direct scattering problem the incident plane wave $u^i(x) = \exp(ikx \cdot \theta)$ and the obstacle $D \subset \mathbb{R}^3$ are again given and the total wave $u \in C^2(\mathbb{R}^3 \setminus \overline{D}) \cap C^1(\mathbb{R}^3 \setminus D)$ has to be determined with

$$\Delta u + k^2 u = 0 \quad \text{outside } D \tag{1.89}$$

and the Neumann boundary condition

$$\frac{\partial u}{\partial v} = 0 \quad \text{on } \Gamma. \tag{1.90}$$

Furthermore, the scattered field $u^s = u - u^i$ satisfies the Sommerfeld radiation condition

$$\frac{\partial u^s}{\partial r} - ik\,u^s = \mathcal{O}\!\left(r^{-2}\right) \quad \text{for } r = |x| \to \infty \tag{1.91}$$

uniformly with respect to $\hat{x} = x/|x|$.

Again, the scattered field u^s satisfies the following exterior boundary value problem for $f = -\partial u^i/\partial v$:

Given $f \in H^{-1/2}(\Gamma)$ find $v \in H^1_{loc}(\mathbb{R}^3 \setminus \overline{D})$ such that

$$\Delta v + k^2 v = 0 \quad \text{outside } D\,, \tag{1.92}$$

$$\frac{\partial v}{\partial v} = f \quad \text{on } \Gamma\,, \tag{1.93}$$

and

$$\frac{\partial v}{\partial r} - ik\,v = \mathcal{O}\!\left(r^{-2}\right) \quad \text{for } r = |x| \to \infty \tag{1.94}$$

uniformly with respect to $\hat{x} = x/|x|$.

The data-to-pattern operator $G : H^{-1/2}(\Gamma) \to L^2(S^2)$ is now defined to map $f \in H^{-1/2}(\Gamma)$ into the far field pattern $v^\infty = Gf$ of the exterior Neumann boundary value problem (1.92), (1.93), and (1.94).

The solution is again understood in the variational sense, i.e., $v \in H^1_{loc}(\mathbb{R}^3 \setminus \overline{D})$ is a variational solution of (1.92) and (1.93) if it satisfies

$$\iint\limits_{\mathbb{R}^3 \setminus \overline{D}} \left[\nabla u \cdot \nabla \phi - k^2 u \phi \right] dx = \langle f, \phi \rangle$$

for all $\phi \in H^1(\mathbb{R}^3 \setminus \overline{D})$ with compact support. Here, $\langle \cdot, \cdot \rangle$ denotes again the dual form in $\langle H^{-1/2}(\Gamma), H^{1/2}(\Gamma) \rangle$. Existence and uniqueness of this exterior Neumann boundary value problem is well established, see [183], Lecture 4, [150], Section 2.6, or Chapter 1, Theorem 2.2.

The far field patterns $u^\infty = u^\infty(\hat{x}, \theta)$ of the scattered fields u^s define again the far field operator $F : L^2(S^2) \to L^2(S^2)$ by (compare (1.36))

$$(Fg)(\hat{x}) = \int\limits_{S^2} u^\infty(\hat{x}, \theta) \, g(\theta) \, ds(\theta) \quad \text{for } \hat{x} \in S^2. \tag{1.95}$$

By literally the same proofs as in Theorems 1.6 and 1.8 one shows reciprocity (1.27) of the far field patterns, normality of F and unitarity of the scattering operator $\mathcal{S} = I + \frac{ik}{8\pi^2} F$. Furthermore, F is one-to-one if k^2 is not a Neumann eigenvalue of $-\Delta$ in D.[8] Also, the uniqueness result of Theorem 1.9 holds.

The analogous results of Lemmas 1.13 and 1.14 and Theorem 1.15 are formulated in the following theorem.

Theorem 1.26 *(a) The far field operator $F : L^2(S^2) \to L^2(S^2)$ from (1.95) has a factorization in the form*

$$F = -G N^* G^* \tag{1.96}$$

where $G : H^{-1/2}(\Gamma) \to L^2(S^2)$ maps $f \in H^{-1/2}(\Gamma)$ into the far field pattern $v^\infty = Gf$ of the exterior Neumann boundary value problem (1.92), (1.93), and (1.94), and $N : H^{1/2}(\Gamma) \to H^{-1/2}(\Gamma)$ is the normal derivative of the double layer potential, defined by

$$(N\varphi)(x) = \frac{\partial}{\partial \nu} \int\limits_{\Gamma} \varphi(y) \frac{\partial \Phi}{\partial \nu(y)}(x, y) \, ds(y), \quad x \in \Gamma, \tag{1.97}$$

for $\varphi \in H^{1/2}(\Gamma)$.

(b) G is compact, one-to-one with dense range in $L^2(S^2)$.

(c) N is an isomorphism from $H^{1/2}(\Gamma)$ onto $H^{-1/2}(\Gamma)$ if k^2 is not a Neumann eigenvalue of $-\Delta$ in D.

(d) $\mathrm{Im}\langle N\varphi, \varphi \rangle > 0$ for all $\varphi \in H^{1/2}(\Gamma)$ with $\varphi \neq 0$ if k^2 is not a Neumann eigenvalue of $-\Delta$ in D. Again, $\langle \cdot, \cdot \rangle$ denotes the duality pairing in $\langle H^{-1/2}(\Gamma), H^{1/2}(\Gamma) \rangle$.

---

[8] k^2 is called a Neumann eigenvalue of $-\Delta$ in D if there exists a non-trivial solution $u \in C^2(D) \cap C^1(\overline{D})$ of the Helmholtz equation in D such that $\partial u / \partial \nu$ vanishes on Γ.

(e) Let N_i be the boundary operator (1.97) corresponding to the wavenumber $k = i$. The operator $-N_i$ is self-adjoint and coercive as an operator from $H^{1/2}(\Gamma)$ onto $H^{-1/2}(\Gamma)$, i.e.,

$$-\langle N_i \varphi, \varphi \rangle \geq c_0 \|\varphi\|^2_{H^{1/2}(\Gamma)} \quad \textit{for all } \varphi \in H^{1/2}(\Gamma). \tag{1.98}$$

(f) The difference $N - N_i$ is compact from $H^{1/2}(\Gamma)$ into $H^{-1/2}(\Gamma)$.

Remark: We note that the classical definition of $N\varphi$ by (1.97) makes only sense for sufficiently smooth densities (i.e., for Hölder continuously differentiable functions on Γ, see [43]). It can be shown that N has a bounded extension to an operator from $H^{1/2}(\Gamma)$ into $H^{-1/2}(\Gamma)$ which we also denote by N (see [144]).

Proof: (a) Analogously to H from (1.51) we define the operator $\partial H : L^2(S^2) \to H^{-1/2}(\Gamma)$ by

$$(\partial H)g(x) := \frac{\partial}{\partial \nu} \int_{S^2} g(\theta) \, e^{ikx\cdot\theta} \, ds(\theta) = ik \int_{S^2} g(\theta) \, (\nu(x) \cdot \theta) \, e^{ikx\cdot\theta} \, ds(\theta), \quad x \in \Gamma.$$

$$\tag{1.99}$$

Its adjoint $(\partial H)^* : H^{1/2}(\Gamma) \to L^2(S^2)$ is now given by

$$(\partial H)^* \varphi(\hat{x}) = -ik \int_{\Gamma} (\nu(y) \cdot \hat{x}) \, \varphi(y) \, e^{-ik\hat{x}\cdot y} \, ds(y) = \int_{\Gamma} \varphi(y) \, \frac{\partial}{\partial \nu(y)} e^{-ik\hat{x}\cdot y} \, ds(y)$$

$$\tag{1.100}$$

for $\hat{x} \in S^2$. We note that $(\partial H)^* \varphi$ is now the far field pattern of the double layer potential

$$v(x) = \int_{\Gamma} \varphi(y) \, \frac{\partial}{\partial \nu(y)} \Phi(x,y) \, ds(y), \quad x \in \mathbb{R}^3 \setminus \overline{D}, \tag{1.101}$$

and $\partial v/\partial \nu = N\varphi$ provided $\varphi \in C^{1,\alpha}(\Gamma)$ (see [42]). This yields $(\partial H)^* \varphi = GN\varphi$ and thus $\partial H = N^* G^*$ by a density argument. Furthermore, we note that $F = -G(\partial H)$ and therefore $F = -GN^*G^*$.

(b) This follows from similar the arguments as in the proof of Lemma 1.13.

(c) For this property we refer again to [144].

(d) Here we proceed exactly as in the proof of Lemma 1.14. Define, for any $\varphi \in H^{1/2}(\Gamma)$ the double layer potential v by

$$v(x) = \int_{\Gamma} \varphi(y) \, \frac{\partial \Phi(x,y)}{\partial \nu(y)} \, ds(y), \quad x \in \mathbb{R}^3 \setminus \Gamma. \tag{1.102}$$

Then $v \in H^1(D) \cap H^1_{loc}(\mathbb{R}^3 \setminus \overline{D})$ is again a solution of the Helmholtz equation in $\mathbb{R}^3 \setminus \Gamma$. The traces $v_{\pm}$ and $\partial v_{\pm}/\partial \nu$ exist in the variational sense with $\varphi = v_+ - v_-$ and $\frac{\partial v_-}{\partial \nu} = \frac{\partial v_+}{\partial \nu} = N\varphi$. Therefore, using Green's formula in D and in $D_R := \{x \in \mathbb{R}^3 \setminus D : |x| < R\}$

we conclude that

$$\langle N\varphi, \varphi \rangle = \left\langle \frac{\partial v}{\partial \nu}, v_+ - v_- \right\rangle \tag{1.103}$$

$$= -\iint\limits_{D \cup D_R} \left(|\nabla v(x)|^2 - k^2 |v(x)|^2 \right) dx + \int\limits_{|x|=R} \overline{v(x)} \frac{\partial v(x)}{\partial r} \, ds \tag{1.104}$$

$$= -\iint\limits_{D \cup D_R} \left(|\nabla v|^2 - k^2 |v|^2 \right) dx + ik \int\limits_{|x|=R} |v|^2 ds + \mathcal{O}\left(\frac{1}{R} \right) \tag{1.105}$$

as R tends to infinity. Taking the imaginary part yields

$$\operatorname{Im}\langle N\varphi, \varphi \rangle = k \lim_{R \to \infty} \int\limits_{|x|=R} |v|^2 ds = \frac{k}{(4\pi)^2} \int\limits_{S^2} |v^\infty|^2 ds \geq 0. \tag{1.106}$$

Let now $\operatorname{Im}\langle N\varphi, \varphi \rangle = 0$ for some $\varphi \in H^{1/2}(\Gamma)$. Again, from (1.106), Rellich's Lemma (see Lemma 1.2) and unique continuation we conclude that v vanishes outside of D. Therefore, $N\varphi = 0$ on Γ by the trace theorem. Since N is an isomorphism φ has to vanish.

(e) For $k = i$ the same arguments as above yield

$$\langle N_i\varphi, \varphi \rangle = -\iint\limits_{D \cup D_R} \left(|\nabla v|^2 + |v|^2 \right) dx - \int\limits_{|x|=R} |v|^2 ds + \mathcal{O}\left(\frac{1}{R} \right), \quad R \to \infty,$$

and thus as $R \to \infty$ (note that v decays exponentially):

$$\langle N_i\varphi, \varphi \rangle = -\iint\limits_{\mathbb{R}^3} \left(|\nabla v(x)|^2 + |v(x)|^2 \right) dx = -\|v\|^2_{H^1(\mathbb{R}^3)}.$$

The trace theorem and the boundedness of N_i^{-1} yields the existence of $c > 0$ and $c_0 > 0$ with

$$-\langle N_i\varphi, \varphi \rangle \geq c \, \|\partial v/\partial \nu\|^2_{H^{-1/2}(\Gamma)} = c \, \|N_i\varphi\|^2_{H^{-1/2}(\Gamma)} \geq c_0 \, \|\varphi\|^2_{H^{1/2}(\Gamma)}.$$

(f) This follows again from the fact that the kernel of $N - N_i$ is smooth. $\square$

Now we proceed exactly as in the case of the Dirichlet problem. Indeed, the functional analytic Theorem 1.23 is applicable where now $X = H^{-1/2}(\Gamma)$ and $X^* = H^{1/2}(\Gamma)$ and $A = -N^*$. Moreover, Theorem 1.12 holds also for the Neumann boundary condition. Therefore, we have the following analogy of Theorem 1.25.

Theorem 1.27 *Assume k^2 is not a Neumann eigenvalue of $-\Delta$ in D. For any $z \in \mathbb{R}^3$ define again $\phi_z \in L^2(S^2)$ by*

$$\phi_z(\hat{x}) := e^{-ik\hat{x} \cdot z}, \quad \hat{x} \in S^2.$$

Then

$$z \in D \quad \Longleftrightarrow \quad \phi_z \in \mathcal{R}\big((F^*F)^{1/4}\big). \qquad (1.107)$$

Let $\lambda_j \in \mathbb{C}$ be the eigenvalues of the normal operator F with corresponding eigenfunctions $\psi_j \in L^2(S^2)$. Then the following characterization holds.

A point $z \in \mathbb{R}^3$ belongs to D if, and only if, the series

$$\sum_{j=1}^{\infty} \frac{\big|(\phi_z, \psi_j)_{L^2(S^2)}\big|^2}{|\lambda_j|}$$

converges, i.e., if, and only if,

$$W(z) := \left[\sum_{j=1}^{\infty} \frac{\big|(\phi_z, \psi_j)_{L^2(S^2)}\big|^2}{|\lambda_j|} \right]^{-1} > 0. \qquad (1.108)$$

Therefore, $\chi_D(z) = \operatorname{sign} W(z)$ is the characteristic function of D.

Remark: Of course, the function W is different from the one defined in Theorem 1.25. We do not indicate this by a different symbol.

The characterizations of Theorems 1.25 and 1.27 imply again uniqueness of the inverse scattering problem if k^2 is not an eigenvalue of the underlying boundary value problem in D. It is remarkable that the characterization of the characteristic function depends only on F and makes no use of the boundary condition. We will exploit this fact further down.

1.7   Additional remarks and numerical examples

Before we turn to the presentation of some numerical simulations we want to add some remarks.

In this chapter we considered the set $\{u^\infty(\cdot, \theta) : \theta \in S^2\}$ of far field patterns as data for the inverse problem or, equivalently, the far field operator $F : L^2(S^2) \to L^2(S^2)$. In many applications the incident fields are point sources $v^i = v^i(\cdot, y)$ given by $v^i(x, y) = \Phi(x, y)$ for y from a surface Σ which contains $\overline{D}$ in its interior. The corresponding scattered fields $v^s = v^s(\cdot, y)$ are measured on the same surface Σ.[9] Therefore, the set $\{v^s(\cdot, y) : y \in \Sigma\}$ are the data for the inverse problem or, equivalently, the near field operator $F_\Sigma : L^2(\Sigma) \to L^2(\Sigma)$ given by

$$(F_\Sigma \varphi)(x) = \int_\Sigma \varphi(y)\, v^s(x, y)\, ds(y), \quad x \in \Sigma.$$

---

[9] Or even a different one.

If we define the "data-to-pattern" operator $G : H^{1/2}(\Gamma) \to L^2(\Sigma)$ by $Gf = v|_\Sigma$ where $v \in H^1_{loc}(\mathbb{R}^3 \setminus \overline{D})$ solves the exterior Dirichlet boundary value problem (1.38), (1.39), and (1.40) then one can show by the same arguments than above that

$$F_\Sigma = -G\,S'\,G'$$

where now $S' : H^{-1/2}(\Gamma) \to H^{1/2}(\Gamma)$ and $G' : L^2(\Sigma) \to H^{-1/2}(\Gamma)$ are the adjoints of S and G, respectively, with respect to the *bilinear forms* $(\psi,\phi) \mapsto \int_\Sigma \psi\,\phi\,ds$ for $\psi,\phi \in L^2(\Sigma)$ and $(\psi,\phi) \mapsto \langle \psi, \overline{\phi} \rangle$ for $\psi \in H^{-1/2}(\Gamma)$ and $\phi \in H^{1/2}(\Gamma)$. It is an open problem how to develop a characterization of D from this type of factorization. In the stationary case, however, i.e., for $k = 0$, G' and S' coincide with G^* and S^*, respectively, and the Factorization Method can be applied. We refer to [81, 131]. For the Helmholtz equation there are several ways to overcome this difficulty with the near field operator. They all construct operators P from $L^2(\Sigma)$ into itself such that PF_Σ or $F_\Sigma P$ can be written in the form BTB^*. We refer to Section 2.4 where we will develop one possibility and also to [153] and [121] where in the latter paper a more complicated case (Maxwell's equations in a layered medium where Σ does not enclose D) is treated.

We have presented the analysis in $\mathbb{R}^3$ only because the fundamental solution Φ is more elementary in three than in two dimensions. In $\mathbb{R}^2$ the fundamental solution is given by $\Phi(x, y) = \frac{i}{4} H_0^{(1)}(k|x - y|)$ where $H_0^{(1)}$ denotes the Hankel function of the first kind and order zero. $H_0^{(1)}$ has a singularity at zero of logarithmic type. From the asymptotic behavior of $H_0^{(1)}$ for large arguments one derives the asymptotics of the fundamental solution as

$$\Phi(x, y) = \frac{i}{4} H_0^{(1)}(k|x - y|) = \frac{\exp(ik|x| + i\pi/4)}{\sqrt{8\pi k\,|x|}}\, e^{-ik\,\hat{x}\cdot y}\left[1 + \mathcal{O}(|x|^{-1})\right], \quad |x| \to \infty,$$

uniformly with respect to $\hat{x} = x/|x|$ and y from compact sets (see [43]).

Sommerfeld's radiation condition (1.20) has now the form

$$\frac{\partial v}{\partial r} - ik\,v = \mathcal{O}\!\left(r^{-3/2}\right) \quad \text{for } r = |x| \to \infty.$$

To be consistent with the presentation in $\mathbb{R}^3$ we define the far field pattern v^∞ by (compare (1.25))

$$v(x) = \frac{\exp(ik|x| + i\pi/4)}{\sqrt{8\pi k\,|x|}}\, v^\infty(\hat{x}) + \mathcal{O}(|x|^{-3/2}), \quad |x| \to \infty.$$

With this normalization formula (1.26) holds, and the far field pattern ϕ_z of the fundamental solution $\Phi(\cdot, z)$ is again given by formula (1.41).

All the results of this chapter (with suitably modified proofs) remain valid except Theorem 1.8 and, of course, Section 1.5. In the definition of the scattering matrix S and in Theorem 1.8 one has to change the factor $k/(8\pi^2)$ into $1/(4\pi)$ as it is easily seen from the proof. The example of the unit disk in $\mathbb{R}^2$ can be carried out analogously to the presentation of Section 1.5.

Now we turn to the presentation of some numerical simulations in two dimensions. Theorems 1.25 and 1.27 suggest a visualization of D in a very natural way. We have computed the following examples for the Dirichlet boundary condition. The simplest version is to choose a grid $\mathcal{G}$ of points $z \in \mathbb{R}^2$ such that the unknown obstacle is in the convex hull of the grid points. We assume that we know approximate values $a_{j\ell}$ of the far field pattern $u^\infty(\theta_j, \theta_\ell)$, $j, \ell = 1, \ldots, M$, at equidistantly distributed directions $\theta_j = 2\pi j/M$. Then we compute a singular value decomposition $A = U \Lambda V^*$ of the matrix $A := (a_{j\ell}) \in \mathbb{C}^{M \times M}$. For each z from the grid $\mathcal{G}$ we compute the expansion coefficients of $r_z = (\exp(-ik\,z \cdot \theta_j))_{j=1,\ldots,M} \in \mathbb{C}^M$ with respect to the columns of V by

$$\rho_\ell^{(z)} = \sum_{j=1}^M V_{j,\ell}\, e^{-ik\,z \cdot \theta_j}\,, \quad \ell = 1, \ldots, M\,,$$

which is a matrix-vector multiplication $\rho^{(z)} = V^\top r_z$ of $V^\top$ and r_z. For each z we compute

$$W(z) := \left[\sum_{\ell=1}^M \frac{|\rho_\ell^{(z)}|^2}{|\lambda_\ell|} \right]^{-1}$$

and plot the contour lines of $z \mapsto W(z)$. The values of $W(z)$ should be much smaller for $z \notin D$ than for those lying within D.

In the *first example* D is given as the union of two disjoint obstacles $D = D_1 \cup D_2$, where D_1 is the ellipse with axes .8, .4, and center $(0,0)$, rotated by 20 degrees while D_2 is the "kite" of [43], parameterized by $x(t) = (\cos t + .65\cos(2t) - .65\,, 1.5\sin t)$, $t \in [0, 2\pi]$, then rotated by 70 degrees and shifted to center $(3, 5)$. For the *second example* we have chosen D to be a triangle with vertices at $(0,0)$, $(1,0)$, and $(0,1)$. For both examples we took two wavenumbers $k = 1$ and $k = 5$, and computed the far field pattern $u^\infty(\theta_j, \theta_\ell), j, \ell = 1, \ldots, M$, for $M = 32$ incident and observed directions. Contour plots of W are shown in Figures 1.7–1.10 without noise (left plot) or 1% white noise added to the data (right plot).

In the *third example* we have taken real data from the Ipswich dataset provided by the Electromagnetics Technology Division at Hanscom Air Force Base. The Ipswich

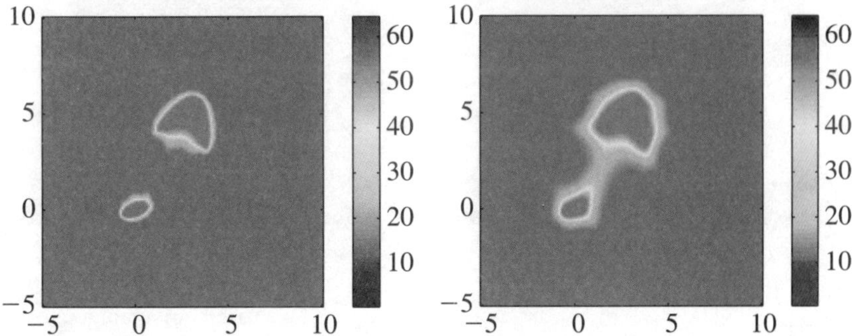

Figure 1.7 Function W for the first example with $k = 1$ and no noise (left) and 1% noise (right)

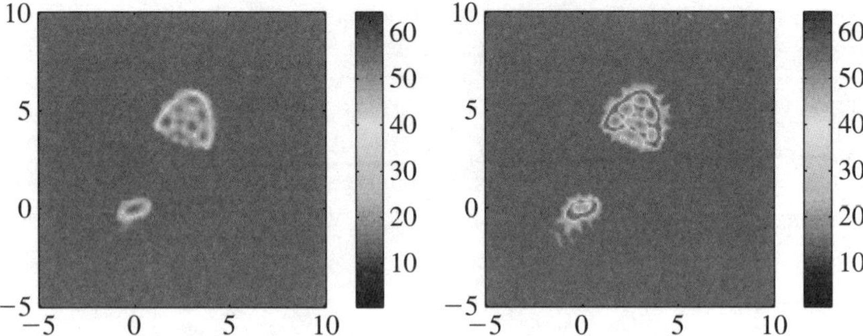

Figure 1.8 Function W for the first example with $k = 5$ and no noise (left) and 1% noise (right)

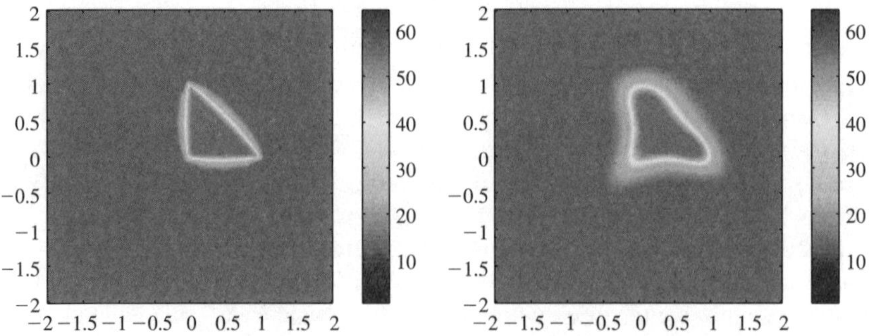

Figure 1.9 Function W for the second example with $k = 1$ and no noise (left) and 1% noise (right)

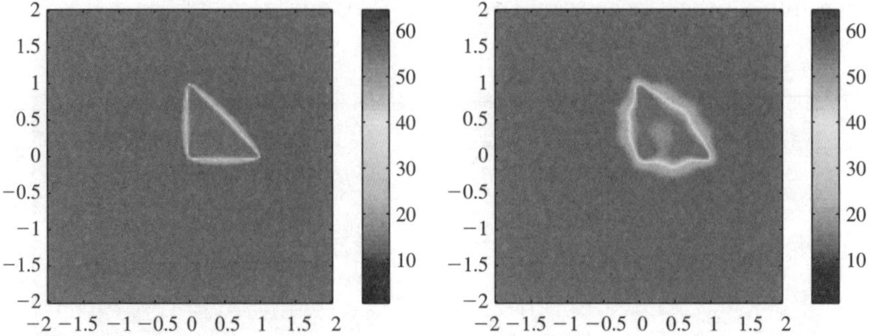

Figure 1.10 Function W for the second example with $k = 5$ and no noise (left) and 1% noise (right)

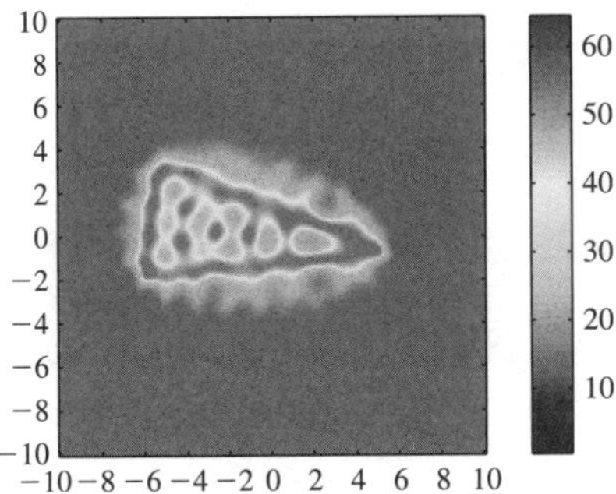

Figure 1.11 Function W for the third example (real data) with wavelength $\lambda = 3$

data is single-frequency electric far field data measured using a multistatic system with multiple views corresponding to different incident angles. A detailed discussion of the data and the measurement process can be found in [143]. The object is supposed to be an aluminum triangle. The wavelength is $\lambda = 3$ which corresponds to wavenumber $k = 2\pi/\lambda \approx 2.0944$. The measured data consist of a matrix $A \in \mathbb{C}^{36 \times 36}$ corresponding to the angles $\theta_j = 0^o, 10^o, \ldots, 350^o$, and for data corresponding to the same angles. We note that the data carry a considerable amount of noise since $\|A^*A - AA^*\| \approx 18.72$. Figure 1.11 shows the result.

Recently, Hyvönen in [90] and Lechleiter in [139] presented rigorous mathematical justifications of the use of finite-dimensional approximations of the data operator for the impedance tomography problem. Lechleiter's approach can easily be applied to the inverse scattering problem. He shows convergence of the noisy $W(z)$ from above to zero for z outside of D if the noise level goes to zero and the dimension of the incorporated data goes to infinity in a suitably regularized way. Under certain assumptions the indicator function remains strictly positive in any compact subset of D.

2

The factorization method for other types of inverse obstacle scattering problems

In this chapter we will investigate some examples of inverse obstacle scattering problems in which the data operator (e.g., the far field operator) fails to be normal. First, we consider the same inverse scattering problem as in the previous chapter except that the Dirichlet or Neumann boundary conditions are replaced by an impedance boundary condition. In physical applications the impedance is complex valued which implies that the far field operator fails to be normal. In Section 2.1 we prove uniqueness and existence of solutions of the direct problem. As we will see in Section 2.2 the inf-criterion of Subsection 1.4.2 is general enough to give a complete characterization of D in terms of the far field operator. In Section 2.3 we consider the case where the far field patterns $u^\infty(\hat{x}, \theta)$ are only given for $\hat{x}, \theta$ from a subset U of the unit sphere S^2. Also in this case the inf-criterion is applicable. In Section 2.4 we investigate the case where our data consist of (near-) field measurements on some surface Σ which correspond to point sources on Σ as incident fields. We show how one can transform the data in such a way that the inf-criterion can be applied. This inf-criterion is not very useful from the numerical point of view as mentioned above already several times. In Section 2.5 we develop a Factorization Method which is based on a range condition of the type $z \in D \iff \phi_z \in \mathcal{R}(F_\#^{1/2})$ rather than a inf-criterion. Here, the operator $F_\#$ can be easily computed from the data operator. In this way we are able to construct a simple and fast method to treat the cases of Sections 2.2, 2.3, and 2.4. We finish this chapter with the investigation of the Factorization Method for a scattering problem where the background medium consists of the half space $\{x \in \mathbb{R}^3 : x_3 > 0\}$.

2.1 The direct scattering problem with impedance boundary conditions

As in previous chapter let $k > 0$ be the wavenumber and $D \subset \mathbb{R}^3$ be an open and bounded domain with C^2-boundary $\Gamma = \partial D$ such that the exterior $\mathbb{R}^3 \setminus \overline{D}$ of $\overline{D}$ is connected. In the direct scattering problem an incident wave u^i is given and one has to find the scattered wave $u^s \in C^2(\mathbb{R}^3 \setminus \overline{D}) \cap C^1(\mathbb{R}^3 \setminus D)$ such that the total wave $u = u^i + u^s$ satisfies the

Helmholtz equation

$$\Delta u + k^2 u = 0 \tag{2.1}$$

in the exterior of $\overline{D}$ and now, different from the previous chapter, the impedance boundary condition

$$\frac{\partial u}{\partial \nu} + \lambda u = 0 \quad \text{on } \Gamma. \tag{2.2}$$

Here, $\nu(x)$ denotes again the exterior unit normal vector at $x \in \Gamma$ and $\lambda \in L^\infty(\Gamma)$ is the given (complex-valued) impedance function. Furthermore, the scattered field u^s satisfies the Sommerfeld radiation condition

$$\frac{\partial u^s}{\partial r} - ik\, u^s = \mathcal{O}\left(r^{-2}\right) \quad \text{for } r = |x| \to \infty \tag{2.3}$$

uniformly with respect to $\hat{x} = x/|x| \in S^2$.

As we mentioned in Section 1.1 the impedance boundary value problem appears in acoustic obstacle scattering if we suppose that the normal velocity is proportional to the excess pressure on the boundary of the impenetrable obstacle. In electromagnetic scattering, the impedance boundary condition describes an obstacle, which is not perfectly conducting, but does not allow the electromagnetic field to penetrate deeply into the scattering domain, see [43, Section 6.1].

Writing the impedance boundary condition (2.2) in the form

$$\frac{\partial u^s}{\partial \nu} + \lambda u^s = -\frac{\partial u^i}{\partial \nu} - \lambda u^i \quad \text{on } \Gamma$$

for the scattered field we observe that the scattering problem is a particular case of the following radiation problem:

Given $g \in H^{-1/2}(\Gamma)$, find a solution $v \in H^1_{loc}(\mathbb{R}^3 \backslash \overline{D})$ of the Helmholtz equation (2.1) in the exterior of $\overline{D}$ which satisfies the radiation condition (2.3) and the boundary condition

$$\frac{\partial v}{\partial \nu} + \lambda v = g \quad \text{on } \Gamma. \tag{2.4}$$

The equations (2.1) and (2.4) are again understood in the variational form, i.e., $v \in H^1_{loc}(\mathbb{R}^3 \setminus \overline{D})$ satisfies the variational equation

$$\iint_{\mathbb{R}^3 \backslash \overline{D}} \left[\nabla v \cdot \nabla \overline{\psi} - k^2 v \, \overline{\psi} \right] dx - \int_{\Gamma} \lambda\, v \, \overline{\psi}\, ds = -\langle g, \psi \rangle \tag{2.5}$$

for all $\psi \in H^1(\mathbb{R}^3 \setminus \overline{D})$ with compact support. Here, again, $\langle \cdot, \cdot \rangle$ denotes the dual form in the dual system $\langle H^{-1/2}(\Gamma), H^{1/2}(\Gamma) \rangle$. A well-known regularity result (see, e.g., [70]) yields that v is analytic outside of $\overline{D}$. Therefore, the radiation condition (2.3) is well defined.

Uniqueness of the scattering problem can be shown by Green's theorem just as in the case of the Dirichlet boundary condition, see below (Theorem 2.2). The proof of existence can be done by variational approaches (cf. [150] or [126] for the Dirichlet boundary condition) or by an integral equation method. We decided to sketch the latter approach since we were not able to find it in any monograph. The boundary integral equation method for L^∞ – impedances λ and boundary data $g \in H^{-1/2}(\Gamma)$ uses the following surface potentials for $\psi \in H^{-1/2}(\Gamma)$ and $\phi \in H^{1/2}(\Gamma)$:

$$v(x) = \int_\Gamma \psi(y)\, \Phi(x, y)\, ds(y)\,, \quad x \notin \Gamma\,, \tag{2.6}$$

$$w(x) = \int_\Gamma \phi(y)\, \frac{\partial}{\partial v(y)} \Phi(x, y)\, ds(y)\,, \quad x \notin \Gamma\,. \tag{2.7}$$

The integral in (2.6) has to be understood in the dual system $\langle H^{-1/2}(\Gamma), H^{1/2}(\Gamma)\rangle$. Here, Φ is again the fundamental solution of the Helmholtz equation, see (1.24). The functions v and w are called single layer potential and double layer potential with density ψ and ϕ, respectively. In the following lemma we collect the most important properties of these potentials. Some of them are used already in the previous chapter (see, e.g., Lemma 1.14 and Theorem 1.26).

Lemma 2.1 *Let v and w be defined in (2.6) and (2.7) for $\psi \in H^{-1/2}(\Gamma)$ and $\phi \in H^{1/2}(\Gamma)$, respectively.*

(a) $v|_D, w|_D \in H^1(D)$ and $v|_{\mathbb{R}^3\setminus\overline{D}}, w|_{\mathbb{R}^3\setminus\overline{D}} \in H^1_{loc}(\mathbb{R}^3 \setminus \overline{D})$, and these restrictions solve the Helmholtz equation (2.1) in D and in $\mathbb{R}^3\setminus\overline{D}$, respectively. Furthermore, the functions v and w satisfy the Sommerfeld radiation condition (2.3). The traces $v_\pm$, $\partial v_\pm/\partial v$, $w_\pm$, and $\partial w_\pm/\partial v$ on Γ exist and are given by

$$v_\pm = S\psi \in H^{1/2}(\Gamma)\,, \tag{2.8}$$

$$\frac{\partial v_\pm}{\partial v} = \mp\frac{1}{2}\psi + D'\psi \in H^{-1/2}(\Gamma)\,, \tag{2.9}$$

$$w_\pm = \pm\frac{1}{2}\phi + D\phi \in H^{1/2}(\Gamma)\,, \tag{2.10}$$

$$\frac{\partial w_\pm}{\partial v} = N\phi \in H^{-1/2}(\Gamma)\,. \tag{2.11}$$

Here, S, D, D' and N are given by the extensions of the boundary integral operators

$$(S\psi)(x) = \int_\Gamma \psi(y)\, \Phi(x, y)\, ds(y)\,, \quad x \in \Gamma\,, \tag{2.12}$$

$$(D\psi)(x) = \int_\Gamma \psi(y)\, \frac{\partial}{\partial v(y)} \Phi(x, y)\, ds(y)\,, \quad x \in \Gamma\,, \tag{2.13}$$

$$(D'\psi)(x) = \int_\Gamma \psi(y)\,\frac{\partial}{\partial\nu(x)}\Phi(x,y)\,ds(y), \quad x \in \Gamma, \tag{2.14}$$

$$(N\phi)(x) = \frac{\partial}{\partial\nu}\int_\Gamma \phi(y)\,\frac{\partial}{\partial\nu(y)}\Phi(x,y)\,ds(y), \quad x \in \Gamma. \tag{2.15}$$

to densities $\psi \in H^{-1/2}(\Gamma)$ *and* $\phi \in H^{1/2}(\Gamma)$.

(b) The operators S, D, D' are bounded from $H^{-1/2}(\Gamma)$ to $H^{1/2}(\Gamma)$ and N is bounded from $H^{1/2}(\Gamma)$ to $H^{-1/2}(\Gamma)$. They satisfy the following commutation relations:

$$NS = (D')^2 - \frac{1}{4}I, \quad SN = D^2 - \frac{1}{4}I, \quad DS = SD', \quad D'S = SD. \tag{2.16}$$

(c) If k^2 is not a Dirichlet eigenvalue of $-\Delta$ in D then S is an isomorphism from $H^{-1/2}(\Gamma)$ onto $H^{1/2}(\Gamma)$. If k^2 is not a Neumann eigenvalue of $-\Delta$ in D then N is an isomorphism from $H^{1/2}(\Gamma)$ onto $H^{-1/2}(\Gamma)$.

For a *proof* we refer to [144].

Theorem 2.2 *Let $k > 0$ be a fixed frequency and $\lambda \in L^\infty(\Gamma)$ complex valued such that $\mathrm{Im}\,\lambda \geq 0$ almost everywhere on Γ. Then there exists a unique radiating solution of the direct scattering problem (2.1), (2.4) in the sense of (2.5).*

Proof: First we prove uniqueness. Let $v \in H^1_{loc}(\mathbb{R}^3 \setminus \overline{D})$ be a solution of (2.5) for $g = 0$. We choose $R > 0$ such that $\overline{D}$ is contained in the ball $B(0, R)$ and fix a real valued function $\varphi \in C^\infty(\mathbb{R}^3)$ with $\varphi = 1$ for $|x| \leq R$ and $\varphi = 0$ for $|x| \geq R + 1$. Substituting $\psi = \varphi v$ in (2.5) yields

$$0 = \iint_{|x|<R} \left[|\nabla v|^2 - k^2|v|^2\right] dx + \iint_{R<|x|<2R} \left[\nabla v \cdot \nabla(\varphi\overline{v}) - k^2\varphi\,|v|^2\right] dx - \int_\Gamma \lambda\,|v|^2\,ds.$$

By well-known regularity results the solution v is smooth in the region $\{x : R \leq |x| \leq R+1\}$. Therefore, the second volume integral can be written as

$$\int_{|x|=R} \overline{v}\,\frac{\partial v}{\partial\nu}\,ds$$

by Green's first theorem. (Note that $\varphi = 1$ for $|x| = R$ and $\varphi = 0$ for $|x| = R+1$.) Taking the imaginary part yields

$$\mathrm{Im}\int_{|x|=R} \overline{v}\,\frac{\partial v}{\partial\nu}\,ds = \int_\Gamma \mathrm{Im}\,\lambda\,|v|^2\,ds \geq 0.$$

Now standard arguments in the proof of uniqueness for classical solutions (see [43, Theorem 10.8]) yield that v vanishes for $|x| \geq R$ and thus, by analytic extension, also in the exterior of D.[1]

---

[1] Here we use again the fact that the exterior is connected.

To prove existence we make the following ansatz for a solution (with density $\psi \in H^{-1/2}(\Gamma)$):

$$u(x) = \int_{\Gamma} \left[\psi(y)\, \Phi(x,y) + i\, (\overline{D}D'\psi)(y)\, \frac{\partial}{\partial \nu(y)} \Phi(x,y) \right] ds(y)\,, \quad x \notin \overline{D}\,. \qquad (2.17)$$

Here, $\overline{D}$ denotes the conjugate of D and is defined by $\overline{D}\psi = \overline{D\overline{\psi}}$. We note that by Lemma 2.1 D' is a bounded operator from $H^{-1/2}(\Gamma)$ into $H^{1/2}(\Gamma)$ and D is compact in $H^{1/2}(\Gamma)$. Therefore, $\overline{D}D'$ is compact from $H^{-1/2}(\Gamma)$ into $H^{1/2}(\Gamma)$. The field u is the sum of the single layer potential with density $\psi \in H^{-1/2}(\Gamma)$ and the double layer potential with density $i\overline{D}D'\psi \in H^{1/2}(\Gamma)$. It follows from Lemma 2.1 that u satisfies the boundary condition (2.4) if, and only if, ψ satisfies $R\psi = g$ where $R : H^{-1/2}(\Gamma) \to H^{-1/2}(\Gamma)$ is defined by

$$R\psi = \left(D' - \frac{1}{2}I \right) \psi + i N \overline{D}D'\psi + \lambda \left[S\psi + i \left(D + \frac{1}{2}I \right) \overline{D}D'\psi \right]. \qquad (2.18)$$

R is a Fredholm operator of index zero because all operators D', $N\overline{D}D'$, S, and $\left(D + \frac{1}{2}I \right) \overline{D}D'$ are compact from $H^{-1/2}(\Gamma)$ into itself.

As a next step we prove injectivity of R. Let $\psi \in H^{-1/2}(\Gamma)$ satisfy $R\psi = 0$, i.e., the potential u, defined by (2.17) in all of $\mathbb{R}^3 \setminus \Gamma$, satisfies the homogeneous impedance boundary value problem (in the sense of (2.5))

$$\frac{\partial u_+}{\partial \nu} + \lambda\, u_+ = 0 \quad \text{on } \Gamma\,. \qquad (2.19)$$

The uniqueness result implies $u \equiv 0$ in the exterior of D. Applying the jump conditions of Lemma 2.1 we get

$$u_- = u_- - u_+ = -i\, \overline{D}D'\psi\,,$$

$$\frac{\partial u_-}{\partial \nu} = \frac{\partial u_-}{\partial \nu} - \frac{\partial u_+}{\partial \nu} = \psi\,, \quad \text{i.e.,}$$

$$u_- = -i\, \overline{D}D' \left(\frac{\partial u_-}{\partial \nu} \right)\,.$$

We note that u satisfies the variational equation

$$\iint_{D} \left[\nabla u \cdot \nabla \overline{\psi} - k^2 u\, \overline{\psi} \right] dx = \left\langle \frac{\partial u_-}{\partial \nu}\,, \psi \right\rangle$$

for all $\psi \in H^1(D)$. Substituting $\psi = u$ and taking the imaginary part yields

$$0 = \operatorname{Im} \left\langle \frac{\partial u_-}{\partial \nu}\,, u_- \right\rangle = \operatorname{Re} \left\langle \frac{\partial u_-}{\partial \nu}\,, \overline{D}D' \frac{\partial u_-}{\partial \nu} \right\rangle = \left\| D' \frac{\partial u_-}{\partial \nu} \right\|_{L^2(\Gamma)}^2$$

where we have used that $\overline{D}$ and D' are L^2 – to each other. This yields $D'\frac{\partial u_-}{\partial \nu} = 0$, thus $D'\psi = 0$ and therefore $u_- = 0$ on Γ. Using the jump conditions again yields

$$\psi = \frac{\partial u_-}{\partial \nu} = \frac{1}{2}\psi + D'\psi = \frac{1}{2}\psi$$

which yields that ψ vanishes on Γ.

Therefore, we have shown that the operator $R : H^{-1/2}(\Gamma) \to H^{-1/2}(\Gamma)$ is one-to-one. Application of the Fredholm alternative to the operator R yields that R is also onto and, moreover, an isomorphism from $H^{-1/2}(\Gamma)$ onto itself. Therefore, the equation (2.18) has exactly one solution v for every $g \in H^{-1/2}(\Gamma)$ and v depends continuously on g. $\square$

We recall from the previous chapter that any radiating solution v of the Helmholtz equation has the asymptotic form

$$v(x) = \frac{\exp(ik|x|)}{4\pi |x|}\, v^\infty(\hat{x}) + \mathcal{O}(|x|^{-2}), \quad |x| \to \infty, \tag{2.20}$$

uniformly with respect to $\hat{x} = x/|x| \in S^2$. The function $v^\infty : S^2 \to \mathbb{C}$ denotes again the far field pattern of v. As in cases of the Dirichlet and Neumann boundary condition, the far field pattern for the scattering problem with plane wave

$$u^i(x) = u^i(x,\theta) = e^{ik\, x\cdot\theta}, \quad x \in \mathbb{R}^3, \tag{2.21}$$

as incident field depends also on $\theta \in S^2$, and we indicate this dependence by writing $u^\infty(\hat{x},\theta)$. The far field operator $F = F_{imp}$ is then defined by

$$(F_{imp}\,\varphi)(\hat{x}) = \int_{S^2} u^\infty(\hat{x},\theta)\,\varphi(\theta)\,ds(\theta), \quad \hat{x} \in S^2. \tag{2.22}$$

The previous theorem guarantees again the existence of the data-to-pattern operator $G = G_{imp} : H^{-1/2}(\Gamma) \to L^2(S^2)$ defined by

$$G_{imp}\, g = v^\infty, \tag{2.23}$$

where v^∞ is a far field pattern of the radiating solution of the exterior impedance boundary value problem (2.1) and (2.4) in the sense of (2.5).

Within this section we will indicate the type of boundary condition by writing F_{imp} and G_{imp} if necessary.[2]

It is convenient to relate the operator G_{imp} to the corresponding data-to-pattern operator G_{Dir} for the Dirichlet boundary condition. To formulate this relationship we have to define the exterior Dirichlet-to-Neumann operator. We use the symbol Λ^{-1} for this operator because we reserve the symbol Λ for the Neumann-to-Dirichlet operator (compare Chapter 6). Therefore, $\Lambda : H^{-1/2}(\Gamma) \to H^{1/2}(\Gamma)$ is defined by $\Lambda g = u|_\Gamma$ where u

---

[2] We will omit this subscript in cases where no misunderstandings can occur.

solves the exterior Neumann boundary value problem with boundary data $g \in H^{-1/2}(\Gamma)$. Then its inverse Λ^{-1} is given by

$$\Lambda^{-1} f = \frac{\partial v_+}{\partial \nu} \quad \text{on } \Gamma \tag{2.24}$$

where v denotes the unique solution of the exterior Dirichlet boundary value problem (2.1) in $\mathbb{R}^3 \setminus \overline{D}$ and (2.3) with $v = f$ on Γ. We refer to the previous chapter and, in particular, to Theorem 1.1. We understand the normal derivative again in the variational sense, i.e.,

$$\langle \Lambda^{-1} f, \psi \rangle = - \iint\limits_{\mathbb{R}^3 \setminus \overline{D}} \left[\nabla v \cdot \nabla \overline{\psi} - k^2 v \, \overline{\psi} \right] dx \tag{2.25}$$

for all $\psi \in H^{1/2}(\Gamma)$. Here we denote by ψ also an arbitrary extension of ψ to a function in $H^1(\mathbb{R}^3 \setminus \overline{D})$ of compact support. It can be shown that the definition of Λ^{-1} does not depend on this extension.

The following properties are well known:

Theorem 2.3 *(a) The Neumann-to-Dirichlet operator Λ is an isomorphism from $H^{-1/2}(\Gamma)$ onto $H^{1/2}(\Gamma)$.*

(b) The operators S, N, Λ, and Λ^{-1} are self-adjoint with respect to the bilinear forms $(\psi, \varphi) \mapsto \int \psi \varphi \, ds$, and D' is the adjoint of D with respect to this bilinear form.

Proof: (a) Existence, injectivity, and surjectivity of Λ^{-1} follow directly from the existence theorem for the exterior Dirichlet and Neumann problems, see Theorems 1.1, and 2.2 (the latter for the case $\lambda = 0$). In particular, from the ansatz (2.17) and equation (2.18) we note that

$$\Lambda = (S + i\overline{D}D' + iD\overline{D}D')R^{-1}$$

where R is given by (2.18) for $\lambda = 0$.

(b) The properties for the boundary operators are just seen by changing the orders of integration. For Λ this is just Green's second formula, combined with the observation that $\int_\Gamma [v \, \partial w/\partial \nu - w \, \partial v/\partial \nu] ds = 0$ for radiating solutions of the Helmholtz equation. $\square$

Now we can express G_{imp} by the corresponding data-to-pattern operator G_{Dir} for the Dirichlet boundary condition.

Theorem 2.4 *Let $\lambda \in L^\infty(\Gamma)$ be a given impedance function with $\text{Im } \lambda \geq 0$.*

(a) The data-to-pattern operators G_{Dir} and G_{imp} satisfy

$$G_{Dir} = G_{imp} (\Lambda^{-1} + \lambda I), \tag{2.26}$$

where $\Lambda^{-1} + \lambda I$ is an isomorphism from $H^{1/2}(\Gamma)$ onto $H^{-1/2}(\Gamma)$. In particular, the ranges of G_{Dir} and G_{imp} coincide.

(b) G_{imp} is compact and one-to-one.

Proof: (a) Consider the exterior Dirichlet boundary value problem with $v = f$ on Γ. Then v solves also the impedance boundary value problem (2.4) with $g = \partial v/\partial v + \lambda f = \Lambda^{-1} f + \lambda f$. Therefore, for the far field pattern v^∞ of v we have

$$v^\infty = G_{Dir} f = G_{imp}\, g = G_{imp}(\Lambda^{-1} + \lambda I)f\,,$$

which implies (2.26).

The inclusion operator $I : H^{1/2}(\Gamma) \to H^{-1/2}(\Gamma)$ is compact which implies by Theorem 2.3 that $\Lambda^{-1} + \lambda I$ is Fredholm with index zero. Also, $\Lambda^{-1} + \lambda I$ is one-to-one, because G_{Dir} is one-to-one, see Lemma 1.13. Therefore, we conclude that $\Lambda^{-1} + \lambda I$ is an isomorphism.

(b) The compactness and the injectivity of G_{imp} follows from part (a) and the compactness and injectivity, respectively, of G_{Dir} by Lemma 1.13. $\square$

We list now some of the main properties of the far field operator (2.22).

Theorem 2.5 *Let $k > 0$ be a fixed frequency and $\lambda \in L^\infty(\Gamma)$ be a given impedance function with $\operatorname{Im} \lambda \geq 0$. Let $F = F_{imp}$ be the far field operator for the scattering problem with impedance boundary condition.*

(a) The far field pattern satisfies the reciprocity relation

$$u^\infty\left(-\hat{x}, \theta\right) = u^\infty\left(-\theta, \hat{x}\right) \quad \text{for all } \hat{x}, \theta \in S^2 \tag{2.27}$$

which is equivalent to the identity

$$F^* = J\,\overline{F}\,J \tag{2.28}$$

where $J : L^2(S^2) \to L^2(S^2)$ is the self-adjoint involution $(Jg)(\theta) = g(-\theta)$, and the conjugate $\overline{F}$ is again defined by $\overline{F}f = \overline{F\bar{f}}$.

(b) Let $v(\cdot, y) = v(x, y) = \Phi(x, y) + v^s(x, y)$ be the total field which corresponds to the incident field $v^i(\cdot, y) = \Phi(\cdot, y)$ at source point $y \notin \overline{D}$. Then there holds the mixed reciprocity principle

$$v^\infty(\theta, y) = u^s(y, -\theta) \quad \text{for all } \theta \in S^2\,,\ y \notin \overline{D}\,. \tag{2.29}$$

(c) F is compact.
(d) F satisfies the relations

$$F - F^* - \frac{ik}{8\pi^2}\, F^*F = 2iR\,, \tag{2.30}$$

$$F^* - F + \frac{ik}{8\pi^2}\, FF^* = -2iJ\,\overline{R}\,J\,, \tag{2.31}$$

where $R : L^2(S^2) \to L^2(S^2)$ is the self-adjoint non-negative operator given by

$$(Rh)(\hat{x}) = \int_{S^2}\left(\int_\Gamma \left(\operatorname{Im}\lambda(y)\right) u(y, \theta)\,\overline{u(y, \hat{x})}, ds(y)\right) h(\theta)\, d\theta\,, \quad \hat{x} \in S^2\,. \tag{2.32}$$

Here again, $u(\cdot, \theta) = u^i(\cdot, \theta) + u^s(\cdot, \theta)$ denotes the total field corresponding to the incident plane wave u^i (compare (2.21)).

(e) The scattering operator

$$S = I + \frac{ik}{8\pi^2} F \tag{2.33}$$

is subunitary and satisfies

$$S^*S = I - \frac{k}{4\pi^2} R \,,$$

where the operator R from (2.32) is non-negative. In particular, if λ is real-valued, the far field operator F is normal and the scattering operator S is unitary.

(f) Assume that there exists no Herglotz wave function which solves the boundary value problem in D with homogeneous impedance boundary condition. Then F is one-to-one and has dense range in $L^2(S^2)$.

(g) Under the assumption of (f) $\mathrm{Im}\, F$ is strictly positive and

$$\mathrm{Im}\, (F\psi, \psi)_{L^2(S^2)} \geq \frac{k}{16\pi^2} \|F\psi\|^2_{L^2(S^2)} > 0 \,, \tag{2.34}$$

$$\mathrm{Im}\, (F\psi, \psi)_{L^2(S^2)} \geq \frac{k}{16\pi^2} \|F^*\psi\|^2_{L^2(S^2)} > 0 \tag{2.35}$$

for any $\psi \neq 0$.

Proof: The reciprocity principles of (a) and (b) hold for all types of boundary conditions. We refer to Theorem 1.6 for the Dirichlet boundary condition and to [43, 160] for other types.

(c) This follows from (2.40) below and the compactness of G_{imp}.

(d), (e) We refer to Section 3.1, Theorem 3.3 below where we will consider a more general relation for the case of mixed boundary conditions.

(f) The proof follows the lines of the corresponding proof of Theorem 1.8, part (d).

(g) The relations (2.30), and (2.31) imply

$$\mathrm{Im}\, (F\varphi, \varphi)_{L^2(S^2)} = \frac{1}{2i} \left((F - F^*)\varphi, \varphi \right)_{L^2(S^2)} = \frac{1}{2i} \left(\left(\frac{ik}{8\pi^2} F^*F + 2iR \right) \varphi, \varphi \right)_{L^2(S^2)}$$

$$= \frac{k}{16\pi^2} \|F\varphi\|^2_{L^2(S^2)} + (R\varphi, \varphi)_{L^2(S^2)} \,,$$

$$\mathrm{Im}\, (F\varphi, \varphi)_{L^2(S^2)} = \frac{1}{2i} \left(\left(\frac{ik}{8\pi^2} FF^* + 2iJ^*RJ \right) \varphi, \varphi \right)_{L^2(S^2)}$$

$$= \frac{k}{16\pi^2} \|F^*\varphi\|^2_{L^2(S^2)} + (RJ\varphi, J\varphi)_{L^2(S^2)} \,.$$

This completes the proof, because the operator R is non-negative. $\square$

2.2   The obstacle reconstruction by the inf-criterion

In the previous chapter we introduced two algorithms for finding the support of a sound-soft or sound-hard scatterer from the far field operator. Both of them are based on a factorization of the kind $F = -GTG^*$ of the far field operator with some coercive operator T which was the adjoint of the single layer potential or of the normal derivative of the double layer potential, respectively. While the inf-criterion (1.66) characterizes the unknown domain by computing the infimum of a certain functional the $(F^*F)^{1/4}$-method (1.80) exploits the normality of the far field operator F and makes use of a second and very natural factorization of F in the form

$$F = \sqrt{|F|}\,(\operatorname{sign} F)\,\sqrt{|F|}\,.$$

Here, $\sqrt{|F|} = (F^*F)^{1/4}$ is self-adjoint and $\operatorname{sign} F$ is given by

$$(\operatorname{sign} F)\,\psi = \sum_{j=1}^{\infty} \frac{\lambda_j}{|\lambda_j|}\,(\psi,\psi_j)_{L^2(S^2)}\,\psi_j\,, \quad \psi \in L^2(S^2)\,,$$

in terms of the eigensystem (λ_j, ψ_j) of F, see (1.75). This second factorization is only possible for normal far field operators which is the case for the Dirichlet or Neumann boundary value problems. For the impedance boundary condition, however, the far field operator fails to be normal in general (except of the case of a real-valued impedance) because the operator R in Theorem 2.5, (d) does not vanish. Hence, we can not expect the square root algorithm to hold in the form (1.80). However, the inf-criterion is applicable in the impedance case as well, see [76]. To prove this criterion we first derive the factorization of F_{imp} in the form $F = -G\,T^*G^*$.

For the proof of the factorization we need again the following two operators $H : L^2(S^2) \rightarrow H^{1/2}(S^2)$ and $\partial H : L^2(S^2) \rightarrow H^{-1/2}(S^2)$, defined by (1.51) and (1.99), respectively, i.e.,

$$(H\psi)(x) = \int_{S^2} \psi(\theta)\,e^{ik\,\theta\cdot x}\,ds(\theta)\,, \quad x \in \Gamma\,, \tag{2.36}$$

$$(\partial H)\psi(x) = \frac{\partial}{\partial \nu}\int_{S^2} \psi(\theta)\,e^{ik\,\theta\cdot x}\,ds(\theta)\,, \quad x \in \Gamma\,. \tag{2.37}$$

Theorem 2.6 *Let again* Λ^{-1} *be the exterior Dirichlet-to-Neumann operator, defined in (2.24) and* $S : H^{-1/2}(\Gamma) \rightarrow H^{1/2}(\Gamma)$ *the single layer boundary operator of Lemma 2.1 (see also (1.43)). Then the following factorization holds:*

$$F_{imp} = -G_{imp}\,T^*_{imp}\,G^*_{imp} \tag{2.38}$$

where $G_{imp} : H^{-1/2}(\Gamma) \rightarrow L^2(S^2)$ *has been defined in (2.23) and* $T_{imp} : H^{1/2}(\Gamma) \rightarrow H^{-1/2}(\Gamma)$ *is given by*

$$T_{imp} = N + i(\operatorname{Im}\lambda)I + D'\overline{\lambda} + \lambda D + \lambda S\overline{\lambda}\,. \tag{2.39}$$

T_{imp} *is a Fredholm operator of index zero.* T_{imp} *is even an isomorphism if* k^2 *is not an eigenvalue of* $-\Delta$ *on D with respect to the impedance boundary condition.*

Proof: First we note that

$$F_{imp} = -G_{imp}\,(\partial H + \lambda H)\,. \tag{2.40}$$

Indeed, from the definition (2.22) of the far field operator we observe that $F_{imp}\,\varphi$ is the far field pattern of the radiating solution v with the impedance boundary condition

$$\frac{\partial v}{\partial \nu} + \lambda v = -\left(\frac{\partial}{\partial \nu} + \lambda\right) \int_{S^2} \varphi(\theta)\, e^{ik\,\theta \cdot} \, ds(\theta) = -(\partial H)\varphi - \lambda H \varphi \quad \text{on } \Gamma$$

which proves (2.40).

Now we consider the adjoint $(\partial H + \lambda H)^*$ of $\partial H + \lambda H$ which is given by

$$(\partial H + \lambda H)^*\varphi(\hat{x}) = \int_{\Gamma} \left[\frac{\partial}{\partial \nu(y)} e^{-ik\,\hat{x}\cdot y} + \overline{\lambda(y)}\, e^{-ik\,\hat{x}\cdot y}\right] \varphi(y)\, ds(y)\,, \quad \hat{x} \in S^2\,.$$

The right-hand side is the far field pattern v^∞ of the potential

$$v(x) = \int_{\Gamma} \left[\frac{\partial}{\partial \nu(y)} \Phi(x,y) + \overline{\lambda(y)}\, \Phi(x,y)\right] \varphi(y)\, ds(y)\,, \quad x \notin \Gamma\,. \tag{2.41}$$

The jump conditions (2.8)–(2.11) yield

$$f = \frac{\partial v_+}{\partial \nu} + \lambda\, v_+ = \left[N\varphi - \frac{1}{2}\overline{\lambda}\varphi + D'(\overline{\lambda}\varphi)\right] + \lambda\left[\frac{1}{2}\varphi + D\varphi + S(\overline{\lambda}\varphi)\right]$$

and thus $v^\infty = G_{imp}\, f$, i.e.,

$$(\partial H + \lambda H)^* = G_{imp}\, T_{imp} \tag{2.42}$$

with T_{imp} given by (2.39). Solving for $(\partial H + \lambda H)$ and substituting this into (2.40) yields (2.38). The difference $T_{imp} - N$ is compact from $H^{1/2}(\Gamma)$ into $H^{-1/2}(\Gamma)$ since S, D, D', and the imbedding I are compact. This proves that T_{imp} is a Fredholm operator of index zero since N has this property. Finally, we show that T_{imp} is one-to-one if k^2 is not an eigenvalue of $-\Delta$ in D with respect to the impedance boundary condition. Let $T_{imp}\,\varphi = 0$ and define v as in (2.41). Then $\partial v_+/\partial \nu + \lambda v_+ = T_{imp}\,\varphi = 0$ on Γ and thus $v = 0$ in the exterior of D by the uniqueness of the exterior impedance boundary value problem. The jump conditions (2.8)–(2.11) yield again

$$v_- = v_- - v_+ = \varphi\,,$$

$$\frac{\partial v_-}{\partial \nu} = \frac{\partial v_-}{\partial \nu} - \frac{\partial v_+}{\partial \nu} = -\overline{\lambda}\,\varphi\,.$$

Eliminating φ yields $\partial \overline{v}_-/\partial \nu + \lambda\,\overline{v}_- = 0$ on Γ. From our assumption on k^2 we conclude that v vanishes also in D and thus also $\varphi = 0$. $\square$

Remark: In the case of the Neumann boundary condition, i.e., where $\lambda \equiv 0$, this coincides with the factorization (1.96), i.e.,

$$F_{Neu} = -G_{Neu} N^* G_{Neu}^* .$$

The next step toward the inf-criterion is the verification of the coercivity of the operator T_{imp} in the factorization (2.38).

Lemma 2.7 *Let $k > 0$ be a fixed frequency and $\lambda \in L^\infty(\Gamma)$ a given impedance with* Im $\lambda \geq 0$ *on* Γ. *Assume that k^2 is not an eigenvalue of $-\Delta$ in D with respect to the impedance boundary condition with impedance λ. Then the operator T_{imp} from (2.38) is coercive, i.e., there exists $c > 0$ with*

$$\left| \langle T_{imp}\, \varphi, \varphi \rangle \right| \geq c \|\varphi\|_{H^{1/2}(\Gamma)}^2 \quad \text{for all } \varphi \in H^{1/2}(\Gamma). \tag{2.43}$$

Proof: Combining the previous theorem with Theorem 1.26 we note that T_{imp} can be written as $T_{imp} = N_i + C$ where C is compact and N_i denotes the normal derivative of the double layer potential (2.15) for wavenumber $k = i$. From Theorem 1.26 we also know that N_i is self-adjoint and coercive. Therefore, in order to apply Lemma 1.17 to the operator $-T_{imp}$ we have to check that $\langle \varphi, T_{imp}\, \varphi \rangle \notin \mathbb{R}$ for $\varphi \neq 0$.

This is a consequence of (2.34), (2.38) and the injectivity of the operators T_{imp} and G_{imp}. Indeed, first we note from (2.34) and (2.38) that

$$0 \leq \frac{k}{16\pi^2} \|F\psi\|_{L^2(S^2)}^2 \leq \mathrm{Im}(F\psi, \psi)_{L^2(S^2)} = -\,\mathrm{Im}\langle T^* G^* \psi, G^* \psi \rangle$$

$$= \mathrm{Im}\langle T\, G^* \psi, G^* \psi \rangle \tag{2.44}$$

where we have dropped the subscript "*imp*" for simpler notation. The range of G^* is dense in $H^{1/2}(\Gamma)$ because of the injectivity of G by Theorem 2.4. Therefore, we conclude that $\mathrm{Im}\langle T\varphi, \varphi \rangle \geq 0$ for all $\varphi \in H^{1/2}(\Gamma)$.

Let now $\varphi \in H^{1/2}(\Gamma)$ with $\mathrm{Im}\langle T\varphi, \varphi \rangle = 0$. Again, since the range of G^* is dense there exists a sequence $\psi_n \in L^2(S^2)$ with $G^* \psi_n \to \varphi$. From (2.44) we conclude that

$$\frac{k}{16\pi^2} \|F\psi_n\|_{L^2(S^2)}^2 \leq \mathrm{Im}\langle T\, G^* \psi_n, G^* \psi_n \rangle .$$

Since the right hand side converges to $\mathrm{Im}\langle T\varphi, \varphi \rangle = 0$ we have shown that $F\psi_n \to 0$ in $L^2(S^2)$ as n tends to infinity. Furthermore, for any $\psi \in L^2(S^2)$ we have that

$$\langle T^* G^* \psi_n, G^* \psi \rangle = -(F\psi_n, \psi)_{L^2(S^2)} \quad \longrightarrow \quad 0 .$$

The left-hand side converges to $\langle T^* \varphi, G^* \psi \rangle$ and thus $\langle T^* \varphi, G^* \psi \rangle = 0$ for all $\psi \in L^2(S^2)$. From the denseness of the range of G^* we arrive at $T^* \varphi = 0$ and thus $\varphi = 0$ by the injectivity of $T^* = T_{imp}^*$. $\square$

Remark: No eigenvalues k^2 with respect to the impedance boundary condition exist if there exists an open set $U \subset \mathbb{R}^3$ such that Im $\lambda > 0$ almost everywhere on $U \cap \Gamma$. This is seen as follows. Let $v \in H^1(D)$ be an eigenfunction, i.e., v satisfies (compare (2.5))

$$\iint_D [\nabla v \cdot \nabla \overline{\psi} - k^2 v\, \overline{\psi}]\, dx - \int_\Gamma \lambda\, v\, \overline{\psi}\, ds = 0 \tag{2.45}$$

for all $\psi \in H^1(D)$. Setting $\psi = v$ and taking the imaginary part yields

$$\int_\Gamma (\operatorname{Im} \lambda) \, |v|^2 ds = 0$$

and thus $v \equiv 0$ on $U \cap \Gamma$. Now we define v by 0 in $U \setminus D$ and show that the extended function v satisfies the Helmholtz equation in $\tilde{D} = D \cup U$. Indeed, for functions $\psi \in H_0^1(\tilde{D})$ we have that

$$\iint\limits_{\tilde{D}} \left[\nabla v \cdot \nabla \overline{\psi} - k^2 v \, \overline{\psi} \right] dx = \iint\limits_{D} \left[\nabla v \cdot \nabla \overline{\psi} - k^2 v \, \overline{\psi} \right] dx = \int_\Gamma \lambda \, v \, \overline{\psi} \, ds = 0 \, .$$

Therefore, this extension of v satisfied the Helmholtz equation $\Delta v + k^2 v = 0$ in $\tilde{D}$. Since it vanishes on $U \setminus \overline{D}$ an analytic continuation argument yields that $v \equiv 0$ in D. Therefore, in this case the interior impedance boundary value problem has no eigenvalues, and the operator T_{imp} is coercive for all $k > 0$.

Now we are ready to formulate and prove the inf-criterion, following the lines of Section 1.4.2, Theorem 1.20.

Theorem 2.8 *Assume that k^2 is not an eigenvalue of $-\Delta$ in D with respect to the impedance boundary condition with impedance $\lambda \in L^\infty(\Gamma)$ with $\operatorname{Im} \lambda \geq 0$. Then for any point $z \in \mathbb{R}^3$ the following inf-criterion holds*

$$z \in D \quad \Longleftrightarrow \quad \inf \left\{ \left| (F_{imp} \psi, \psi)_{L^2(S^2)} \right| : \psi \in L^2(S^2), \, (\phi_z, \psi)_{L^2(S^2)} = 1 \right\} > 0 \, .$$

$$\tag{2.46}$$

Here again, $\phi_z \in L^2(S^2)$ is given by

$$\phi_z(\theta) = e^{-ik \, z \cdot \theta}, \quad \theta \in S^2 \, . \tag{2.47}$$

Proof: From Theorem 2.4 we conclude from that the ranges $\mathcal{R}(G_{Dir})$ and $\mathcal{R}(G_{imp})$ of G_{Dir} and G_{imp}, respectively, coincide. According to Theorem 1.12 a point z belongs to D if and only if $\phi_z \in \mathcal{R}(G_{Dir})$. Therefore, the function ϕ_z belongs to the range $\mathcal{R}(G_{imp})$ if and only if $z \in D$.

Now we apply again Theorem 1.16 to the factorization (2.38) and note that T_{imp} is coercive by Lemma 2.7. $\square$

Remark: As already mentioned, in the case where $\operatorname{Im} \lambda(x) > 0$ on $U \cap \Gamma$ for some open set $U \subset \mathbb{R}^3$ the interior impedance boundary value problem in D has no real eigenvalues (see the previous remark), and the inf-criterion (2.46) holds for all $k > 0$.

2.3   Reconstruction from limited data

Now we consider the case when only partial data are available, i.e., $u^\infty(\hat{x}, \theta)$ for $\hat{x}, \theta \in U$. Here, U denotes some subset of S^2 with nonempty interior (relative to S^2). For convenience, we assume that $U \subset S^2$ itself is already an open set. We define the limited aperture far field operator $F_{la} : L^2(U) \to L^2(U)$ as the integral operator

$$(F_{la}\varphi)(\hat{x}) = \int_U u^\infty(\hat{x}, \theta)\, \varphi(\theta)\, d\theta\,, \quad \hat{x} \in U,\ \varphi \in L^2(U)\,, \tag{2.48}$$

where u^∞ is the far field pattern corresponding to one of the boundary conditions considered so far, i.e., either the Dirichlet boundary condition or the impedance boundary condition which includes the Neumann condition as a special case. The definition of F_{la} can be rewritten as

$$F_{la} = P_U F P_U^*, \tag{2.49}$$

where $F = F_{Dir}$ or F_{imp} is the far field operator and $P_U : L^2(S^2) \to L^2(U)$ is the restriction operator $P_U \varphi = \varphi|_U$. The adjoint $P_U^* : L^2(U) \to L^2(S^2)$ is given by the extension, i.e., $(P_U^* \psi)(\theta) = \psi(\theta)$ for $\theta \in U$ and $(P_U^* \psi)(\theta) = 0$ otherwise.

Numerical simulations indicate that the far field operator (2.48) is not normal, even in the case of Dirichlet or Neumann boundary conditions, unless $U = S^2$. Therefore, the square root characterization (1.80) is not expected to hold. The inf-criterion (2.46), however, is general enough to give a complete solution of the inverse problem from the limited aperture and even from discrete data.

We define the data-to-pattern operator G_{la} by

$$G_{la} = P_U\, G\,, \tag{2.50}$$

where G denotes one of the operators $G_{Dir} : H^{1/2}(\Gamma) \to L^2(S^2)$ or $G_{imp} : H^{-1/2}(\Gamma) \to L^2(S^2)$, respectively. In the following theorem we list the main properties of F_{la} and G_{la} which are essential for the validity of the inf-criterion. For the sake of simplicity, we omit the subscripts which indicate the boundary condition because the assertions hold for all types, i.e., Dirichlet or impedance boundary condition.

Theorem 2.9 *Let $k > 0$ be not an eigenvalue of $-\Delta$ in D with respect to the corresponding boundary condition. For the impedance boundary condition we assume also that $\operatorname{Im}\lambda \geq 0$ on Γ. Then the following assertions hold.*

(a) G_{la} and F_{la} are compact, injective, and have dense ranges in $L^2(U)$.

(b) The factorization $F_{la} = -G_{la} T G_{la}^*$ holds with the same T as for the corresponding complete data operator F. In particular, T is coercive.

Proof: (a) The compactness of G_{la} and F_{la} follows from the compactness of G and F, respectively. Injectivity of G_{la} and F_{la} follows from the injectivity of G and F, respectively, and the fact that the far field patterns are analytic functions on the sphere. Indeed, $P_U Gf = 0$ implies that $v^\infty = 0$ on U and therefore, since U is open relative to S^2, $v^\infty = 0$ on all of S^2 which means that $Gf = 0$. Denseness of the ranges of G_{la} and F_{la} follows directly from the corresponding denseness properties of G and F, respectively, since P_U is onto.

The factorization (b) follows immediately from the corresponding results for the complete data by the definitions (2.50) and (2.49), respectively. $\square$

The following lemma relates the unknown obstacle D to the range of the operator G_{al} and corresponds to Theorem 1.12.

Lemma 2.10 *Let $\phi_z \in L^2(S^2)$ be defined by (2.47). Then the restriction $\phi_z|_U = P_U\phi_z$ belongs to the range $\mathcal{R}(G_{la})$ of G_{la} if and only if $z \in D$.*

Proof: As we know already, $z \in D$ if and only if $\phi_z \in \mathcal{R}(G)$. This implies immediately that $P_U\phi_z \in \mathcal{R}(G_{la})$ if $z \in D$. The reverse statement follows again from the analyticity of Gf on S^2. Indeed, $P_U\phi_z \in \mathcal{R}(G_{la})$ implies $\phi_z = v^\infty$ on U for some far field pattern $v^\infty = Gf$ and thus by analyticity $\phi_z = Gf$ which implies that $z \in D$. $\square$

Now we can prove the main result.

Theorem 2.11 *Under the assumptions of Theorem 2.9 the following inf-criterion holds (for any of the boundary conditions):*

$$z \in D \quad \Longleftrightarrow \quad \inf\left\{\left|(F_{la}\psi, \psi)_{L^2(U)}\right| : \psi \in L^2(U), \, (\phi_z, \psi)_{L^2(U)} = 1\right\} > 0.$$

Proof: We combine the previous Lemma 2.10 with Theorem 2.9 and Theorem 1.16 and take for X the same space as for the corresponding complete data problem (i.e., $H^{1/2}(\Gamma)$ for the Dirichlet boundary condition and $H^{-1/2}(\Gamma)$ for the Neumann and the impedance boundary condition). Furthermore, we set $Y = L^2(U)$, $B = G$ and $A = T$. $\square$

2.4   Reconstruction from near field data

In this section we will investigate the following scattering problem. Let Ω denote an open bounded connected domain which contains $\overline{D}$. Assume that k^2 is not a Dirichlet eigenvalue for the operator $-\Delta$ in Ω. Consider again the scattering problem with a point source $u^i(\cdot, y) = \Phi(\cdot, y)$ as incident field for any of the boundary conditions of Dirichlet type, Neumann type, or impedance type. Therefore, for the *direct scattering problem* an incident point source $v^i(\cdot, y)$ is given where y is a some point on the boundary $\Sigma = \partial\Omega$ of Ω, find a radiating scattered wave $v^s(\cdot, y) \in C^2(\mathbb{R}^3 \setminus \overline{D}) \cap C(\mathbb{R}^3 \setminus D)$ such that the total wave $v = v^i + v^s$ satisfies the Helmholtz equation (2.1) and one of the boundary conditions (1.16), (1.90), or (2.2).

Definition 2.12 *(a) We call the set $\{v^s(x, y) : x, y \in \partial\Omega\}$ the* point source data. *Recalling the notion $\Sigma = \partial\Omega$, the* near field operator $F_\Sigma : H^{-1/2}(\Sigma) \longrightarrow H^{1/2}(\Sigma)$ *is given by*

$$(F_\Sigma h)(x) = \int_\Sigma v^s(x, y)\, h(y)\, ds(y), \quad x \in \Sigma. \tag{2.51}$$

The (near field) *inverse problem* is to reconstruct D from F_Σ.

We note that the kernel of the integral operator F_Σ is smooth. For $h \in H^{-1/2}(\Sigma)$ we understand the integral again as the dual form $\langle h, \overline{v^s(x, \cdot)}\rangle$ in $\langle H^{-1/2}(\Sigma), H^{1/2}(\Sigma)\rangle$.

Our method of solving this problem is to convert the near field data into far field data and thus to reduce the near field inverse problem to the visualization problem considered in the previous section. This can be done by using the operators $S_\Sigma : H^{-1/2}(\Sigma) \to$

$H^{1/2}(\Sigma)$ and $H_\Sigma : L^2(S^2) \to H^{1/2}(\Sigma)$ defined in the previous chapter, Section 1.4.1 with respect to the domain Ω. We recall the definitions for the convenience of the reader:

$$(S_\Sigma \varphi)(x) = \int_\Sigma \varphi(y)\, \Phi(x,y)\, ds(y)\,, \quad x \in \Sigma\,,$$

$$(H_\Sigma g)(x) = \int_{S^2} g(\theta)\, e^{ik\, x\cdot\theta}\, ds(\theta)\,, \quad x \in \Sigma\,.$$

With these operators we can show:

Lemma 2.13 *Assume as before that k^2 is not a Dirichlet eigenvalue of $-\Delta$ in Ω. Let F be the far field operator with respect to the boundary condition under consideration. Then*

$$F = H_\Sigma^* \, S_\Sigma^{-1} \, F_\Sigma \, S_\Sigma^{-1} \, H_\Sigma\,. \tag{2.52}$$

Proof: From the proof of Theorem 1.15 we note that $H_\Sigma^* S_\Sigma^{-1} = G_{Dir,\Sigma}$ where $G_{Dir,\Sigma} : H^{1/2}(\Sigma) \to L^2(S^2)$ is the data-to-pattern operator for Ω, i.e., $G_{Dir,\Sigma} f = w^\infty$ where w solves the exterior Dirichlet problem for the Helmholtz equation in the exterior of Ω with boundary condition $w = f$ on Σ. From the definition of F_Σ and the mixed reciprocity principle (2.29), we observe that for $h \in H^{-1/2}(\Sigma)$ and fixed $\hat{x} \in S^2$:

$$(H_\Sigma^* S_\Sigma^{-1} F_\Sigma h)(\hat{x}) = (G_{Dir,\Sigma} F_\Sigma h)(\hat{x}) = \int_\Sigma v^\infty(\hat{x}, y)\, h(y)\, ds(y)$$

$$= \int_\Sigma u^s(y, -\hat{x})\, h(y)\, ds(y) = \left\langle h, \overline{u^s(\cdot, -\hat{x})} \right\rangle\,.$$

Now we set $h = S_\Sigma^{-1} H_\Sigma g$ for some $g \in L^2(S^2)$ which yields

$$(H_\Sigma^* S_\Sigma^{-1} F_\Sigma S_\Sigma^{-1} H_\Sigma g)(\hat{x}) = \left\langle S_\Sigma^{-1} H_\Sigma g\,, \overline{u^s(\cdot, -\hat{x})} \right\rangle$$

$$= \left(g\,, H_\Sigma^*(S_\Sigma^*)^{-1} \overline{u^s(\cdot, -\hat{x})} \right)_{L^2(S^2)} = \left(g, H_\Sigma^* \overline{f} \right)_{L^2(S^2)}$$

where $f = S_\Sigma^{-1} u^s(\cdot, -\hat{x})$ on Σ. (Note that $S_\Sigma^* = \overline{S}_\Sigma$.) We define the single layer potential w by

$$w(x) = \int_\Sigma f(y)\, \Phi(x,y)\, ds(y) \quad \text{for } x \in \mathbb{R}^3\,.$$

Then $w_\pm = S_\Sigma f = u^s(\cdot, -\hat{x})$ on Σ and thus $w = u^s(\cdot, -\hat{x})$ in the exterior of Σ by the uniqueness of the exterior Dirichlet problem. This implies that the far field pattern coincide, i.e., $w^\infty(-\theta) = u^\infty(-\theta, -\hat{x}) = u^\infty(\hat{x}, \theta)$ for all $\theta \in S^2$ where we used the reciprocity principle (2.27). Furthermore, from the jump conditions of the normal derivative of the single layer potential we have that $f = \partial w_-/\partial \nu - \partial w_+/\partial \nu$. Substituting this for f yields

$$(\overline{H}_\Sigma^* f)(\theta) = \int_\Sigma f(y)\, e^{ik\,\theta\cdot y}\, ds(y)$$

$$= \int_\Sigma \left[\left(\frac{\partial w_-(y)}{\partial \nu} - \frac{\partial w_+(y)}{\partial \nu} \right) e^{ik\,\theta\cdot y} - \underbrace{(w_-(y) - w_+(y))}_{=0}\, \frac{\partial}{\partial \nu(y)} e^{ik\,\theta\cdot y} \right] ds(y)$$

$$= \int_\Sigma \left[w_+(y)\, \frac{\partial}{\partial \nu(y)} e^{ik\,\theta\cdot y} - \frac{\partial w_+(y)}{\partial \nu}\, e^{ik\,\theta\cdot y} \right] ds(y)$$

$$= w^\infty(-\theta) = u^\infty(\hat{x},\theta)$$

by the representation (1.26). Therefore, we have shown that

$$(H_\Sigma^* S_\sigma^{-1} F_\Sigma S_\Sigma^{-1} H_\Sigma g)(\hat{x}) = \left(g, \overline{u^\infty(\hat{x},\cdot)} \right)_{L^2(S^2)} = (Fg)(\hat{x})$$

which proves the assertion. $\square$

Now we can give a characterization of D in terms of the operator F_Σ.

Theorem 2.14 *Assume that k^2 is not a eigenvalue of the operator $-\Delta$ in the domain D with respect to the boundary condition under consideration. Furthermore, assume that k^2 is not a Dirichlet eigenvalue of $-\Delta$ in the domain Ω. Then, for any point $z \in \Omega$, the following inf-criterion holds:*

$$z \in D \iff \inf\left\{ \left| \langle \varphi, F_\Sigma S_\Sigma^{-1} S_\Sigma^* \varphi \rangle \right| : \varphi \in H^{-1/2}(\Sigma),\ \langle \varphi, \Phi(\cdot,z) \rangle = 1 \right\} > 0.$$

Proof: We recall the characterizations (1.66) for the Dirichlet boundary condition and (2.46) for the impedance boundary condition, i.e.,

$$z \in D \iff \inf\left\{ \left| (\psi, F\psi)_{L^2(S^2)} \right| : \psi \in L^2(S^2),\ (\psi, \phi_z)_{L^2(S^2)} = 1 \right\} > 0 \quad (2.53)$$

where $\phi_z(\hat{x}) = \exp(-ikz \cdot \hat{x})$. By (2.52), we can rewrite

$$(\psi, F\psi)_{L^2(S^2)} = \left(\psi, H_\Sigma^* S_\Sigma^{-1} F_\Sigma S_\Sigma^{-1} H_\Sigma \psi \right)_{L^2(S^2)} = \langle (S_\Sigma^{-1})^* H_\Sigma \psi,\ F_\Sigma S_\Sigma^{-1} H_\Sigma \psi \rangle$$

$$= \langle \varphi, F_\Sigma S_\Sigma^{-1} S_\Sigma^* \varphi \rangle$$

where we have set $\varphi = (S_\Sigma^{-1})^* H_\Sigma \psi = G_{Dir,\Sigma}^* \psi$. Furthermore, we observe again, that ϕ_z is the far field pattern of $\Phi(\cdot,z)$, hence $\phi_z = G_{Dir,\Sigma}\, \Phi(\cdot,z)$. So we can write for $\psi \in L^2(S^2)$:

$$(\psi, \phi_z)_{L^2(S^2)} = \left(\psi, G_{Dir,\Sigma}\, \Phi(\cdot,z) \right)_{L^2(S^2)} = \langle \varphi, \Phi(\cdot,z) \rangle.$$

The theorem follows now from the inf-criterion (2.53). $\square$

2.5   The $F_\#$ – factorization method

As we mentioned already in the previous chapter (remark at the end of Subsection 1.4.2) the inf-criterion (2.46) is very time consuming due to the non-convexity of the functional $\varphi \mapsto \left|(\varphi, F\varphi)_{L^2(S^2)}\right|$. This was the reason for the development of a criterion which uses an equation rather than a minimization problem. The fast square root algorithm, however, is valid (to our knowledge) only for far field operators which are normal. In this section we modify the approach in order to handle cases where the far field operator fails to be normal. Following [73], [75], and [77], we suggest to factorize the auxiliary self-adjoint positive operator

$$F_\# = |\operatorname{Re} F| + |\operatorname{Im} F| \tag{2.54}$$

instead of F. Here, the real and the imaginary parts of F are self-adjoint operators given by

$$\operatorname{Re} F = \frac{1}{2}(F + F^*) \quad \text{and} \quad \operatorname{Im} F = \frac{1}{2i}(F - F^*).$$

As already mentioned at the end of Subsection 1.4.3, for sound-soft and sound-hard obstacles the following criterion holds: A test point z belongs to D if and only if the integral equation

$$F_\#^{1/2}\varphi = \phi_z$$

has a solution in $L^2(S^2)$. We will show that this characterization keeps valid also in the case of an impedance boundary condition for arbitrary impedances $\lambda \in L^\infty(\Gamma)$ with $\operatorname{Im} \lambda \geq 0$. The operator $F_\#$ defined by (2.54) can be used in a number of other inverse elliptic boundary problems (see [120]) as well as in the MUSIC-algorithm (see [118]). We will come back to it in Chapters 3–6.

2.5.1   The functional analytic background

In this subsection we take a more general point of view and consider the following functional analytic situation.

Theorem 2.15 *Let $X \subset U \subset X^*$ be a Gelfand triple with a Hilbert space U and a reflexive Banach space X such that the imbedding is dense. Furthermore, let Y be a second Hilbert space and let $F : Y \to Y$, $G : X \to Y$, and $T : X^* \to X$ be linear bounded operators such that*

$$F = GTG^*. \tag{2.55}$$

We make the following assumptions:

(A1) G is compact with dense range.

(A2) There exists $t \in [0, 2\pi]$ such that $\operatorname{Re}[\exp(it)T]$ has the form $\operatorname{Re}[\exp(it)T] = C + K$ with some compact operator K and some self-adjoint and coercive operator $C : X^* \to X$, i.e., there exists $c > 0$ with

$$\langle \varphi, C\varphi \rangle \geq c\|\varphi\|^2 \quad \text{for all } \varphi \in X^*.$$

(A3) Im T *is compact and non-negative on* $\mathcal{R}(G^*) \subset X^*$, *i.e.,* $\langle \varphi, (\operatorname{Im} T)\varphi \rangle \geq 0$ *for all* $\varphi \in \mathcal{R}(G^*)$.

(A4) Re[exp$(it)T$] *is one-to-one or* Im T *is strictly positive on the closure* $\overline{\mathcal{R}(G^*)}$ *of* $\mathcal{R}(G^*)$, *i.e.,* $\langle \varphi, (\operatorname{Im} T)\varphi \rangle > 0$ *for all* $\varphi \in \overline{\mathcal{R}(G^*)}$ *with* $\varphi \neq 0$.

Then the operator $F_{\#} = \left| \operatorname{Re}[\exp(it)F] \right| + \operatorname{Im} F$ *is positive, and the ranges of* $G :$ $X \to Y$ *and* $F_{\#}^{1/2} : Y \to Y$ *coincide. Furthermore, the operator* $F_{\#}^{-1/2}G$ *is a bounded isomorphism from* X *onto* Y.

Proof: The proof is divided into six parts.

Part A (Reduction): In the first part we show that we can restrict ourselves to the case $X = U$ and $C = I$ and, furthermore, to the case where G is one-to-one.

Proof of Part A: The operator C is positively coercive, hence there exists a bounded isomorphism W from X^* onto U such that $C = W^*W$. We sketch the proof of this fact. First, the restriction $C|_U : U \to U$ is self-adjoint and positive. Therefore, there exists a bounded, self-adjoint and positive square root $\tilde{W} : U \to U$. From

$$\|\tilde{W}\varphi\|_U^2 = (\tilde{W}\varphi, \tilde{W}\varphi)_U = (\varphi, C\varphi)_U \leq \|C\|\|\varphi\|_{X^*}^2$$

we conclude that $\tilde{W}$ has a bounded extension W from X^* to U. It is easy to check that $W^* = \tilde{W}$. Therefore, we have that $C = W^*W$. The estimate $c\|\varphi\|_{X^*}^2 \leq \langle \varphi, C\varphi \rangle = \|W\varphi\|_U^2$ yields that W is an isomorphism from X^* onto U.

The factorization (2.55) can be rewritten as

$$F = [GW^*]\left[(W^*)^{-1}TW^{-1}\right][GW^*]^* = \tilde{G}\tilde{T}\tilde{G}^*,$$

with obvious meanings of $\tilde{G} : U \to Y$ and $\tilde{T} : U \to U$. The operator $\operatorname{Re}[\exp(it)\tilde{T}]$ has the decomposition as

$$\operatorname{Re}[\exp(it)\tilde{T}] = (W^*)^{-1}(C + K)W^{-1} = I + \tilde{K}.$$

The new operators $\tilde{G}, \tilde{T}$, and $\tilde{C} = I$ satisfy all the assumptions (A1)–(A4). Furthermore, $\mathcal{R}(\tilde{G}) = \mathcal{R}(G)$ holds since W^* is an isomorphism. Therefore, we can restrict ourselves to the case $X = U$ and $C = I$.

Next, we show that we can restrict ourselves to the case where G is one-to-one. Indeed, let $P : U \to \overline{\mathcal{R}(G^*)}$ be the orthogonal projector onto $\hat{U} = \overline{\mathcal{R}(G^*)} = \mathcal{N}(G)^{\perp} \subset U$. Then $PG^* = G^*$ and thus $G = GP$. Therefore, we can rewrite the factorization (2.55) in the form

$$F = GPTPG^* = \hat{G}\hat{T}\hat{G}^*$$

with $\hat{G} = G|_{\hat{U}}$ and $\hat{T} = PT|_{\hat{U}} : \hat{U} \to \hat{U}$. Again, all of the assumptions (A1)–(A4) are fulfilled.

Part B (Decomposition of U): We set $F_r = \operatorname{Re}[\exp(it)F] : Y \to Y$ for abbreviation. Then F_r is self-adjoint compact and satisfies

$$F_r = G T_r G^* \quad \text{with} \quad T_r = \operatorname{Re}[\exp(it)T]. \tag{2.56}$$

F_r has a complete orthonormal eigensystem $\{(\lambda_j, \psi_j) : j \in \mathbb{N}\}$. We split the space Y into the two closed orthogonal subspaces $Y = Y^- \oplus Y^+$ where

$$Y^- = \mathrm{span}\{\psi_j : \lambda_j \leq 0\} \quad \text{and} \quad Y^+ = \mathrm{span}\{\psi_j : \lambda_j > 0\}. \tag{2.57}$$

Therefore, the form $(F_r\varphi, \varphi)_Y$ is positive on Y^+ and non-positive on Y^-. Both subspaces are invariant under F_r.

We show that the spaces Y^- and $G^*(Y^-)$ are finite-dimensional, and

$$U = G^*(Y^-) + \overline{G^*(Y^+)}. \tag{2.58}$$

Proof: Let $\{\kappa_j, \phi_j\}$ be an eigensystem of the self-adjoint compact operator $K : U \to U$. Define

$$V^- = \mathrm{span}\{\phi_j : \kappa_j \leq -1\} \quad \text{and} \quad V^+ = \mathrm{span}\{\phi_j : \kappa_j > -1\}.$$

Then V^- is finite-dimensional since $\kappa_j \to 0$ by the compactness of K. Setting

$$c_1 = 1 + \min\{\kappa_j : \kappa_j > -1\}$$

we note that $c_1 > 0$ and

$$\big((I + K)\phi, \phi\big)_U \geq c_1 \|\phi\|_U^2 \quad \text{for all } \phi \in V^+.$$

We denote by $R^\pm : U \to V^\pm$ the orthogonal projector onto $V^\pm$. For any $\phi \in U$ we have $\phi = R^+\phi + R^-\phi$. In particular, for $\phi = G^*\psi^-$ with $\psi^- \in Y^-$ we conclude that

$$\begin{aligned}
0 \geq \big(F_r\psi^-, \psi^-\big)_Y &= \big((I + K)\phi, \phi\big)_U \\
&= \big((I + K)R^+\phi, R^+\phi\big)_U + \big((I + K)R^-\phi, R^-\phi\big)_U \\
&\geq c_1 \|R^+\phi\|_U^2 - c_2 \|R^-\phi\|_U^2 = c_1 \|\phi\|_U^2 - (c_1 + c_2)\|R^-\phi\|_U^2,
\end{aligned}$$

with $c_2 = \|I + K\|$. Therefore, we arrive at

$$\|\phi\|_U \leq \left(1 + \frac{c_2}{c_1}\right)^{1/2} \|R^-\phi\|_U$$

for all $\phi \in G^*(Y^-)$. Therefore, $R^-|_{G^*(Y^-)} : G^*(Y^-) \to V^-$ is one-to-one and thus, since V^- is finite-dimensional, also $G^*(Y^-)$ has finite dimension. From the injectivity of G^* we conclude that also Y^- is finite-dimensional. Furthermore, the sum $G^*(Y^-) + \overline{G^*(Y^+)}$ is dense in U, and, by a well-known result from functional analysis, this sum is also closed as the sum of a closed space and a finite-dimensional space. Therefore, $G^*(Y^-) + \overline{G^*(Y^+)} = U$ which ends the proof of Part B.

Part C (Construction of projectors):

(a) The intersection $U^0 = \overline{G^*(Y^+)} \cap G^*(Y^-)$ is finite-dimensional and $U^0 \subset \mathcal{N}(T_r)$.

(b) Let $U^+ \subset \overline{G^*(Y^+)}$ be any closed complementary subspace to U^0, i.e., $U^+ \oplus U^0 = \overline{G^*(Y^+)}$. With $U^- = G^*(Y^-)$ we have the splitting of U into a direct sum as

$$U = U^+ \oplus U^-. \tag{2.59}$$

The projectors $Q^\pm : U \rightarrow U^\pm$ onto $U^\pm$ parallel to $U^\mp$ are well defined and bounded.

(c) The difference $Q = Q^+ - Q^-$ is an involution, i.e., $Q^2 = I$, and hence an isomorphism from U onto itself.

Proof of Part C: We recall that $T_r = \mathrm{Re}[\exp(it)T]$.

(a) From Part B it follows that the intersection U^0 is finite-dimensional. Also, from (2.56) we conclude that $(T_r\varphi, \varphi)_U \geq 0$ for $\varphi \in \overline{G^*(Y^+)}$ and $(T_r\varphi, \varphi)_U \leq 0$ for $\varphi \in \overline{G^*(Y^-)}$. Therefore, $(T_r\varphi, \varphi)_U = 0$ for $\varphi \in U^0$. Now consider any $\varphi \in U^0$ and $\varphi^+ \in \overline{G^*(Y^+)}$. Then $\varphi_t = \varphi^+ + t\varphi$ for any $t \in \mathbb{C}$ belongs to $\overline{G^*(Y^+)}$ as well, and it holds that

$$0 \leq (T_r\varphi_t, \varphi_t)_U = (T_r\varphi^+, \varphi^+)_U + \bar{t}\,(T_r\varphi^+, \varphi)_U + t\,(T_r\varphi, \varphi^+)_U$$

$$= (T_r\varphi^+, \varphi^+)_U + 2\,\mathrm{Re}\big[t\,(T_r\varphi, \varphi^+)_U\big].$$

This is possible for all $t \in \mathbb{C}$ only if $(T_r\varphi, \varphi^+)_U = 0$. The same argument for $\varphi^- \in G^*(Y^-)$ shows that $(T_r\varphi, \varphi^-)_U = 0$ for any $\varphi^- \in G^*(Y^-)$. This implies $(T_r\varphi, \widetilde{\varphi})_U = 0$ for all $\widetilde{\varphi} \in U$. Hence $T_r\varphi = 0$ holds which proves that $U^0 \subset \mathcal{N}(T_r)$.

(b) By definition, the spaces $U^\pm$ are closed and satisfy $U^+ \cap U^- = \{0\}$ and $U^+ \oplus U^- = \overline{G^*(Y^+)} + \overline{G^*(Y^-)} = U$. It is well known that the projectors $Q^\pm$ are bounded, but we repeat the argument for the convenience of the reader. Consider the Banach space $U^+ \times U^-$ with norm $\|(\phi^+, \phi^-)\| = \|\phi^+\| + \|\phi^-\|$. The operator $A : U^+ \times U^- \rightarrow U$, defined by $A(\phi^+, \phi^-) = \phi^+ + \phi^-$ is a bounded bijection by (2.59) with $A^{-1}\phi = (Q^+\phi, Q^-\phi)$ and $\|A\| = 1$. According to the Open Mapping Theorem, A^{-1} is bounded which yields boundedness of Q^+ and Q^-.

(c) This follows immediately from the relation

$$Q^2 = (Q^+ - Q^-)(Q^+ - Q^-) = Q^+ + Q^- = I,$$

and completes the proof of Part C.

Part D (Factorization): Let $P^\pm : Y \rightarrow Y^\pm$ denote the orthogonal projectors onto $Y^\pm$. Then we can write

$$|F_r| = (P^+ - P^-)\,F_r = F_r\,(P^+ - P^-) = G\,T_r\,G^*(P^+ - P^-).$$

With

$$F_\# = |F_r| + \mathrm{Im}\,F = (P^+ - P^-)\,F_r + \mathrm{Im}\,F, \tag{2.60}$$

$$T_\# = T_r\,(Q^+ - Q^-) + \mathrm{Im}\,T. \tag{2.61}$$

we can prove the following factorization:

$$F_\# = G\,T_\#\,G^*. \tag{2.62}$$

Proof of Part D: First, we show the following commutation rules where we set again $T_r = \text{Re}[\exp(it)T]$.

$$T_r\, G^* P^\pm = T_r\, Q^\pm G^*\,. \tag{2.63}$$

Let $Q_0 : U \to U^0 = \overline{G^*(Y^+)} \cap G^*(Y^-)$ be any projector onto U^0 such that $Q_0|_{U^+} = 0$. For each $\varphi \in U$ we have the following relations which are easily seen from the definitions of $Q^\pm$ and Q_0.

$$Q^- G^* P^- \varphi = G^* P^- \varphi\,,$$

$$Q^- G^* P^+ \varphi = Q_0 G^* P^+ \varphi\,,$$

$$Q^+ G^* P^- \varphi = 0\,,$$

$$Q^+ G^* P^+ \varphi = G^* P^+ \varphi - Q_0 G^* P^+ \varphi\,.$$

Therefore, by these identities and by Part C, (c) we conclude that

$$T_r\, Q^- G^* = T_r\, Q^- G^* (P^+ + P^-) = T_r (Q_0 G^* P^+ + G^* P^-) = T_r\, G^* P^-$$

since $U^0 \subset \mathcal{N}(T_r)$ and

$$T_r\, Q^+ G^* = T_r\, Q^+ G^* (P^+ + P^-) = T_r (G^* P^+ - Q_0 G^* P^+) = T_r\, G^* P^+\,,$$

which proves (2.63). Using these relations and (2.55) we have:

$$F_\# = F_r\,(P^+ - P^-) + \text{Im}\, F = G\, T_r\, G^* (P^+ - P^-) + G(\text{Im}\, T)G^*$$

$$= G\, T_r (Q^+ - Q^-)G^* + G(\text{Im}\, T)G^* = G\, T_\#\, G^*$$

which ends the proof of Part D.

Part E (Coercivity): The operator $T_\#$ given by (2.61) is self-adjoint and positively coercive, i.e., there exists $c > 0$ with

$$(T_\# \varphi, \varphi)_U \geq c\|\varphi\|_U^2 \quad \text{for all } \varphi \in U\,. \tag{2.64}$$

Proof of Part E: The operator $T_\# : U \to X$ is self-adjoint, since

$$(T_\# G^* \psi_1, G^* \psi_2)_U = (F_\# \psi_1, \psi_2)_Y = (\psi_1, F_\# \psi_2)_Y = (G^* \psi_1, T_\# G^* \psi_2)_U$$

and $\mathcal{R}(G^*)$ is dense in U (note that G is one-to-one). It follows from Assumptions (A2)–(A4) that T_r and $T_\#$ are Fredholm operators of index 0, because $Q^+ - Q^-$ is an isomorphism and $\text{Im}\, T$ is compact. The operator $T_\#$ is strictly positive, since $T_r\,(Q^+ - Q^-)$ is non-negative and $\text{Im}\, T$ or $T_r\,(Q^+ - Q^-)$ is strictly positive by Assumption (A4). Hence $T_\#$ has a bounded inverse $T_\#^{-1}$. From the positivity of the form

$$\big(T_\#(\varphi + t\psi), \varphi + t\psi\big)_U = (T_\# \varphi, \varphi)_U + 2\,\text{Re}\big[t(\psi, T_\# \varphi)_U\big] + |t|^2 (T_\# \psi, \psi)_U$$

for any $\varphi, \psi \in U$ and any $t \in \mathbb{C}$ we conclude that

$$\big|(T_\# \varphi, \psi)\big|^2 \leq (T_\# \varphi, \varphi)_U (T_\# \psi, \psi)_U\,.$$

For $\psi = T_\#^{-1}\varphi$ this yields

$$\|\varphi\|_U^4 = \left|(T_\#\varphi, T_\#^{-1}\varphi)_U\right|^2 \le (T_\#\varphi, \varphi)_U (\varphi, T_\#^{-1}\varphi)_U \le (T_\#\varphi, \varphi)_U \|T_\#^{-1}\| \|\varphi\|_U^2,$$

i.e.,

$$\|T_\#^{-1}\|^{-1} \|\varphi\|_U^2 \le (T_\#\varphi, \varphi)_U,$$

which proves the coercivity of $T_\#$ with $c = \|T_\#^{-1}\|_U^{-1}$.

Part F (Final part): Now we can complete the proof of Theorem 2.15. The operator $F_\#$ has two factorizations:

$$F_\# = G\,T_\#\,G^* \quad \text{and also} \quad F_\# = F_\#^{1/2}(F_\#^{1/2})^*.$$

Application of Theorem 1.21 yields

$$\mathcal{R}(G) = \mathcal{R}(F_\#^{1/2}) \tag{2.65}$$

and that the operators $F_\#^{-1/2}G$ and $G^{-1}F_\#^{1/2}$ are isomorphisms. $\quad\square$

Remark: In the case where both of the subspaces U_1 and U_2 are infinite-dimensional the condition that $U_1 + U_2$ is dense in U does not guarantee that $\overline{U_1} + \overline{U_2} = U$. We give a simple example. Let

$$U = \ell^2 = \left\{ \{c_k\} : \sum_{k=-\infty}^{\infty} |c_k|^2 < \infty \right\},$$

$$U_1 = \left\{ \{c_k\} \in \ell^2 : c_k = 0 \text{ for all } k < 0 \right\},$$

$$U_2 = \left\{ \{c_k\} \in \ell^2 : c_k = k\,c_{-k} \text{ for all } k > 0 \right\}.$$

Both subspaces U_1 and U_2 are closed. The sum $U_1 + U_2$ is dense in U because it contains the subspace ℓ_0^2 of all finite sequences $\{c_k\}$ (i.e., sequences which vanishing members for sufficiently large k). On the other hand any $\{c_k\} \in U_1 + U_2$ satisfies $c_k = \mathcal{O}(|k|^{-1})$ for $k \to -\infty$. However, it is easy to find an element $\{c_k\} \in \ell^2$ for which this condition does not hold, for example, $c_k = (-k)^{-2/3}$ for $k < 0$ and $c_k = 0$ for $k \ge 0$.

Remark: In the case where the operator $\operatorname{Im} T$ is negative, one sets

$$F_\# = |\operatorname{Re}[\exp(it)F]| - \operatorname{Im} F = |\operatorname{Re}[\exp(it)F]| + |\operatorname{Im} F|.$$

2.5.2 Applications to some inverse scattering problems

The abstract Theorem 2.15 can immediately be applied to the inverse scattering problem with Dirichlet, Neumann, or impedance boundary condition. Indeed, in the Dirichlet case we have $T = -S^*$ and $C = -S_i$ which is negatively coercive by Lemma 1.14, part (c). Therefore, we set $t = \pi$. For the impedance boundary condition (or, as a special case, the Neumann boundary condition) we have $T = -T_{imp}^*$ and $C = -N_i$, which is positively coercive by Theorem 1.26, part (e). For this case we set $t = 0$. Furthermore, assumption (A3) and (A4) are satisfied by Lemma 1.14 again for the case of Dirichlet boundary conditions and by the arguments in the proof of Lemma 2.7 for the case of impedance boundary conditions. Therefore, we can characterize the range of G by $F_\# = |\operatorname{Re} F| + \operatorname{Im} F$. Furthermore, from Theorem 1.12 we know that a given point $z \in \mathbb{R}^3$ belongs to D if, and only if, $\phi_z \in \mathcal{R}(G_{Dir})$ which coincides with $\mathcal{R}(G_{imp})$ by the proof of Theorem 2.8. Here ϕ_z is given again by (2.47). We conclude:

Corollary 2.16 *Consider the inverse scattering problem with respect to the Dirichlet boundary condition or the impedance boundary condition. In the latter case we assume that $\operatorname{Im} \lambda \geq 0$ almost everywhere on Γ. Furthermore, we assume that k^2 is not an eigenvalue of $-\Delta$ in D with respect to the boundary condition under consideration. Then the obstacle D is characterized by those points $z \in \mathbb{R}^3$ for which*

$$\sum_{j=1}^{\infty} \frac{\left|(\phi_z, \psi_j)_{L^2(S^2)}\right|^2}{\lambda_j} < \infty. \tag{2.66}$$

Here, $\phi_z(\hat{x}) = \exp(-ik\, z \cdot \hat{x})$ for $\hat{x} \in S^2$ and $\{\lambda_j, \psi_j\}$ is an eigensystem of the (positive) operator $F_\# : L^2(S^2) \to L^2(S^2)$ given by $F_\# = |\operatorname{Re} F| + |\operatorname{Im} F|$.

The same technique is also applicable for the reconstruction from the limited aperture data. Let the data set contain all of the far field patterns $\{u^\infty(\hat{x}, \theta) : \hat{x}, \theta \in U\}$. Here again, $U \subset S^2$ is an open subdomain of the unit sphere. We have shown in Section 2.3 that the far field operator F_{la} from (2.3) is factorized as $F_{la} = -G_{la} T^* G_{la}^*$ with the same T as for the case of complete data, see Theorem 2.9. We apply Theorem 2.15, Lemma 2.10 and summarize the result in the following corollary.

Corollary 2.17 *Consider the same situation as in the previous Corollary 2.16. Assume again that k^2 is not an eigenvalue of $-\Delta$ in D with respect to the boundary condition under consideration. Then the obstacle D is characterized by those points $z \in \mathbb{R}^3$ for which*

$$\sum_{j=1}^{\infty} \frac{\left|(\phi_z, \psi_j)_{L^2(U)}\right|^2}{\lambda_j} < \infty. \tag{2.67}$$

Now, $\phi_z(\hat{x}) = \exp(-ik\, z \cdot \hat{x})$ for $\hat{x} \in U$ and $\{\lambda_j, \psi_j\}$ is an eigensystem of the (positive) operator $F_\# = |\operatorname{Re} F_{la}| + |\operatorname{Im} F_{la}| : L^2(U) \to L^2(U)$.

2.6 Obstacle scattering in a half-space

In this section we study the scattering of an incident plane wave by an obstacle D which is imbedded in the homogeneous half-space $\mathbb{R}^3_+ = \{x \in \mathbb{R}^3 : x_3 > 0\}$. The physical background is the same as in the case of whole space, but an additional boundary condition is posed on the plane $\mathbb{R}^3_0 = \{x \in \mathbb{R}^3 : x_3 = 0\}$.

The boundary conditions on $\mathbb{R}_0^3$ produce reflected parts of the incident field $x \mapsto \exp(ik\,\theta \cdot x)$. We will restrict ourselves to the Dirichlet boundary condition on $\mathbb{R}_0^3$, i.e., the total field for the case where no obstacle D is present is given by

$$u^i(x, \theta) = e^{ik\,\theta \cdot x} - e^{ik\,\theta' \cdot x}, \quad x \in \mathbb{R}_+^3. \tag{2.68}$$

Here, y' denotes the reflection of $y \in \mathbb{R}^3$ with respect to the plane $\mathbb{R}_0^3$, i.e.,

$$y' = (y_1, y_2, y_3)' = (y_1, y_2, -y_3), \quad y \in \mathbb{R}^3. \tag{2.69}$$

We can equally well pose a Neumann boundary condition on $\mathbb{R}_0^3$. In this case the total field is given by

$$u^i(x, \theta) = e^{ik\,\theta \cdot x} + e^{ik\,\theta' \cdot x}, \quad x \in \mathbb{R}_+^3.$$

As in the case of whole space we suppose that the scattered field u^s satisfies the Helmholtz equation

$$\Delta u^s + k^2 u^s = 0 \quad \text{in } \mathbb{R}_+^3 \setminus \overline{D} \tag{2.70}$$

and the Sommerfeld radiation condition (2.3) for $\hat{x} \in S_+^2 = S^2 \cap \mathbb{R}_+^3$, i.e.,

$$\frac{\partial u^s}{\partial r} - ik\,u^s = \mathcal{O}\left(r^{-2}\right) \quad \text{for } r \to \infty \tag{2.71}$$

uniformly with respect to $\hat{x} = x/|x| \in S_+^2$. In addition, a boundary condition on Γ has to be assumed which is of either Dirichlet type or impedance type, i.e., either

$$u^i + u^s = 0 \quad \text{on } \Gamma$$

or

$$\frac{\partial(u^i + u^s)}{\partial \nu} + \lambda\,(u^i + u^s) = 0 \quad \text{on } \Gamma. \tag{2.72}$$

This model is used in various areas like geophysics, underwater acoustics or electromagnetic imaging, see [165]. As the first possible application of the corresponding inverse problem we mention obstacle imaging in ocean environments, see e.g., [185]. For the mathematical modeling techniques and methods of underwater acoustics we refer to [109], for recent results in inverse underwater acoustics see the special issue of the *Inverse Problems* [184].

A related – but different – problem of the parameter reconstruction of an inhomogeneous elastic half-space (or half-plane) has been studied in [31]. In [136] and [64] the uniqueness of the reconstruction of the compactly disturbed medium was proven under certain conditions. In contrast to the problem to identify the complete parameter we are only interested in the determination of the obstacle, i.e., in the reconstruction of the shape and the position of the scatterer.

2.6.1 The direct scattering problem

In the remaining part of this chapter we restrict ourselves to the case of the Dirichlet boundary condition on $\mathbb{R}_0^3$ and a impedance boundary condition on Γ. The direct scattering problem is formulated as follows, see [71].

For a given incident plane wave $u^i(\cdot, \theta)$ of the form (2.68) for $\theta \in S_+^2$ find the scattered wave $u^s(\cdot, \theta) \in C^2(\mathbb{R}_+^3 \setminus \overline{D}) \cap C^1((\mathbb{R}_+^3 \cup \mathbb{R}_0^3) \setminus D)$ which satisfies the Helmholtz equation (2.70) and the Sommerfeld radiation condition (2.71) in $\mathbb{R}_+^3$. The total wave $u = u^i + u^s$ has to satisfy the homogeneous Dirichlet boundary condition on $\mathbb{R}_0^3$ and the homogeneous impedance boundary condition (2.72) on Γ.

In the following theorem we reduce the question of existence and uniqueness of the exterior boundary value problem for arbitrary right-hand sides $g \in H^{-1/2}(\Gamma)$ to a boundary value problem in the whole space by a reflection argument.

Theorem 2.18 *Let $k > 0$ be a fixed wavenumber, $g \in H^{-1/2}(\Gamma)$ and $\lambda \in L^\infty(\Gamma)$ with $\operatorname{Im} \lambda \geq 0$ on Γ. We define $\widetilde{D} = D \cup D'$ and $\widetilde{\Gamma} = \Gamma \cup \Gamma'$ where we define D' and Γ' as the reflected sets, i.e., $V' = \{x' : x \in V\}$ for $V = D$ or $V = \Gamma$.*

We extend g and λ to odd and even functions, respectively, on $\widetilde{\Gamma}$, i.e., $\tilde{g} = g$ on Γ and $\tilde{g}(x) = -g(x')$ on Γ' and $\tilde{\lambda} = \lambda$ on Γ and $\tilde{\lambda}(x) = \lambda(x')$ on Γ'.

(a) Let $v \in H_{loc}^1(\mathbb{R}_+^3 \setminus \overline{D})$ be a solution of the Helmholtz equation (2.70) in $\mathbb{R}_+^3$ which satisfies the homogeneous Dirichlet boundary condition $v = 0$ on $\mathbb{R}_0^3$, the inhomogeneous impedance boundary condition

$$\frac{\partial v}{\partial \nu} + \lambda v = g \quad on \ \Gamma, \tag{2.73}$$

and the Sommerfeld radiation condition (2.71). Then the odd extension $\tilde{v}$ of v (i.e., $\tilde{v} = v$ on $\mathbb{R}_+^3 \setminus D$ and $\tilde{v}(x) = -v(x')$ on $\mathbb{R}_-^3 \setminus D'$) satisfies the Helmholtz equation in $\mathbb{R}^3 \setminus (\overline{D} \cup \overline{D}')$, the Sommerfeld radiation condition (2.3), and the impedance boundary condition

$$\frac{\partial \tilde{v}}{\partial \nu} + \tilde{\lambda}\tilde{v} = \tilde{g} \quad on \ \widetilde{\Gamma}. \tag{2.74}$$

The solutions are again understood in the variational sense as in Section 2.1.

(b) Let, on the other hand, $\tilde{v} \in H_{loc}^1(\mathbb{R}^3)$ be a solution of the Helmholtz equation in $\mathbb{R}^3 \setminus (\overline{D} \cup \overline{D}')$, the Sommerfeld radiation condition (2.3), and the impedance boundary condition (2.74). Then the restriction $v = \tilde{v}|_{\mathbb{R}_+^3 \setminus D}$ solves the corresponding boundary value problem in $\mathbb{R}_+^3 \setminus D$.

(c) The boundary value problem (2.70), (2.71), (2.73) has a unique solution.

Proof: (a) First we note that $\tilde{v} \in H_{loc}^1(\mathbb{R}^3)$ because of the boundary condition $v = 0$ on $\mathbb{R}_0^3$ (cf. [167], Section 6.4). Second, the extended function $\tilde{v}$ satisfies $\Delta\tilde{v}(x') = \Delta v(x)$ and $\partial\tilde{v}(x')/\partial\nu = -\partial v(x)/\partial\nu$ for $x \in \Gamma$ which proves part (a).

(b) We set $\hat{v}(x) = -\tilde{v}(x')$ for $x \notin \widetilde{D}$. Then $\hat{v}$ solves the same exterior boundary value problem as $\tilde{v}$ since $\widetilde{D}$ and $\tilde{\lambda}$ are symmetric with respect to $\mathbb{R}_0^3$ and $\tilde{g}$ is an odd function.

By the uniqueness of the exterior boundary value problem (compare Theorem 2.2) we conclude that $\hat{v} = \tilde{v}$ i.e., in particular, that $\tilde{v}$ vanishes on $\mathbb{R}^3_0$.

(c) This follows again from Theorem 2.2. $\square$

As a consequence of this theorem we note that the far field pattern $v^\infty = \tilde{v}^\infty|_{S^2_+}$ is well defined and analytic on S^2_+. The solution v has the asymptotic behavior

$$v(x) = \frac{\exp(ik\,|x|)}{4\pi\,|x|}\,v^\infty(\hat{x}) + \mathcal{O}(|x|^{-2}) \quad \text{as } |x| \to \infty, \tag{2.75}$$

uniformly with respect to $\hat{x} \in S^2_+$.

The scattered field $u^s(\cdot, \theta)$ corresponding to the incident plane wave $u^i(\cdot, \theta)$ is well defined as a special case of the previous theorem. Also the far field pattern $u^\infty = u^\infty(\hat{x}, \theta)$ for $\hat{x}, \theta \in S^2_+$ is well defined. We define the far field operator F from $L^2(S^2_+)$ into itself by

$$(F\varphi)(\hat{x}) = \int_{S^2_+} u^\infty(\hat{x}, \theta)\,\varphi(\theta)\,ds(\theta), \quad \hat{x} \in S^2_+. \tag{2.76}$$

Also, we define the data-to-pattern operator $G : H^{-1/2}(\Gamma) \to L^2(S^2_+)$ by

$$Gg = v^\infty|_{S^2_+}, \tag{2.77}$$

where v^∞ is the far field pattern of the solution v of (2.70), (2.71), and (2.73).

As in the previous theorem we can relate the operator $\tilde{G} : H^{-1/2}(\tilde{\Gamma}) \to L^2(S^2)$ to G which can be considered as the odd extensions of G and is the data-to-pattern operator for the exterior boundary value problem in $\mathbb{R}^3 \setminus \tilde{D}$ with impedance $\tilde{\lambda}$ on $\tilde{\Gamma}$. Analogously, $\tilde{F} : L^2(S^2) \to L^2(S^2)$ is the corresponding far field operator.

We want to reformulate the previous theorem in terms of two operators. First, let J be the operator defined by $(J\psi)(x) = \psi(x')$. We consider J as an operator from $L^2(S^2)$ into itself as well as from $H^{-1/2}(\tilde{\Gamma})$ into itself and indicate this by writing J_{S^2} or $J_{\tilde{\Gamma}}$, respectively. Second, let R be the restriction operator, either from $L^2(S^2)$ onto $L^2(S^2_+)$ or from $H^{-1/2}(\tilde{\Gamma})$ onto $H^{-1/2}(\Gamma)$, denoted by $R_{S^2_+}$ or R_Γ, respectively. The adjoint R^* is then the extension by zero.

Lemma 2.19 *The following relations hold:*

(a) $G R_\Gamma\,(I - J_{\tilde{\Gamma}}) = R_{S^2_+}\,\tilde{G}\,(I - J_{\tilde{\Gamma}}).$

(b) $\tilde{G} J_{\tilde{\Gamma}} = J_{S^2}\,\tilde{G}$ *and* $\tilde{F} J_{S^2} = J_{S^2}\,\tilde{F}.$

(c) $F = R_{S^2_+}\,\tilde{F}\,(I - J_{S^2})\,R^*_{S^2_+}$ *and* $G = R_\Gamma\,\tilde{G}\,(I - J_{\tilde{\Gamma}})\,R^*_\Gamma.$

Proof: (a) For $g \in H^{-1/2}(\tilde{\Gamma})$ we set $\tilde{g} = (I - J_{\tilde{\Gamma}})g$. Then $\tilde{g}$ is an odd function in $H^{-1/2}(\tilde{\Gamma})$. Furthermore, $GR_\Gamma(I - J_{\tilde{\Gamma}})g = GR_\Gamma\tilde{g} = v^\infty$ where v^∞ is the far field pattern corresponding to v with boundary condition $v = R_\Gamma\tilde{g} = \tilde{g}|_\Gamma$ on Γ. Theorem 2.18 yields (using the same notations) that $v|_\Gamma = \tilde{v}|_\Gamma$ and thus $v^\infty = \tilde{v}^\infty|_{S^2_+} = R_{S^2_+}\,\tilde{G}\,\tilde{g}$.

(b) This follows directly from Theorem 2.18.

(c) This time, let $\tilde{f} = (I - J_{S^2}) R^*_{S^2_+} f$ for $f \in L^2(S^2_+)$. Then it is easily seen that $\tilde{f}$ is the odd extension of f to S^2 and thus $R_{S^2_+} \tilde{F}(I - J_{\tilde{\Gamma}}) R^*_{S^2_+} f = R_{S^2_+} \tilde{F} \tilde{f} = Ff$ by Theorem 2.18. $\square$

2.6.2 The factorization method for the inverse problem

We assume again that the total wave $u = u^i + u^s$ satisfies the homogeneous Dirichlet boundary condition on $\mathbb{R}^3_0$ and the homogeneous impedance boundary condition (2.2) on Γ. The knowledge of the range $\mathcal{R}(G)$ of the data-to-pattern operator G from (2.77) allows us to characterize the domain D by the following lemma.

Lemma 2.20 *For $z \in \mathbb{R}^3_+$ we define the function*

$$\phi_z(\theta) = e^{-ik\, z\cdot\theta} - e^{-ik\, z\cdot\theta'}, \quad \theta \in S^2_+ . \tag{2.78}$$

Then ϕ_z belongs to $\mathcal{R}(G)$ if and only if $z \in D$.

Proof: This follows from the corresponding result for the whole space, i.e., Theorem 2.4 combined with Theorem 1.12. $\square$

Now we turn to the factorization of the far field operator as we have done it for the other boundary value problems treated so far. Let, again, $\tilde{G}$ and $\tilde{F}$ correspond to the scattering problem with impedance boundary conditions on $\tilde{\Gamma} = \Gamma \cup \Gamma'$ where the impedance λ has been extended to an even function $\tilde{\lambda} \in L^\infty(\tilde{\Gamma})$. We denote by $\tilde{T} : H^{1/2}(\tilde{\Gamma}) \to H^{-1/2}(\tilde{\Gamma})$ the operator T_{imp} of (2.39) for the extended problem.

With Lemma 2.19 and $(I - J_{S^2})^2 = 2\,(I - J_{S^2})$ and the factorization (2.38) we have:

$$F = R_{S^2_+} \tilde{F}(I - J_{S^2})\, R^*_{S^2_+} = \frac{1}{2}\, R_{S^2_+} \tilde{F}(I - J_{S^2})^2\, R^*_{S^2_+}$$

$$= \frac{1}{2}\, R_{S^2_+}\, (I - J_{S^2}) \tilde{F}(I - J_{S^2})\, R^*_{S^2_+}$$

$$= -\frac{1}{2}\, R_{S^2_+}\, (I - J_{S^2}) \tilde{G} \tilde{T} \tilde{G}^*(I - J_{S^2})\, R^*_{S^2_+}$$

$$= -\frac{1}{2}\, \big[R_{S^2_+} \tilde{G}\,(I - J_{\tilde{\Gamma}})\big] \tilde{T} \big[R_{S^2_+} \tilde{G}\,(I - J_{\tilde{\Gamma}})\big]^*$$

$$= -\frac{1}{2}\, \big[GR_{\Gamma}\,(I - J_{\tilde{\Gamma}})\big] \tilde{T} \big[GR_{\Gamma}\,(I - J_{\tilde{\Gamma}})\big]^*$$

$$= -\frac{1}{2}\, GR_{\Gamma}\,(I - J_{\tilde{\Gamma}}) \tilde{T}(I - J_{\tilde{\Gamma}})\, R^*_{\Gamma}\, G^*$$

i.e.,

$$F = -G\,T\,G^* \tag{2.79}$$

with

$$T = \frac{1}{2}\, R_{\Gamma}\,(I - J_{\tilde{\Gamma}}) \tilde{T}(I - J_{\tilde{\Gamma}})\, R^*_{\Gamma} . \tag{2.80}$$

Lemma 2.21 *Assume that k^2 is not an eigenvalue of $-\Delta$ in D with respect to the impedance boundary condition where again* $\operatorname{Im}\lambda \geq 0$. *Then the operator* $T : H^{1/2}(\Gamma) \to H^{-1/2}(\Gamma)$ *defined by (2.80) is coercive.*

Proof: The operator $\tilde{T}$ is coercive according to Proposition 2.7. Therefore, for any $\psi \in H^{1/2}(\Gamma)$ we have that

$$
\left| \langle T\psi, \psi \rangle \right| = \frac{1}{2} \left| \langle \tilde{T}(I - J_{\tilde{\Gamma}})R_\Gamma^* \psi, (I - J_{\tilde{\Gamma}})R_\Gamma^* \psi \rangle \right|
$$
$$
\geq \frac{c}{2} \left\| (I - J_{\tilde{\Gamma}})R_\Gamma^* \psi \right\|_{H^{1/2}(\tilde{\Gamma})}^2 = c \left\| \psi \right\|_{H^{1/2}(\Gamma)}^2 ,
$$

since $(I - J_{\tilde{\Gamma}})R_\Gamma^* \psi$ is just the odd extension of ψ to $\tilde{\Gamma}$. This proves coercivity of T with the same constant as for $\tilde{T}$. $\square$

Now again, we can apply the Theorem 1.16 and characterize D by the following inf-criterion.

Theorem 2.22 *Assume that k^2 is not an eigenvalue of $-\Delta$ in D with respect to the impedance boundary condition where again* $\operatorname{Im}\lambda \geq 0$. *Then, for any point $z \in \mathbb{R}^3_+$,*

$$
z \in D \iff \inf\left\{ \left| (F\psi, \psi)_{L^2(S_+^2)} \right| : \psi \in L^2(S_+^2), \ (\phi_z, \psi)_{L^2(S_+^2)} = 1 \right\} > 0 ,
$$

where ϕ_z is again given by (2.78).

Proof: We combine Lemma 2.20 and Theorem 1.16. $\square$

Finally, we want to apply the $F_\#$ – factorization method of Subsection 2.5.1. In order to apply Theorem 2.15 we have to check the assumptions (A1)–(A4) for the operator T given by (2.80). Assumption (A1) follows from Lemma 2.19 Part (c) since $\tilde{G}$ is compact. For (A2) we recall from the proof of Corollary 2.16 that $\operatorname{Re}\tilde{T}$ has the required form $\operatorname{Re}\tilde{T} = C + K$ with compact K and coercive C. Then $\operatorname{Re}T$ has the same form with C and K replaced by $-\frac{1}{2} R_\Gamma (I - J_{\tilde{\Gamma}}) C (I - J_{\tilde{\Gamma}}) R_\Gamma^*$ and $-\frac{1}{2} R_\Gamma (I - J_{\tilde{\Gamma}}) K (I - J_{\tilde{\Gamma}}) R_\Gamma^*$, respectively. The same argument yields (A3). For part (A4) we have that

$$
\operatorname{Im}\langle T\psi, \psi \rangle = -\frac{1}{2} \operatorname{Im}\langle \tilde{T}(I - J_{\tilde{\Gamma}})R_\Gamma^* \psi, (I - J_{\tilde{\Gamma}})R_\Gamma^* \psi \rangle
$$

and this is positive by the corresponding property of $\tilde{T}$ provided $\tilde{\psi} = (I - J_{\tilde{\Gamma}}) R_\Gamma^* \psi$ does not vanish for $\psi \neq 0$ and belongs to $\overline{\mathcal{R}(\tilde{G}_{imp}^*)}$. But this is the case since $\tilde{\psi}$ is just the odd continuation of ψ on $\tilde{\Gamma}$. Therefore, we have shown:

Corollary 2.23 *Assume that k^2 is not an eigenvalue of $-\Delta$ in D with respect to the impedance boundary condition where we again assume that* $\operatorname{Im}\lambda \geq 0$ *on Γ. Then the following criterion holds for any point $z \in \mathbb{R}^3_+$:*

$$
z \in D \iff \sum_{j=1}^{\infty} \frac{\left| (\phi_z, \psi_j)_{L^2(S_+^2)} \right|^2}{\lambda_j} < \infty , \tag{2.81}
$$

where $\phi_z \in L^2(S_+^2)$ is again given by (2.78), i.e.,

$$\phi_z(\theta) = e^{-ik\,z\cdot\theta} - e^{-ik\,z\cdot\theta'}, \quad \theta \in S_+^2\,,$$

and $\{\lambda_j, \psi_j : j \in \mathbb{N}\}$ is an eigensystem of the positive and self-adjoint operator $F_\# = |\operatorname{Re} F| + \operatorname{Im} F : L^2(S_+^2) \to L^2(S_+^2)$.

We finish this section with some remarks.

As mentioned above, we can equally well pose a homogeneous Neumann boundary condition on $\mathbb{R}_0^3$. The odd extensions of g and v have to be replaced by the even extensions, and in (2.78) the minus sign has to be replaced by the plus sign. Furthermore, also the Dirichlet boundary condition can be assumed on Γ. For all these cases the characterizations of Theorem 2.22 and Corollary 2.23 remain valid. In particular, the characterization is independent on the type of the boundary condition on Γ.

Also, one can treat the case of limited data. Let the far field patterns $u^\infty(\hat{x}, \theta)$ be given for all $\hat{x}, \theta \in U$ where U is some open (relative to S_+^2) subset of S_+^2. As in Section 2.3 we introduce the restriction operator $P_U : L^2(S_+^2) \to L^2(U)$ by $P_U\varphi = \varphi|_U$ and observe that its adjoint operator $P_U^* : L^2(U) \to L^2(S_+^2)$ is just the zero-extension onto S_+^2. Then the operator

$$(F_{la}\varphi)(\hat{x}) = \int_U u^\infty(\hat{x}, \theta)\, \varphi(\theta)\, ds(\theta)\,, \quad \hat{x} \in U\,,$$

can be written as $F_{la} = P_U F P_U^*$. Combining this with the factorization (2.79) yields, by the same arguments as in Section 2.3, the characterizations of Theorem 2.22 and Corollary 2.23.

3

The mixed boundary value problem

So far, we assumed always one type of boundary condition on the components of the obstacle D which was either of Dirichlet type or of impedance type (which included the Neumann boundary condition as a special case). In this chapter we will consider mixed boundary conditions, i.e., D consists of two open and bounded domains $D = D_1 \cup D_2$ such that $D_1, D_2 \subset \mathbb{R}^3$ satisfy $\overline{D_2} \cap \overline{D_2} = \emptyset$. We do not assume that each D_j has to be connected, it may consist of several (finitely many) connected components. However, we assume that the exterior $\mathbb{R}^3 \setminus \overline{D}$ is connected. We will pose Dirichlet boundary conditions on $\Gamma_1 = \partial D_1$ and impedance boundary conditions on $\Gamma_2 = \partial D_2$. In our presentation we follow mainly the work published in [73], [72], and [74].

The direct scattering problem is formulated as follows.

For a given incident plane wave $u^i(\cdot, \theta)$ of the form (2.21), i.e.,

$$u^i(x) = e^{ik\, x \cdot \theta}, \quad x \in \mathbb{R}^3,$$

find a scattered wave $u^s = u^s(\cdot, \theta) \in C^2(\mathbb{R}^3 \setminus \overline{D}) \cap C^1(\mathbb{R}^3 \setminus D)$ which satisfies the Helmholtz equation

$$\Delta u^s + k^2 u^s = 0$$

in the exterior of D and the Sommerfeld radiation condition (2.3). The total wave $u = u^i + u^s$ has to satisfy the boundary conditions

$$u = 0 \qquad \text{on } \Gamma_1 = \partial D_1, \tag{3.1}$$

$$\frac{\partial u}{\partial \nu} + \lambda u = 0 \qquad \text{on } \Gamma_2 = \partial D_2. \tag{3.2}$$

Here again, $\nu = \nu(x)$ denotes the unit normal vector at $x \in \partial D$ which is directed into the exterior of D. In this setting we can expect a classical solution only for continuous impedances λ. In order to treat the case $\lambda \in L^\infty(\Gamma_2)$ we have again to weaken the notion of a solution. In the next section we will derive some of the basic properties of the solution of the direct scattering problem before we proceed with the inverse scattering problem.

3.1 The direct scattering problem

The basis is again the following existence and uniqueness theorem of the corresponding exterior boundary value problem.

Theorem 3.1 *Let the wavenumber $k > 0$ and a function $\lambda \in L^\infty(\Gamma_2)$ be fixed such that $\operatorname{Im}\lambda \geq 0$ on Γ_2. For any $f \in H^{1/2}(\Gamma_1)$ and $g \in H^{-1/2}(\Gamma_2)$ the following exterior boundary value problem has a unique solution $v \in H^1_{loc}(\mathbb{R}^3 \setminus \overline{D})$:*

$$\Delta v + k^2 v = 0 \quad in \ \mathbb{R}^3 \setminus \overline{D}, \tag{3.3}$$

$$\frac{\partial v}{\partial r} - ik\,v = \mathcal{O}\left(r^{-2}\right) \quad for \ r = |x| \to \infty \tag{3.4}$$

uniformly with respect to $\hat{x} = x/|x| \in S^2$,

$$v = f \quad on \ \Gamma_1 = \partial D_1, \tag{3.5}$$

$$\frac{\partial v}{\partial \nu} + \lambda v = g \quad on \ \Gamma_2 = \partial D_2. \tag{3.6}$$

The solution has to be understood in the variational sense, compare (1.22) and (2.5) of the previous chapters, i.e., $v \in H^1_{loc}(\mathbb{R}^3 \setminus \overline{D})$ solves (3.3), (3.5), and (3.6) if $v|_{\Gamma_1} = f$ in the sense of traces and

$$\iint_{\mathbb{R}^3 \setminus \overline{D}} \left[\nabla v \cdot \nabla \overline{\psi} - k^2 v\,\overline{\psi}\right] dx - \int_{\Gamma_2} v\,\overline{\psi}\,ds = \langle g, \psi \rangle$$

for all $\psi \in H^1(\mathbb{R}^3 \setminus \overline{D})$ with compact support.

Proof: This can be proved in a standard way as in the case of only one type of Dirichlet or impedance boundary condition. We refer to Theorem 2.2 or, for a detailed proof of this theorem, to [144], Theorem 7.15 (iii). $\square$

Applying this theorem to the particular choice

$$f(x) = -\exp(ik\theta \cdot x)|_{\Gamma_1} \quad \text{and} \quad g(x) = -\left(\frac{\partial \exp(ik\theta \cdot x)}{\partial \nu} + \lambda(x)\exp(ik\theta \cdot x)\right)\Bigg|_{\Gamma_2}$$

proves existence and uniqueness for the scattering problem. Interior regularity results yield that the solution is analytic in the exterior of $\overline{D}$. The scattered field is denoted by $u^s(\cdot, \theta) \in C^2(\mathbb{R}^3 \setminus \overline{D})$ and depends on the angle $\theta \in S^2$ of the incident plane wave u^i.

The far field pattern of v is again denoted by v^∞. In the particular case of the scattering problem by plane incident fields we denote the far field pattern again by $u^\infty = u^\infty(\hat{x}, \theta)$. Furthermore, the far field operator $F = F_{mix}$ is defined as in (2.22), i.e.,

$$(F_{mix}\psi)(\hat{x}) = \int_{S^2} u^\infty(\hat{x}, \theta)\,\psi(\theta)\,ds(\theta), \quad \hat{x} \in S^2. \tag{3.7}$$

Also, we define the data-to-pattern operator $G_{mix} : H^{1/2}(\Gamma_1) \times H^{-1/2}(\Gamma_2) \to L^2(S^2)$ by

$$G_{mix} \begin{pmatrix} f \\ g \end{pmatrix} = v^\infty,$$

where v^∞ is a far field pattern of the solution to (3.3), (3.4), (3.5), and (3.6).

It is helpful to relate the operators G_{mix} and F_{mix} to the corresponding operators G_{Dir} and F_{Dir}, respectively, for the pure Dirichlet boundary condition on both components $\Gamma = \Gamma_1 \cup \Gamma_2$ – just as in Theorem 2.4. From the previous chapter we recall the exterior Dirichlet-to-Neumann operator $\Lambda^{-1} : H^{1/2}(\Gamma) \to H^{-1/2}(\Gamma)$. This operator has two components $\Lambda_j^{-1} : H^{1/2}(\Gamma) \to H^{-1/2}(\Gamma_j)$ for $j = 1, 2$.

Furthermore, we recall the operators $H_j : L^2(S^2) \to H^{1/2}(\Gamma_j)$ and $(\partial H)_j : L^2(S^2) \to H^{-1/2}(\Gamma_j)$ for $j = 1, 2$, defined by (2.36) and (2.37), respectively, i.e.,

$$(H_j \psi)(x) = \int_{S^2} \psi(\theta)\, e^{ik\,\theta \cdot x}\, ds(\theta), \quad x \in \Gamma_j, \tag{3.8}$$

$$(\partial H)_j \psi(x) = \frac{\partial}{\partial \nu} \int_{S^2} \psi(\theta)\, e^{ik\,\theta \cdot x}\, ds(\theta), \quad x \in \Gamma_j. \tag{3.9}$$

Theorem 3.2 *Let $k > 0$ and $\lambda \in L^\infty(\Gamma_2)$ with $\operatorname{Im}\lambda \geq 0$.*

(a) The data-to-pattern operator G_{mix} is compact, one-to-one, and satisfies the relation

$$G_{Dir} = G_{mix} \begin{pmatrix} R_1 \\ \Lambda_{2,+}^{-1} + \lambda R_2 \end{pmatrix}, \tag{3.10}$$

where $R_1 : H^{1/2}(\Gamma) \to H^{1/2}(\Gamma_1)$ and $R_2 : H^{1/2}(\Gamma) \to H^{-1/2}(\Gamma_2)$ are the restriction operators to Γ_j for $j = 1, 2$. Furthermore, the operator

$$\begin{pmatrix} R_1 \\ \Lambda_{2,+}^{-1} + \lambda R_2 \end{pmatrix} : H^{1/2}(\Gamma) \longrightarrow H^{1/2}(\Gamma_1) \times H^{-1/2}(\Gamma_2) \tag{3.11}$$

is an isomorphism from $H^{1/2}(\Gamma)$ onto $H^{1/2}(\Gamma_1) \times H^{-1/2}(\Gamma_2)$.

(b) F_{mix} and G_{mix} are related as follows:

$$F_{mix} = -G_{mix} \begin{pmatrix} H_1 \\ (\partial H)_2 + \lambda H_2 \end{pmatrix}. \tag{3.12}$$

Proof: (a) Let v be the solution of the exterior Dirichlet boundary value problem for some $f \in H^{1/2}(\Gamma)$, i.e., $v = f$ on $\Gamma = \Gamma_1 \cup \Gamma_2$. Then v also solves the mixed boundary value problem (3.3), (3.4), (3.5), and (3.6) with

$$f_{mix} = R_1 f \quad \text{and} \quad g_{mix} = \left(\frac{\partial v}{\partial \nu} + \lambda v \right)\bigg|_{\Gamma_2} = \Lambda_{2,+}^{-1} f + \lambda R_2 f.$$

The far field pattern of v is $v^\infty = G_{Dir}f$ (considered as the Dirichlet boundary value problem). On the other hand, the same far field pattern is given by

$$v^\infty = G_{mix}\begin{pmatrix} f_{mix} \\ g_{mix} \end{pmatrix} = G_{mix}\begin{pmatrix} R_1 f \\ \Lambda_{2,+}^{-1} f + \lambda R_2 f \end{pmatrix}$$

which implies (3.10). To prove injectivity of the operator (3.11) let $f = 0$ on Γ_1 and $\Lambda_{2,+}^{-1} f + \lambda R_2 f = 0$ on Γ_2. By definition of $\Lambda_{2,+}^{-1}$ we have $\Lambda_{2,+}^{-1} f = \partial v_+ / \partial \nu$ on Γ_2 where $v_+ = f$ on $\Gamma_1 \cup \Gamma_2$. In particular, $v_+ = f = 0$ on Γ_1 and $\partial v_+ / \partial \nu + \lambda v_+ = 0$ on Γ_2. The uniqueness result of Theorem 3.1 yields that v vanishes identically, thus $f = 0$ also on Γ_2.

The proof of surjectivity follows the same lines. The proofs of injectivity and compactness of G_{mix} is now a consequence of (3.10) and the injectivity and compactness, respectively, of G_{Dir}.

(b) It follows from the definition (3.7) that $F_{mix}\psi$ is the far field pattern v^∞ of the radiating solution v corresponding to the bounday data

$$f(x) = -\int_{S^2} e^{ik\,x\cdot\theta}\psi(\theta)\,ds(\theta), \quad x \in \Gamma_1,$$

$$g(x) = -\left(\frac{\partial}{\partial \nu} + \lambda(x)\right)\int_{S^2} e^{ik\,x\cdot\theta}\psi(\theta)\,ds(\theta), \quad x \in \Gamma_2.$$

Thus, we have

$$F_{mix}\psi = -G_{mix}\begin{pmatrix} H_1\psi \\ (\partial H)_2\psi + \lambda H_2\psi \end{pmatrix},$$

which proves (3.12). $\square$

Theorem 3.3 *Let $k > 0$ be a fixed frequency and $\lambda \in L^\infty(\Gamma_2)$ a given impedance function with $\mathrm{Im}\,\lambda \geq 0$ on Γ_2.*

(a) The far field pattern u^∞ satisfies the reciprocity relations (2.27) and (2.29).

(b) F_{mix} is compact.

(c) $F = F_{mix}$ satisfies the relations

$$F - F^* - \frac{ik}{8\pi^2}F^*F = 2i\,R_{mix}, \tag{3.13}$$

$$F^* - F + \frac{ik}{8\pi^2}FF^* = -2iJ\,\overline{R}_{mix}\,J, \tag{3.14}$$

where $R_{mix} : L^2(S^2) \to L^2(S^2)$ is the self-adjoint non-negative operator given by

$$(R_{mix}\psi)(\eta) = \int_{S^2}\left(\int_{\Gamma_2} u(x,\theta)\overline{u(x,\eta)}\,\mathrm{Im}\,\lambda(x)\,ds(x)\right)\psi(\theta)\,ds(\theta), \quad \eta \in S^2, \tag{3.15}$$

and $u(\cdot,\theta) = u^i(\cdot,\theta) + u^s(\cdot,\theta)$ *is the total scattered field on* Γ_2 *corresponding to the incident plane wave* $u^i(\cdot,\theta)$ *of direction* θ.

(d) The scattering operator

$$S = S_{mix} = I + \frac{ik}{8\pi^2} F_{mix} \tag{3.16}$$

satisfies

$$S^*S = I - \frac{k}{4\pi^2} R_{mix}.$$

If λ *is real-valued then the far field operator* F_{mix} *is normal, and the scattering operator* S_{mix} *is unitary.*

(e) Assume that there exists no Herglotz wave function which solves the homogeneous mixed boundary value problem in D. Then the operator $F = F_{mix}$ *is one-to-one and has a dense range in* $L^2(S^2)$.

In this case $\mathrm{Im}\, F = (F - F^*)/(2i)$ *is strictly positive and*

$$\mathrm{Im}(F\psi, \psi)_{L^2(S^2)} \geq \frac{k}{16\pi^2} \|F\psi\|^2_{L^2(S^2)} > 0 \qquad \text{for all } \psi \neq 0. \tag{3.17}$$

Proof: The proof of (a) is standard and follows the lines of the proof of Theorems 1.6 and 1.7.

(b) This follows from (3.12) and the compactness of G_{mix} or directly from the smoothness of the kernel with respect to both arguments.

(c) We modify the corresponding proof for the Dirichlet boundary condition from Theorem 1.8, part (a). Because of the different boundary conditions, formula (1.34) has to be replaced by

$$2i \int_{\Gamma_2} v\,\overline{w}\, \mathrm{Im}\,\lambda\, ds = \int_{\Gamma_2} \left[v\, \frac{\partial \overline{w}}{\partial \nu} - \overline{w}\, \frac{\partial v}{\partial \nu} \right] ds = \int_{|x|=R} \left[v\, \frac{\partial \overline{w}}{\partial \nu} - \overline{w}\, \frac{\partial v}{\partial \nu} \right] ds \tag{3.18}$$

which is derived by applying Green's formula to v and $\overline{w}$ in the region $\{x \in \mathbb{R}^3 \setminus \overline{D} : |x| < R\}$. We recall that v and w are the total fields of the scattering problem with incident fields

$$v^i(x) = \int_{S^2} e^{ikx\cdot\theta}\, g(\theta)\, ds(\theta), \quad x \in \mathbb{R}^3,$$

$$w^i(x) = \int_{S^2} e^{ikx\cdot\theta}\, h(\theta)\, ds(\theta), \quad x \in \mathbb{R}^3.$$

By superposition, we conclude that

$$v(x) = \int_{S^2} u(x,\theta)\, g(\theta)\, ds(\theta), \quad x \notin D,$$

$$w(x) = \int_{S^2} u(x,\theta)\, h(\theta)\, ds(\theta), \quad x \notin D.$$

Substituting this into the left-hand side of (3.18) and interchanging the orders of integration yields

$$2i \int_{\Gamma_2} v\,\overline{w}\, \operatorname{Im}\lambda\, ds = 2i \int_{S^2}\int_{S^2} \left[\int_{\Gamma_2} u(x,\theta)\, \overline{u(x,\eta)}\, \operatorname{Im}\lambda(x)\, ds(x) \right] g(\theta)\, \overline{h(\eta)}\, ds(\theta)\, ds(\eta)$$

$$= 2i\, (R_{mix}g, h)_{L^2(S^2)}.$$

From this we observe that R_{mix} is non-negative because of $\operatorname{Im}\lambda \geq 0$ on Γ_2. Furthermore, equation (1.35) of Chapter 1 takes the form

$$2i\, (R_{mix}g, h)_{L^2(S^2)} = -\frac{ik}{8\pi^2}\, (Fg, Fh)_{L^2(S^2)} - (g, Fh)_{L^2(S^2)} + (Fg, h)_{L^2(S^2)}$$

$$= -\frac{ik}{8\pi^2}\, (F^*Fg, h)_{L^2(S^2)} - (F^*g, h)_{L^2(S^2)} + (Fg, h)_{L^2(S^2)},$$

from which (3.13) follows. The second form (3.14) follows from the reciprocity principle in the form (2.28).

(d) We just compute

$$S^*S = (I - \frac{ik}{8\pi^2}\, F^*)\,(I + \frac{ik}{8\pi^2}\, F) = I + \frac{ik}{8\pi^2}\left[(F - F^*) - \frac{ik}{8\pi^2}\, F^*F\right]$$

$$= I + \frac{ik}{8\pi^2}\, 2i\, R_{mix} = I - \frac{k}{4\pi^2}\, R_{mix}$$

by part (c).

(e) By (3.12) and the injectivity of G_{mix} it suffices to show injectivity of

$$\begin{pmatrix} H_1 \\ (\partial H)_2 + \lambda H_2 \end{pmatrix}.$$

Let $\psi \in L^2(S^2)$ satisfy $H_1\psi = 0$ on Γ_1 and $(\partial H)_2\psi + \lambda H_2\psi = 0$ on Γ_2. With $v(x) = \int_{S^2} \psi(\theta)\, \exp(ikx \cdot \theta)\, ds(\theta)$ we conclude that $v = 0$ on Γ_1 and $\partial v/\partial \nu + \lambda v = 0$ on Γ_2. The uniqueness of the exterior mixed boundary value problem yields $v = 0$ in the exterior of D and thus also $\psi = 0$ by the injectivity of H (which follows, e.g., from (1.54) and the injectivity of F by Theorem 1.8).

The estimate (3.17) follows from (3.13), (3.14), and the observation that the operator R_{mix} is non-negative. $\square$

3.2 Factorization of the far field operator

We start our inverse scattering problem analysis with the following factorization result, see [73].

Theorem 3.4 *(a) The far field operator $F : L^2(S^2) \to L^2(S^2)$ can be factorized as*

$$F_{mix} = -G_{mix}\, T^*_{mix}\, G^*_{mix} \tag{3.19}$$

where the operator $T_{mix} : H^{-1/2}(\Gamma_1) \times H^{1/2}(\Gamma_2) \to H^{1/2}(\Gamma_1) \times H^{-1/2}(\Gamma_2)$ is of the form

$$T_{mix} = \begin{pmatrix} S_1 & 0 \\ 0 & N_2 \end{pmatrix} + K_{mix}. \tag{3.20}$$

Here, $S_1 : H^{-1/2}(\Gamma_1) \to H^{1/2}(\Gamma_1)$ and $N_2 : H^{1/2}(\Gamma_2) \to H^{-1/2}(\Gamma_2)$ are the single layer boundary operator on Γ_1 and the normal derivative of the double layer operator on Γ_2 which appear in the factorizations (1.50) and (1.96) for the pure Dirichlet and Neumann boundary conditions on the boundaries Γ_1 and Γ_2, respectively. The operator $K_{mix} : H^{-1/2}(\Gamma_1) \times H^{1/2}(\Gamma_2) \to H^{1/2}(\Gamma_1) \times H^{-1/2}(\Gamma_2)$ is compact.

(b) Assume that k^2 is neither a Dirichlet eigenvalue of $-\Delta$ in the region D_1 nor an eigenvalue of $-\Delta$ in the region D_2 with respect to the impedance boundary condition. Then the operators T_{mix} and T^*_{mix} are isomorphisms.

(c) Under the assumptions of (b) the operator $\mathrm{Im}\, T_{mix}$ is strictly positive, i.e.,

$$\mathrm{Im}\langle T_{mix}\varphi, \varphi \rangle > 0 \quad \text{for all } \varphi \in H^{-1/2}(\Gamma_1) \times H^{1/2}(\Gamma_2) \text{ with } \varphi \neq 0.$$

Remark: The dual form $\langle \cdot, \cdot \rangle$ is defined by

$$\left\langle \begin{pmatrix} \psi_1 \\ \psi_2 \end{pmatrix}, \begin{pmatrix} \varphi_1 \\ \varphi_2 \end{pmatrix} \right\rangle = \overline{\langle \varphi_1, \psi_1 \rangle_1} + \langle \psi_2, \varphi_2 \rangle_2$$

for all $\begin{pmatrix} \psi_1 \\ \psi_2 \end{pmatrix} \in H^{1/2}(\Gamma_1) \times H^{-1/2}(\Gamma_2)$ and $\begin{pmatrix} \varphi_1 \\ \varphi_2 \end{pmatrix} \in H^{-1/2}(\Gamma_1) \times H^{1/2}(\Gamma_2)$.

Here, $\langle \cdot, \cdot \rangle_j$ are the dual forms in $\langle H^{-1/2}(\Gamma_j), H^{1/2}(\Gamma_j) \rangle$ for $j = 1, 2$.

Proof of Theorem 3.4: (a) From Theorem 3.2 we have

$$F_{mix} = -G_{mix} \begin{pmatrix} H_1 \\ (\partial H)_2 + \lambda H_2 \end{pmatrix}. \tag{3.21}$$

From the definitions of H_1 and $(\partial H)_2$ we compute its adjoint for smooth functions as

$$\begin{pmatrix} H_1 \\ (\partial H)_2 + \lambda H_2 \end{pmatrix}^* \begin{pmatrix} \varphi_1 \\ \varphi_2 \end{pmatrix} (\hat{x}) = \int_{\Gamma_1} e^{-ik\,\hat{x}\cdot y}\varphi_1(y)\, ds(y)$$

$$+ \int_{\Gamma_2} \left[\frac{\partial}{\partial \nu(y)} e^{-ik\,\hat{x}\cdot y} + \overline{\lambda(y)}\, e^{-ik\,\hat{x}\cdot y} \right] \varphi_2(y)\, ds(y)$$

which is equal to the far field pattern v^∞ of

$$v(x) = \int_{\Gamma_1} \Phi(x,y)\,\varphi_1(y)\,ds(y) + \int_{\Gamma_2} \left[\frac{\partial}{\partial\nu(y)}\Phi(x,y) + \overline{\lambda(y)}\,\Phi(x,y) \right] \varphi_2(y)\,ds(y)$$

(3.22)

for $x \in \mathbb{R}^3 \setminus (\Gamma_1 \cup \Gamma_2)$. Using the jump conditions of the single and double layer potentials we compute the Dirichlet boundary values of v on Γ_1 and the impedance boundary values on Γ_2 (both from the exterior) as

$$v|_{\Gamma_1} = S_1\varphi_1 + D_{2\to1}\varphi_2 + S_{2\to1}(\overline{\lambda}\varphi_2),$$

$$\left.\frac{\partial v}{\partial\nu}\right|_{\Gamma_2} + \lambda\,v|_{\Gamma_2} = (D'_{1\to2} + \lambda S_{1\to2})\varphi_1 + \left(N_2 - \frac{1}{2}\overline{\lambda}I_2 + D'_2\overline{\lambda}\right)\varphi_2$$

$$+ \lambda\left(\frac{1}{2}I_2 + D_2 + S_2\overline{\lambda}\right)\varphi_2$$

where the operators S_1, S_2, N_2, D_2, D'_2 are the boundary operators defined in (2.12)–(2.15) for the surfaces Γ_1 and Γ_2, respectively. Furthermore, the operators $S_{2\to1}$ and $D_{2\to1}$ from $H^{-1/2}(\Gamma_2) \to H^{1/2}(\Gamma_1)$ are defined by

$$(S_{2\to1}\psi)(x) = \int_{\Gamma_2} \Phi(x,y)\,\psi(y)\,ds(y), \quad x \in \Gamma_1,$$

$$(D_{2\to1}\psi)(x) = \int_{\Gamma_2} \frac{\partial}{\partial\nu(y)}\Phi(x,y)\,\psi(y)\,ds(y), \quad x \in \Gamma_1.$$

The operators $S_{1\to2}$ and $D_{1\to2}$ are analogously defined by interchanging the roles of Γ_1 and Γ_2. All of these operators $S_{1\to2}$, $D_{1\to2}$, $S_{2\to1}$, and $D_{2\to1}$ are smoothing and thus compact. Therefore, we have that

$$v^\infty = G_{mix}\,T_{mix}\begin{pmatrix}\varphi_1 \\ \varphi_2\end{pmatrix}$$

with

$$T_{mix} = \begin{pmatrix} S_1 & D_{2\to1} + S_{2\to1}\overline{\lambda} \\ D'_{1\to2} + \lambda S_{1\to2} & \left(N_2 - \frac{1}{2}\overline{\lambda}I_2 + D'_2\overline{\lambda}\right) + \lambda\left(\frac{1}{2}I_2 + D_2 + S_2\overline{\lambda}\right) \end{pmatrix}$$

$$= \begin{pmatrix} S_1 & 0 \\ 0 & N_2 \end{pmatrix} + K_{mix}$$

(3.23)

and obvious form of K_{mix}. We have therefore shown that

$$\begin{pmatrix} H_1 \\ (\partial H)_2 + \lambda H_2 \end{pmatrix}^* = G_{mix}\,T_{mix}$$

(3.24)

and thus

$$\begin{pmatrix} H_1 \\ (\partial H)_2 + \lambda H_2 \end{pmatrix} = T_{mix}^* \, G_{mix}^*.$$

Substituting this into (3.21) yields the factorization (3.19). The operator K_{mix} is obviously compact.

(b) From the decomposition (3.20) and the facts that K_{mix} is compact and S_1 and N_2 are Fredholm operators of index zero we conclude that also T_{mix} is a Fredholm operator of index zero. Therefore, it is sufficient to prove injectivity of T_{mix}. Let $T_{mix}\varphi = 0$ and denote again by φ_j the restrictions of φ on Γ_j for $j = 1, 2$. Define the potential v by (3.22) in $\mathbb{R}^3 \setminus (\Gamma_1 \cup \Gamma_2)$. By the jump conditions of the single and double layer potentials we conclude again that $v_+ = T_1\varphi = 0$ on Γ_1 and $\partial v_+/\partial \nu + \lambda v_+ = T_2\varphi = 0$ on Γ_2 where $T_1 : H^{-1/2}(\Gamma_1) \times H^{1/2}(\Gamma_2) \to H^{1/2}(\Gamma_1)$ and $T_2 : H^{-1/2}(\Gamma_1) \times H^{1/2}(\Gamma_2) \to H^{-1/2}(\Gamma_2)$ denote the components of the operator T_{mix}. The uniqueness result of Theorem 3.1 yields that v vanishes in the exterior of $D_1 \cup D_2$. Now we use the fact that $v_- = v_+ = 0$ on Γ_1 which yields that v vanishes in D_1 since k^2 is not a Dirichlet eigenvalue of $-\Delta$ in D_1. We conclude that $\varphi_1 = \partial v_+/\partial \nu - \partial v_-/\partial \nu = 0$ on Γ_1. Analogously, we have on Γ_2 that

$$v_- = v_- - v_+ = \varphi_2 \quad \text{and}$$

$$\frac{\partial v_-}{\partial \nu} = \frac{\partial v_-}{\partial \nu} - \frac{\partial v_+}{\partial \nu} = -\bar{\lambda}\,\varphi_2 = -\bar{\lambda}\,v_-,$$

i.e., $\partial \bar{v}_-/\partial \nu + \lambda \bar{v}_- = 0$ on Γ_2. Now we use that k^2 is not an eigenvalue of $-\Delta$ in D_2 with respect to the impedance boundary condition which implies that v vanishes also in D_2. From the above jump conditions also φ_2 vanishes which ends the proof of this part.

(c) This follows by exactly the same arguments as in the proof of Lemma 2.7 of the previous chapter. We repeat the arguments for the convenience of the reader. We note from (3.17) and the factorization (3.19) that

$$0 \leq \frac{k}{16\pi^2} \|F\psi\|_{L^2(S^2)}^2 \leq \mathrm{Im}(F\psi, \psi)_{L^2(S^2)} = -\,\mathrm{Im}\langle T^*G^*\psi, G^*\psi\rangle$$

$$= \mathrm{Im}\langle T\,G^*\psi, G^*\psi\rangle \qquad (3.25)$$

where we have dropped again the subscript "*mix*" for simpler notation. The range of G^* is dense in $H^{-1/2}(\Gamma_1) \times H^{1/2}(\Gamma_2)$ because of the injectivity of G by Theorem 3.2. Therefore, we conclude that $\mathrm{Im}\langle T\varphi, \varphi\rangle \geq 0$ for all φ.

Let now $\varphi \in H^{-1/2}(\Gamma_1) \times H^{1/2}(\Gamma_2)$ with $\mathrm{Im}\langle T\varphi, \varphi\rangle = 0$. Again, since the range of G^* is dense there exists a sequence $\psi_n \in L^2(S^2)$ with $G^*\psi_n \to \varphi$. From (3.25) we conclude that

$$\frac{k}{16\pi^2} \|F\psi_n\|_{L^2(S^2)}^2 \leq \mathrm{Im}\langle T\,G^*\psi_n, G^*\psi_n\rangle.$$

Since the right hand side converges to $\mathrm{Im}\langle T\varphi, \varphi\rangle = 0$ we have shown that $F\psi_n \to 0$ in $L^2(S^2)$ as n tends to infinity. Furthermore, for any $\psi \in L^2(S^2)$ we have that

$$\langle T^*\,G^*\psi_n, G^*\psi\rangle = -(F\psi_n, \psi)_{L^2(S^2)} \longrightarrow 0.$$

The left hand side converges to $\langle T^*\varphi, G^*\psi\rangle$ and thus $\langle T^*\varphi, G^*\psi\rangle = 0$ for all $\psi \in L^2(S^2)$. From the denseness of the range of G^* we arrive at $T^*\varphi = 0$ and thus $\varphi = 0$ by part (b). $\square$

The representation $T_{mix} = \mathrm{diag}(S_1, N_2) + K_{mix}$ of (3.20) explains immediately the difficulties arising in the application of the standard factorization method to the mixed boundary value problem. The real part of the coupling operator T_{mix} fails to be the sum of a coercive and a compact operator – in contrast to the pure Dirichlet or impedance boundary condition! Indeed, according to Lemma 1.14 and Theorem 1.26 the real part $\mathrm{Re}\, S_1$ of S_1 is the sum of a positively coercive and a compact part while $\mathrm{Re}\, N_2$ is the sum of a negatively and a compact part. No one of these two terms dominates unless we restrict ourselves to a special class of functions φ subject to some additional constraints. In the following we will derive a modified version of the square root algorithm under the geometric a priori assumption that D_1 or D_2 lies in a known region Ω_1 or Ω_2, respectively.

For the further investigation we recall the following factorization which has been derived in the proof of Theorem 3.4 when we eliminate G_{mix} in (3.19) by using (3.24):

$$F_{mix} = -\begin{pmatrix} H_1 \\ (\partial H)_2 + \lambda H_2 \end{pmatrix}^* T_{mix}^{-1} \begin{pmatrix} H_1 \\ (\partial H)_2 + \lambda H_2 \end{pmatrix}. \tag{3.26}$$

3.3 Application of the $F_\#$ – factorization method

In this section we will apply the $F_\#$ – Factorization Method from the previous chapter. With some modifications we will essentially follow the approach of [77]. The idea of this fast algorithm is the same as for the Dirichlet, Neumann or impedance boundary condition, namely to study the auxiliary operator $F_\# = |\mathrm{Re}\, F_{mix}| + \mathrm{Im}\, F_{mix}$ from (2.54) instead of F_{mix} itself. However, in the case of the mixed Dirichlet–impedance boundary condition, we are not able to apply Theorem 2.15, because the assumption (A2) is violated as mentioned at the end of the previous section. Therefore, we make the **assumption** that we a priori know open and bounded domains Ω_1 and Ω_2 with C^2–boundaries $\widetilde{\Gamma}_1$ and $\widetilde{\Gamma}_2$, respectively, such that

$$\overline{D_j} \subset \Omega_j \quad \text{for } j = 1, 2.$$

We assume that the closures of the domains Ω_j, $j = 1, 2$, are disjoint and their complements are connected, see Figure 3.1. We call Ω_1 the Dirichlet separator and Ω_2 the impedance separator, see again Figure 3.1. For any real and positive parameter ρ we introduce two modified data operators F_I and F_D by

$$F_I = F_{mix} + \rho\, \widetilde{H}_1^* \widetilde{H}_1, \qquad F_D = F_{mix} - \rho\, \widetilde{H}_2^* \widetilde{H}_2. \tag{3.27}$$

where $\widetilde{H}_j : L^2(S^2) \to L^2(\widetilde{\Gamma}_j), j = 1, 2$, denote the Herglotz operators for the boundaries $\widetilde{\Gamma}_j = \partial\Omega_j$, i.e.,

$$(\widetilde{H}_j \psi)(x) = \int_{S^2} \psi(\theta)\, e^{ik\, x\cdot\theta}\, ds(\theta), \quad x \in \widetilde{\Gamma}_j, \ j = 1, 2. \tag{3.28}$$

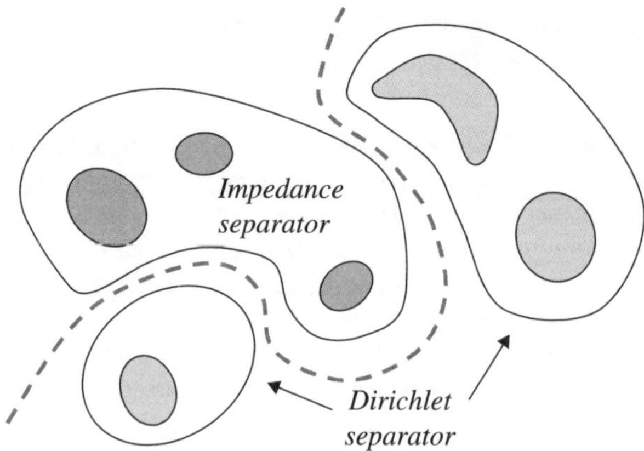

Figure 3.1 Separators with respect to Dirichlet and impedance boundary conditions

We note that $\widetilde{H}_1$ and $\widetilde{H}_2$ are known by our assumption. In contrast to the operators H_j we consider $\widetilde{H}_j$ as operators into $L^2(\widetilde{\Gamma}_j)$. Then we can show:

Lemma 3.5 *Assume that k^2 is neither a Dirichlet eigenvalue of $-\Delta$ in D_1, Ω_1, Ω_2 nor an eigenvalue of $-\Delta$ in D_2 with respect to the impedance boundary condition.*

(a) Then the operators F_D and F_I from (3.27) satisfy the factorizations

$$F_D = -\begin{pmatrix} H_1 \\ \widetilde{H}_2 \end{pmatrix}^{*} T_D \begin{pmatrix} H_1 \\ \widetilde{H}_2 \end{pmatrix}, \qquad F_I = -\begin{pmatrix} \widetilde{H}_1 \\ (\partial H)_2 + \lambda H_2 \end{pmatrix}^{*} T_I \begin{pmatrix} \widetilde{H}_1 \\ (\partial H)_2 + \lambda H_2 \end{pmatrix}$$

$$(3.29)$$

with some Fredholm operators $T_D : H^{1/2}(\Gamma_1) \times L^2(\widetilde{\Gamma}_2) \to H^{-1/2}(\Gamma_1) \times L^2(\widetilde{\Gamma}_2)$ and $T_I : L^2(\widetilde{\Gamma}_1) \times H^{-1/2}(\Gamma_2) \to L^2(\widetilde{\Gamma}_1) \times H^{1/2}(\Gamma_2)$. Here, T_D and T_I have the following forms:

$$T_D = \begin{pmatrix} S_1^{-1} & 0 \\ 0 & \rho I_2 \end{pmatrix} + K_D, \qquad T_I = \begin{pmatrix} -\rho I_1 & 0 \\ 0 & N_2^{-1} \end{pmatrix} + K_I \qquad (3.30)$$

with some compact operators K_D and K_I. The operators S_1 and N_2 are as in Theorem 3.4, part (a).

(b) The operators $\operatorname{Im} T_D$ and $\operatorname{Im} T_I$ are strictly negative, i.e.,

$$\operatorname{Im}\langle T_D\varphi, \varphi\rangle < 0 \text{ and } \operatorname{Im}\langle T_I\varphi, \varphi\rangle < 0 \text{ for any } \varphi \neq 0.$$

(c) The real parts $\operatorname{Re} T_D = (T_D + T_D^{*})/2$ and $\operatorname{Re} T_I = (T_I + T_I^{*})/2$ have decompositions in the form

$$\operatorname{Re} T_D = C_D + \widetilde{K}_D \quad \text{and} \quad \operatorname{Re} T_I = -C_I + \widetilde{K}_I$$

with self-adjoint and (positively) coercive operators C_D and C_I and compact operators $\widetilde{K}_D$ and $\widetilde{K}_I$.

Proof: (a) We first study F_D and recall the factorization of F_{mix} in the form (3.26). We note that $(\partial H)_2 + \lambda H_2 = R_2 \widetilde{H}_2$, where $R_2 : L^2(\widetilde{\Gamma}_2) \to H^{-1/2}(\Gamma_2)$ is defined by $R_2 f = \left(\partial v/\partial \nu + \lambda v\right)\big|_{\Gamma_2}$ and where v solves the interior Dirichlet boundary value problem in Ω_2 with boundary data $v = f$ on $\widetilde{\Gamma}_2$. The operator R_2 is certainly compact because of interior regularity results.[1] R_2 is also one-to-one. Indeed, let $R_2 f = 0$. Then the corresponding solution v solves the homogeneous interior impedance boundary value problem in D_2 and, by the assumption on k, has to vanish in all of D_2. Analytic continuation yields $v = 0$ in Ω_2 and thus also $f = 0$.

We write the factorization (3.26) in the form

$$F_{mix} = -\begin{pmatrix} H_1 \\ \widetilde{H}_2 \end{pmatrix}^* \begin{pmatrix} I_1 & 0 \\ 0 & R_2^* \end{pmatrix} T_{mix}^{-1} \begin{pmatrix} I_1 & 0 \\ 0 & R_2 \end{pmatrix} \begin{pmatrix} H_1 \\ \widetilde{H}_2 \end{pmatrix}$$

and thus

$$F_D = -\begin{pmatrix} H_1 \\ \widetilde{H}_2 \end{pmatrix}^* \left[\begin{pmatrix} I_1 & 0 \\ 0 & R_2^* \end{pmatrix} T_{mix}^{-1} \begin{pmatrix} I_1 & 0 \\ 0 & R_2 \end{pmatrix} + \begin{pmatrix} 0 & 0 \\ 0 & \rho I_2 \end{pmatrix} \right] \begin{pmatrix} H_1 \\ \widetilde{H}_2 \end{pmatrix}. \tag{3.31}$$

Using (3.20) the operator T_{mix}^{-1} can be written as

$$T_{mix}^{-1} = \begin{pmatrix} S_1^{-1} & 0 \\ 0 & N_2^{-1} \end{pmatrix} - T_{mix}^{-1} K_{mix} \begin{pmatrix} S_1^{-1} & 0 \\ 0 & N_2^{-1} \end{pmatrix}. \tag{3.32}$$

Substituting this into (3.31) yields

$$F_D = -\begin{pmatrix} H_1 \\ \widetilde{H}_2 \end{pmatrix}^* \left[\begin{pmatrix} S_1^{-1} & 0 \\ 0 & \rho I_2 \end{pmatrix} + K_D \right] \begin{pmatrix} H_1 \\ \widetilde{H}_2 \end{pmatrix} \tag{3.33}$$

where we collect all compact parts in the operator K_D. This has the form of the factorization (3.29) for F_D with

$$T_D = \begin{pmatrix} I_1 & 0 \\ 0 & R_2^* \end{pmatrix} T_{mix}^{-1} \begin{pmatrix} I_1 & 0 \\ 0 & R_2 \end{pmatrix} + \begin{pmatrix} 0 & 0 \\ 0 & \rho I_2 \end{pmatrix}.$$

This proves the factorization of T_D. For T_I we proceed quite similarly. It is $H_1 = R_1 \widetilde{H}_1$ where now $R_1 : L^2(\widetilde{\Gamma}_1) \to H^{-1/2}(\Gamma_1)$ is given by $R_1 f = v|_{\Gamma_1}$ where v solves the interior Dirichlet boundary value problem in Ω_1 with boundary data $v = f$ on $\widetilde{\Gamma}_1$. Then we follow the lines of the proof for F_D.

(b) We certainly have for any $\varphi = (\varphi_1, \varphi_2)^\top \in H^{1/2}(\Gamma_1) \times L^2(\widetilde{\Gamma}_2)$ with $\varphi \neq 0$ that

$$\text{Im}\langle T_D \varphi, \varphi \rangle = \text{Im}\langle \widetilde{\varphi}, T_{mix}\widetilde{\varphi} \rangle + \rho \, \text{Im} \, \|\varphi_2\|^2_{L^2(\widetilde{\Gamma}_2)} = \text{Im}\langle \widetilde{\varphi}, T_{mix}\widetilde{\varphi} \rangle < 0$$

by Theorem 3.4 since $\widetilde{\varphi} = T_{mix}^{-1} \begin{pmatrix} I_1 & 0 \\ 0 & R_2 \end{pmatrix} \varphi$ does not vanish by the injectivity of T_{mix}^{-1} and R_2. For T_I this is seen analogously.

---

[1] The existence of solutions with L^2–boundary data can be shown, for example, by boundary integral equation methods. An ansatz for v as a double layer potential with L^2–density leads to an integral equation of the second kind in $L^2(\widetilde{\Gamma}_2)$. The compactness of R_2 is immediately seen by the smoothness of the kernel. We refer to [110, 5, 4] for detailed treatments of boundary value problems with L^2–boundary data.

(c) This follows from the decomposition (3.30) and the fact that $\mathrm{Re}\, S_1$ is a compact perturbation of the coercive operator $S_{1,i}$ (which is the single layer operator on Γ_1 for wavenumber $k = i$. The same argument holds for $\mathrm{Re}\, N_2$. $\square$

Now we proceed to the solution of the inverse obstacle problem by the Factorization Method. As in the previous section, we first characterize the parts of the scatterer by the corresponding data-to-pattern operators or, in our setting, by the operators in (3.29) which enclose T_D and T_I.

Lemma 3.6 *Assume that k^2 is neither a Dirichlet eigenvalue of $-\Delta$ in D_1, Ω_1, Ω_2 nor an eigenvalue of $-\Delta$ in D_2 with respect to the impedance boundary condition.*
(a) A point $z \in \mathbb{R}^3 \setminus \overline{\Omega_2}$ belongs to D_1 if, and only if, the function ϕ_z, defined by

$$\phi_z(\hat{x}) = e^{-ik\, z\cdot\hat{x}}, \quad \hat{x} \in S^2, \tag{3.34}$$

belongs to the range of the operator

$$\left(H_1^*, \tilde{H}_2^*\right) : H^{-1/2}(\Gamma_1) \times L^2(\tilde{\Gamma}_2) \longrightarrow L^2(S^2). \tag{3.35}$$

(b) A point $z \in \mathbb{R}^3 \setminus \overline{\Omega_1}$ belongs to D_2 if, and only if, the function ϕ_z belongs to the range of the operator

$$\left(\tilde{H}_1^*, (\partial H)_2^* + H_2^*\overline{\lambda}\right) : L^2(\tilde{\Gamma}_1) \times H^{1/2}(\Gamma_2) \longrightarrow L^2(S^2). \tag{3.36}$$

Proof: We restrict ourselves to the proof of part (b) and leave the (simpler) proof of part (a) to the reader.

(b) Let $G_{2,imp}$, $G_{2,Dir}$, and $T_{2,imp}$ be the operators G_{imp}, G_{Dir}, and T_{imp}, respectively, from Chapters 1 and 2 for the domain D_2 when no domain D_1 is present (see Definition 1.11, (2.23), and (2.39)). Theorem 1.12 applied to this situation yields that $z \in D_2$ if, and only if, ϕ_z belongs to the range of $G_{2,Dir}$ which coincides with the range of $G_{2,imp}$ by Theorem 2.4 and, furthermore, with the range of $(\partial H)_2^* + H_2^*\overline{\lambda}$ by (2.42) and the invertibility of $T_{2,imp}$.

Let first $z \in D_2$. Therefore, ϕ_z belongs to the range of $(\partial H)_2^* + H_2^*\overline{\lambda}$, i.e., there exists $\varphi_2 \in H^{1/2}(\Gamma_2)$ with $\phi_z = (\partial H)_2^*\varphi_2 + H_2^*(\overline{\lambda}\varphi_2)$. Setting $\varphi_1 = 0$ yields that ϕ_z belongs to the range of $\left(\tilde{H}_1^*, (\partial H)_2^* + H_2^*\overline{\lambda}\right)$.

Second, let $z \notin D_2$ and assume on the contrary that $\phi_z = \tilde{H}_1^*\varphi_1 + (\partial H)_2^*\varphi_2 + H_2^*(\overline{\lambda}\varphi_2)$ for some $\varphi_1 \in L^2(\tilde{\Gamma}_1)$ and $\varphi_2 \in H^{1/2}(\Gamma_2)$. Since both sides are far field patterns the Lemma 1.2 of Rellich yields that

$$\Phi(x,z) = \int_{\tilde{\Gamma}_1} \varphi_1(y)\, \Phi(x,y)\, ds(y) + \int_{\Gamma_2} \left[\varphi_2(y)\, \frac{\partial}{\partial\nu(y)}\Phi(x,y) + \overline{\lambda(y)}\, \varphi_2(y)\, \Phi(x.y) \right] ds(y)$$

for $x \notin \left(\overline{D_2} \cup \overline{\Omega_1} \cup \{z\}\right)$. For $x \to z$ this leads to a contradiction since the right-hand side belongs to $H^1(B \setminus \overline{D_2})$ for any small ball B with center z in contrast to the left-hand side. $\square$

The last step in the characterization of D is again to relate the ranges of the operators in (3.35) and (3.36) to the far field operators F_D and F_I, respectively. It follows from Lemma 3.5, that the triples

$$F_D, \quad \begin{pmatrix} H_1 \\ \widetilde{H}_2 \end{pmatrix}^*, \quad -T_D \qquad \text{and} \qquad F_I, \quad \begin{pmatrix} \widetilde{H}_1 \\ (\partial H)_2 + \lambda H_2 \end{pmatrix}^*, \quad -T_I$$

satisfy all the conditions of Theorem 2.15. Application of (2.65) yields the final main result of this section.

Theorem 3.7 *Assume that k^2 is neither a Dirichlet eigenvalue of $-\Delta$ in D_1, Ω_1, Ω_2, nor an eigenvalue of $-\Delta$ in D_2 with respect to the impedance boundary condition. Let $\rho > 0$ be an arbitrary parameter and let $\phi_z \in L^2(S^2)$ be defined by (3.34), i.e., $\phi_z(\hat{x}) = \exp(-ikz \cdot \hat{x})$, $\hat{x} \in S^2$. Then the following characterization holds:*
(a) For any point $z \notin \overline{\Omega_2}$ we have:

$$z \in D_1 \quad \Longleftrightarrow \quad \phi_z \in \mathcal{R}\big(F_\#^{1/2}\big),$$

where $F_\# = |\operatorname{Re} F_D| + \operatorname{Im} F_D$ and $F_D = F_{mix} - \rho\, \widetilde{H}_2^* \widetilde{H}_2$.
(b) For any point $z \notin \overline{\Omega_1}$ we have:

$$z \in D_2 \quad \Longleftrightarrow \quad \phi_z \in \mathcal{R}\big(F_\#^{1/2}\big),$$

where now $F_\# = |\operatorname{Re} F_I| + \operatorname{Im} F_I$ and $F_I = F_{mix} + \rho\, \widetilde{H}_1^* \widetilde{H}_1$.
The operators $\widetilde{H}_j : L^2(S^2) \to L^2(\widetilde{\Gamma}_j)$ are defined in (3.28).
We can reformulate this by using Picard's criterion: A point $z \notin \overline{\Omega}_{3-j}$ belongs to D_j if, and only if, the series

$$\sum_{n=1}^{\infty} \frac{\big|(\phi_z, \psi_n)_{L^2(S^2)}\big|^2}{\lambda_n}$$

converges where $\{\lambda_n, \psi_n\}_{n\in\mathbb{N}}$ is an eigensystem of the self-adjoint and positive definite operator $F_\#$.

As we indicated at the end of Chapter 1 the same characterization holds also in the two-dimensional case, i.e., where $D \subset \mathbb{R}^2$. The fundamental solution $\Phi = \Phi(x, y)$ has to be replaced by $\frac{i}{4} H_0^{(1)}(k|x - y|)$ where $H_0^{(1)}$ denotes the Hankel function of first kind and order zero. In order to achieve $\operatorname{Im} F \geq 0$ we have to normalize the far field pattern again as

$$u^s(x) = \frac{\exp(ik|x| + i\pi/4)}{\sqrt{8\pi k\,|x|}} \, u^\infty(\hat{x}) + \mathcal{O}\big(|x|^{-3/2}\big), \quad |x| \to \infty,$$

uniformly with respect to $\hat{x} = x/|x|$. We present a numerical example of the obstacle reconstruction in the two dimensional case. We choose a region which consists of two components, both being ellipses. The plot of the region is shown in Figure 3.2. On the first ellipse (the one in the south-west corner) we impose the Dirichlet boundary condition, on the second one the Neumann boundary conditions. Furthermore, we assume a priori

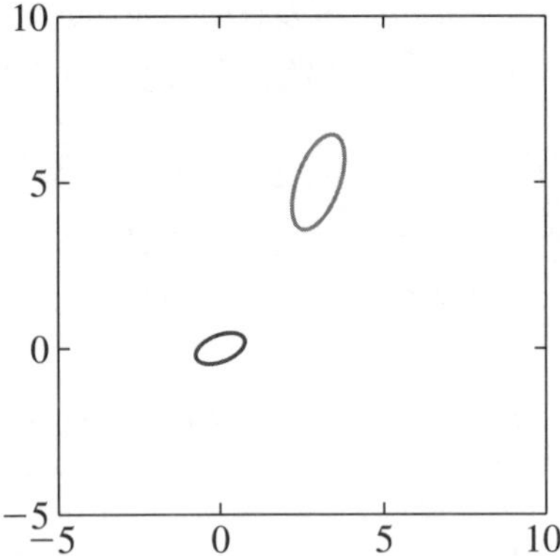

Figure 3.2 The original domain

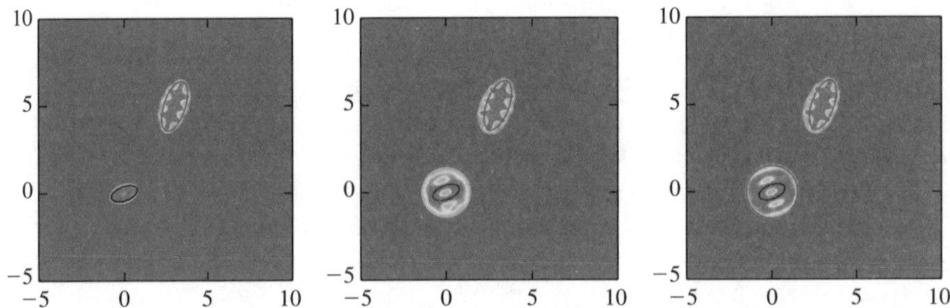

Figure 3.3 Reconstruction for $\rho = 0$, $\rho = 5$, $\rho = 10$ and no noise

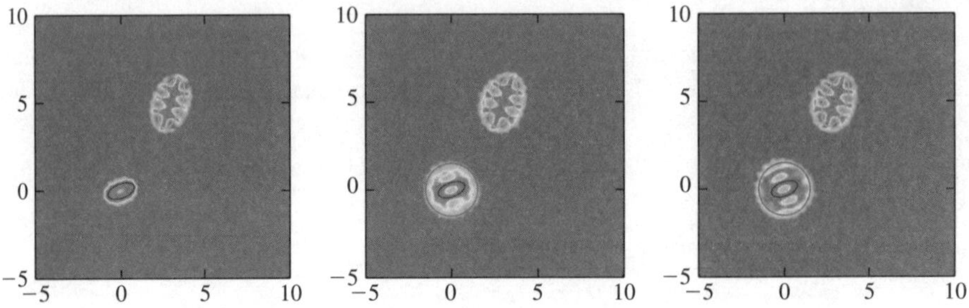

Figure 3.4 Reconstruction for $\rho = 0$, $\rho = 5$, $\rho = 10$ and 1% noise

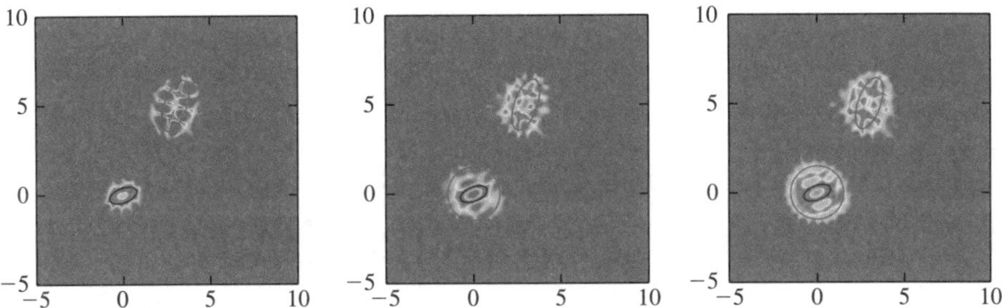

Figure 3.5 Reconstruction for $\rho = 0$, $\rho = 5$, $\rho = 10$ and 5% noise

that the first ellipse is contained in a disk Ω_1 of radius 1.5. For the case of a disk with center 0 and radius R the operator $\widetilde{H}^*\widetilde{H}$ can be computed explicitly and is given by

$$(\widetilde{H}^*\widetilde{H}\psi)(\hat{x}) = 2\pi R \int_{|\hat{y}|=1} J_0\big(kR|\hat{x} - \hat{y}|\big)\,\psi(\hat{y})\,ds(\hat{y}), \quad |\hat{x}| = 1,$$

where J_0 denotes again the Bessel function of order zero. We choose 64 incident plane waves with directions θ_j distributed equidistantly in $[0, 2\pi]$. For each incident field we computed the direct scattering problem by a boundary integral equation method and evaluated the far field patterns at the same 64 directions θ_j. The wavenumber k is chosen to be 5, i.e., the wavelength is $2\pi/5 \approx 1.257$.

The following plots show contour lines of the functions

$$W(z) = \left[\sum_{n=1}^{64} \frac{\big|(\phi_z, \psi_n)_{L^2(S^2)}\big|^2}{\lambda_n}\right]^{-1}, \quad z \in \mathbb{R}^2,$$

for the choices $\rho = 0$, $\rho = 5$, and $\rho = 10$, respectively. We also indicated the shapes of the true ellipses and the circle $\widetilde{\Gamma}_1 = \partial\Omega_1$.

Figures 3.3, 3.4, and 3.5 below show the reconstruction without noise, with 1%, and with 5% extra noise added to the data, respectively.

We observe that the choice $\rho = 0$ seems to be the best in all cases. However, as we noted above, there is no general theory for this case because we can not apply Theorem 2.15. The numerical experiments indicate, however, that the statement of this theorem holds also for $\rho = 0$. Also we note – and this is in agreement with the theory – that the function W more and more emphasizes the disk Ω_1 rather than the first obstacle when enlarging ρ.

4

The MUSIC algorithm and scattering by an inhomogeneous medium

The factorization method presented in the previous chapter has a discrete analogue which is known as the MUSIC algorithm (standing for **MU**ltiple–**SI**gnal–**C**lassification). This algorithm is well known for signal-processing applications. As Devenay pointed out in [60] it could also been used for imaging, i.e., it provides a method to determine one or more unknown targets (point scatterers) from the so-called *multistatic response matrix* **F**. This is the complex $N \times N$ matrix where $F_{j\ell}$ is the measured far field pattern at receiver number j for source number ℓ. In general, **F** is symmetric but fails to be hermitean. Nevertheless, under generic assumptions on the geometry of the targets and for exact data, the rank of **F** coincides with the number M of targets provided M does not exceed N. Even more, as in the previous chapter one can explicitly define vectors $\phi_z \in \mathbb{C}^N$ depending on an arbitrary point z of space such that ϕ_z belongs to the range $\mathcal{R}(\mathbf{F})$ of **F** if and only if z coincides with one of the targets. A description of this algorithm can be found in, e.g., [60] or [30].

4.1 The MUSIC algorithm

In this section we recall the MUSIC algorithm in inverse scattering theory. The underlying scattering model is very simple. We consider an array of M point scatterers at locations $y_1, \ldots, y_M \in \mathbb{R}^3$ in the homogeneous, isotropic space $\mathbb{R}^3$. Incident plane waves of the form

$$u^i(x, \theta) = e^{ikx \cdot \theta}, \quad x \in \mathbb{R}^3,$$

are scattered by the targets at y_m. In this model we neglect all multiple scattering between the scatterers. Then the scattered field u^s is given by

$$u^s(x, \theta) = \sum_{m=1}^{M} \tau_m u^i(y_m, \theta) \, \Phi(x, y_m).$$

Here, $\tau_m \in \mathbb{C} \setminus \{0\}$ is the scattering strength of the m-th target, $m = 1, \ldots, M$, and Φ is again the fundamental solution of the Helmholtz equation in $\mathbb{R}^3$ given by (1.24), i.e.,

$$\Phi(x, y) = \frac{\exp(ik|x - y|)}{4\pi |x - y|}, \quad x \neq y. \tag{4.1}$$

Again, from the asymptotic behavior of $\Phi(x, y)$ we conclude that

$$u^s(x, \theta) = \frac{\exp(ik|x|)}{4\pi |x|} \sum_{m=1}^{M} \tau_m \, u^i(y_m, \theta) \, e^{-ik\hat{x}\cdot y_m} + \mathcal{O}\left(|x|^{-2}\right) \quad |x| \to \infty.$$

Therefore, the far field pattern is given by

$$u^\infty(\hat{x}, \theta) = \sum_{m=1}^{M} \tau_m \, u^i(y_m, \theta) \, e^{-ik\hat{x}\cdot y_m}, \quad \hat{x} \in S^2.$$

The *inverse scattering problem* is to determine the locations $y_1, \ldots, y_M$ from the response $u^\infty(\hat{x}, \theta)$ for all $\hat{x}, \theta \in S^2$ – or for all $\hat{x}, \theta$ from a finite subset $\{\theta_j : j = 1, \ldots, N\} \subset S^2$. In this finite case we assume $N \geq M$ and define the *multistatic response matrix* $\mathbf{F} \in \mathbb{C}^{N \times N}$ by

$$F_{j\ell} := u^\infty(\theta_j, \theta_\ell) = \sum_{m=1}^{M} \tau_m \, u^i(y_m, \theta_\ell) \, e^{-ik\theta_j \cdot y_m} = \sum_{m=1}^{M} \tau_m \, e^{ik \, y_m \cdot (\theta_\ell - \theta_j)}, \quad j, \ell = 1, \ldots, N.$$

Defining the matrices $\mathbf{H} \in \mathbb{C}^{M \times N}$ and $\mathbf{T} \in \mathbb{C}^{M \times M}$ by

$$H_{m\ell} = \sqrt{|\tau_m|} \, e^{ik \, \theta_\ell \cdot y_m}, \quad \ell = 1, \ldots, N, \; m = 1, \ldots, M, \quad \text{and} \quad \mathbf{T} = \mathrm{diag}(\mathrm{sign}\,\tau_m),$$

where $\mathrm{sign}\,\tau_m = \tau_m/|\tau_m|$ we observe that $\mathbf{F}$ has a factorization in the form

$$\mathbf{F} = \mathbf{H}^* \, \mathbf{T} \, \mathbf{H} \tag{4.2}$$

with adjoint $\mathbf{H}^* \in \mathbb{C}^{N \times M}$ of $\mathbf{H}$. If $N \geq M$ and if the locations y_m are such that $\mathbf{H}$ has maximal rank M then, by a standard result from linear algebra, the ranges $\mathcal{R}(\mathbf{H}^*)$ and $\mathcal{R}(\mathbf{F})$ of $\mathbf{H}^*$ and $\mathbf{F}$, respectively, coincide. Here we note the similarity – and the difference – to the factorization of the far field operator F from Subsection 1.4.1. In particular, we refer to the factorization of the form (1.55) where the matrix $\mathbf{H}$ corresponds to the integral operator H. But, as we have seen in Subsection 1.4.3, the range of the operator H^* does not coincide with the range of the operator F in contrast to the finite dimensional case.

For any point $z \in \mathbb{R}^3$ we define the vector $\phi_z \in \mathbb{C}^N$ by

$$\phi_z = \left(e^{-ik\theta_1 \cdot z}, \ldots, e^{-ik\theta_N \cdot z} \right)^\top. \tag{4.3}$$

Then we are able to show the following result:

Theorem 4.1 *Let $\{\theta_n : n \in \mathbb{N}\} \subset S^2$ be a countable set of directions such that any analytic function on S^2 that vanishes in θ_n for all $n \in \mathbb{N}$ vanishes identically. Let K be* *a compact subset of $\mathbb{R}^3$ containing all y_m. Then there exists $N_0 \in \mathbb{N}$ such that for any $N \geq N_0$ the following characterization holds for every $z \in K$:*

$$z \in \{y_1, \ldots, y_M\} \iff \phi_z \in \mathcal{R}(\mathbf{H}^*). \tag{4.4}$$

Furthermore, the ranges of $\mathbf{H}^*$ and $\mathbf{F}$ coincide and thus

$$z \in \{y_1, \ldots, y_M\} \iff \phi_z \in \mathcal{R}(\mathbf{F}) \iff \mathbf{P}\phi_z = 0, \tag{4.5}$$

where $\mathbf{P} : \mathbb{C}^N \to \mathcal{R}(\mathbf{F})^\perp = \mathcal{N}(\mathbf{F}^*)$ is the orthogonal projection onto the null space $\mathcal{N}(\mathbf{F}^*)$ of $\mathbf{F}^*$. Here, $W^\perp = \{u \in \mathbb{C}^N : u^*w = 0 \text{ for all } w \in W\}$ denotes the orthogonal complement of W.

Proof: The lengthy proof corrects the insufficient arguments in [118].

First we note that $\phi_z \in \mathcal{R}(\mathbf{H}^*)$ if $z \in \{y_1, \ldots, y_M\}$ because $\sqrt{|\tau_m|}\,\phi_{y_m}, m = 1, \ldots, M$, are the columns of the matrix $\mathbf{H}^* \in \mathbb{C}^{N \times M}$.

We show now that there exists $N_0 \in \mathbb{N}$ such that the vectors $\{\phi_{y_1} \ldots, \phi_{y_M}, \phi_z\}$ are linearly independent for all $N \geq N_0$ and all points $z \in K \setminus \{y_1, \ldots, y_M\}$. In particular, this would imply that $\mathbf{H}^*$ has maximal rank M and that $\phi_z \notin \mathcal{R}(\mathbf{H}^*)$ for all $z \in K \setminus \{y_1, \ldots, y_M\}$.

Assume on the contrary that this is not the case. Then there exist sequences $N_\ell \to \infty$ and $\{z^{(\ell)}\} \subset K \setminus \{y_1, \ldots, y_M\}$ and $\{\lambda^{(\ell)}\} \subset \mathbb{C}^M$ and $\{\mu^{(\ell)}\} \subset \mathbb{C}$ such that

$$\left|\mu^{(\ell)}\right| + \sum_{m=1}^{M} \left|\lambda_m^{(\ell)}\right| = 1$$

and

$$\mu^{(\ell)}\, e^{-ik\, z^{(\ell)} \cdot \theta_j} + \sum_{m=1}^{M} \lambda_m^{(\ell)}\, e^{-ik\, y_m \cdot \theta_j} = 0 \qquad \text{for all } j = 1, \ldots, N_\ell. \tag{4.6}$$

Since all of the sequences are bounded there exist converging subsequences $z^{(\ell)} \to z \in K$ and $\lambda^{(\ell)} \to \lambda \in \mathbb{C}^M$ and $\mu^{(\ell)} \to \mu \in \mathbb{C}$ as ℓ tends to infinity. We fix $j \in \mathbb{N}$ and let ℓ tend to infinity. Then

$$|\mu| + \sum_{m=1}^{M} |\lambda_m| = 1 \quad \text{and} \quad \mu\, e^{-ik\, z \cdot \theta_j} + \sum_{m=1}^{M} \lambda_m\, e^{-ik\, y_m \cdot \theta_j} = 0. \tag{4.7}$$

Since this holds for every $j \in \mathbb{N}$ we conclude from the assumption on the "richness" of the set $\{\theta_j : j \in \mathbb{N}\}$ that

$$\mu\, e^{-ik\, z \cdot \theta} + \sum_{m=1}^{M} \lambda_m\, e^{-ik\, y_m \cdot \theta} = 0 \qquad \text{for all } \theta \in S^2.$$

The left-hand side is the far field pattern of the function

$$x \mapsto \mu\, \Phi(x, z) + \sum_{m=1}^{M} \lambda_m\, \Phi(x, y_m).$$

Here, Φ is again the fundamental solution of the Helmholtz equation as defined in (4.1). Therefore, by Rellich's Lemma and unique continuation,

$$\mu\,\Phi(x,z) + \sum_{m=1}^{M} \lambda_m\,\Phi(x,y_m) = 0 \qquad \text{for all } x \notin \{z, y_1, \ldots, y_M\}\,.$$

Now we distinguish between two cases:

(A) Let $z \notin \{y_1, \ldots, y_M\}$. By letting x tend to z and to y_m for $m = 1, \ldots, M$ we conclude that all coefficients μ and λ_m for $m = 1, \ldots, M$ have to vanish. This contradicts the first equation of (4.7).

(B) Let now $z \in \{y_1, \ldots, y_M\}$. Without loss of generality we assume that $z = y_1$. By the same arguments as in part (A) we conclude that

$$\mu + \lambda_1 = 0 \quad \text{and} \quad \lambda_m = 0 \qquad \text{for } m = 2, \ldots, M\,. \tag{4.8}$$

Now we write (4.6) in the form

$$\left[\mu^{(\ell)} + \lambda_1^{(\ell)}\right] e^{-ik\,y_1\cdot\theta_j} + \mu^{(\ell)} \left[e^{-ik\,z^{(\ell)}\cdot\theta_j} - e^{-ik\,y_1\cdot\theta_j} \right] + \sum_{m=2}^{M} \lambda_m^{(\ell)}\, e^{-iky_m\cdot\theta_j} = 0 \tag{4.9}$$

for all $j = 1, \ldots, N_\ell$. The quantity

$$\rho_\ell = \left|\mu^{(\ell)} + \lambda_1^{(\ell)}\right| + \sum_{m=2}^{M} \left|\lambda_m^{(\ell)}\right| + \left|z^{(\ell)} - y_1\right|$$

converges to zero as ℓ tends to infinity. By Taylor's formula we have that

$$e^{-ik\,z^{(\ell)}\cdot\theta_j} - e^{-ik\,y_1\cdot\theta_j} = -ik\,\theta_j \cdot \left(z^{(\ell)} - y_1\right) e^{-ik\,y_1\cdot\theta_j} + \mathcal{O}\!\left(\left|z^{(\ell)} - y_1\right|^2\right)$$

as ℓ tends to infinity. Division of (4.9) by ρ_ℓ yields

$$\left[\tilde{\lambda}_1^{(\ell)} - ik\,\mu^{(\ell)}\,\theta_j \cdot a^{(\ell)}\right] e^{-ik\,y_1\cdot\theta_j} + \sum_{m=2}^{M} \tilde{\lambda}_m^{(\ell)}\, e^{-iky_m\cdot\theta_j} = \mathcal{O}\!\left(\left|z^{(\ell)} - y_1\right|\right)$$

for all $j = 1, \ldots, N_\ell$, where

$$\tilde{\lambda}_1^{(\ell)} = \frac{\mu^{(\ell)} + \lambda_1^{(\ell)}}{\rho_\ell}, \quad \tilde{\lambda}_m^{(\ell)} = \frac{\lambda_m^{(\ell)}}{\rho_\ell}, \quad m = 2, \ldots, M, \quad a^{(\ell)} = \frac{z^{(\ell)} - y_1}{\rho_\ell}\,.$$

These sequences are all bounded as well, i.e., we can extract further subsequences $\tilde{\lambda}_m^{(\ell)} \to \tilde{\lambda}_m$ for $m = 1, \ldots, M$, and $a^{(\ell)} \to a \in \mathbb{R}^3$ as ℓ tends to infinity. We have that

$$\sum_{m=1}^{M} \left|\tilde{\lambda}_m\right| + |a| = 1 \tag{4.10}$$

and

$$\left[\tilde{\lambda}_1 - ik\,\mu\,\theta_j \cdot a\right] e^{-ik\,y_1\cdot\theta_j} + \sum_{m=2}^{M} \tilde{\lambda}_m\, e^{-iky_m\cdot\theta_j} = 0$$

for all $j \in \mathbb{N}$. Again, by the assumption on the set $\{\theta_j : j \in \mathbb{N}\}$ we conclude that this equation holds for all $\theta \in S^2$. The left-hand side is now the far field pattern of the function

$$x \mapsto \tilde{\lambda}_1 \, \Phi(x, y_1) + \mu \, a \cdot \nabla_y \Phi(x, y_1) + \sum_{m=2}^{M} \tilde{\lambda}_m \, \Phi(x, y_m) \, ,$$

i.e., by Rellich's Lemma and unique continuation again,

$$\tilde{\lambda}_1 \, \Phi(x, y_1) + \mu \, a \cdot \nabla_y \Phi(x, y_1) + \sum_{m=2}^{M} \tilde{\lambda}_m \, \Phi(x, y_m) = 0$$

for all $x \notin \{y_1, \ldots, y_M\}$. By letting x tend to y_1 and to y_m for $m = 2, \ldots, M$ we conclude that all coefficients $\tilde{\lambda}_m$ for $m = 2, \ldots, M$ have to vanish and also the term

$$\tilde{\lambda}_1 \, \Phi(x, y_1) + \mu \, a \cdot \nabla_y \Phi(x, y_1) = \left[\tilde{\lambda}_1 + \mu \, \frac{a \cdot (y_1 - x)}{|y_1 - x|} \left(ik - \frac{1}{|y_1 - x|} \right) \right]$$
$$\times \, \frac{\exp(ik|y_1 - x|)}{4\pi |y_1 - x|}$$

vanishes for $x \neq y_1$. From this easily $\tilde{\lambda}_1 = 0$ and $\mu \, a = 0$ follows. We recall from (4.7) and (4.8) that $|\mu| = 1/2$ and thus $a = 0$. This, finally, contradicts (4.10). $\square$

Therefore, plotting the function

$$W(z) = \frac{1}{|\mathbf{P}\phi_z|} \, , \quad z \in \mathbb{R}^3 \, ,$$

should result in sharp peaks at $y_1, \ldots, y_M$. The plots of Figure 4.1 show the result for the example where $d = 2$, $M = 3$, $N = 10$, wavelength $\lambda = 1$, i.e., $k = 2\pi$, and θ_j, $j = 1, \ldots, 10$, are equidistantly chosen directions. The values of τ are 1, $1.5 + i$, and 2, respectively. To the data 10% relative noise has been added. On the left side we show a contour plot of W while on the right side the same function is shown in a 3D-plot.

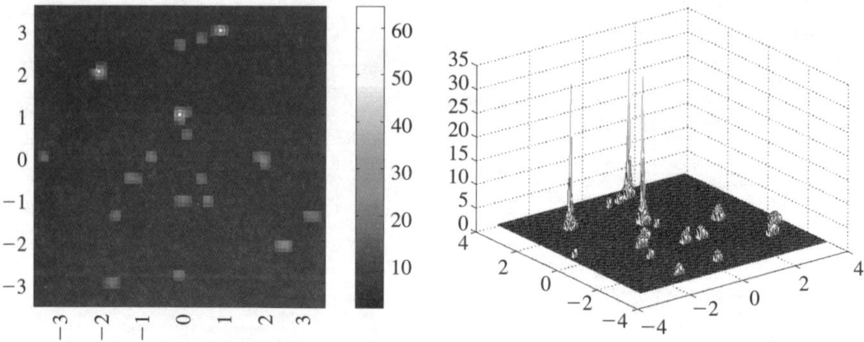

Figure 4.1 Plots of function W

4.2 Scattering by an inhomogeneous medium

As shown by [2], the simple model of the previous section can be considered as the limiting case of the scattering by an inhomogeneous medium, i.e., a medium with continuously distributed scatterers which are modeled by a "contrast function" $q \in L^\infty(D)$. In this section we will formulate the model and will state uniqueness and existence results. A popular approximation of the true model is given by the Born approximation. It will lead us to a simple factorization which is completely analogous to (4.2).

For this and the following sections we make the following general assumptions

Assumption 4.2 *Let $D \subset \mathbb{R}^3$ be open and bounded such that the complement $\mathbb{R}^3 \setminus \overline{D}$ is connected. Furthermore, we assume that the boundary ∂D of D is smooth enough such that the imbedding of $H^1(D)$ into $L^2(D)$ is compact. Let $q \in L^\infty(D)$ satisfy*

(1) $\operatorname{Im} q \geq 0$ *on D.*

(2) There exists $c_0 > 0$ with $1 + \operatorname{Re} q(x) \geq c_0$ for almost all $x \in D$.

(3) $|q|$ is locally bounded below, i.e., for every compact subset $M \subset D$ there exists $c > 0$ (depending on M) such that

$$|q(x)| \geq c \quad \text{for almost all } x \in M . \tag{4.11}$$

We extend q by zero outside of D.

We note that part (3) of this assumption is satisfied for continuous contrasts q. It must be guaranteed that q vanishes at most on the boundary of D. In the physical situation q is related to the index of refraction n by $q = n^2 - 1$ (compare equations (1.12) and (4.12) below).

By $k > 0$ we denote again the wavenumber. For smooth functions q, i.e., $q \in C(\overline{D})$ and incident field $u^i(x, \theta) = \exp(ik\theta \cdot x)$ with some $\theta \in S^2$, the classical formulation of the direct scattering problem is to determine $u^s = u^s(\cdot, \theta) \in C^1(\mathbb{R}^3) \cap C^2(\mathbb{R}^3 \setminus \partial D)$ such that the total field[1] $u := u^s + u^i$ satisfies the Helmholtz equation

$$\Delta u(x) + k^2 \big(1 + q(x)\big) u(x) = 0 \quad \text{in } \mathbb{R}^3 \setminus \partial D , \tag{4.12}$$

and u^s satisfies the Sommerfeld radiation condition

$$\frac{\partial u^s(x)}{\partial r} - ik u^s(x) = \mathcal{O}\big(r^{-2}\big), \quad r = |x| \to \infty . \tag{4.13}$$

Writing (4.12) for the scattered field u^s we get

$$\Delta u^s + k^2(1 + q)\, u^s = -k^2 q\, u^i \quad \text{in } \mathbb{R}^3 \setminus \partial D . \tag{4.14}$$

---

[1] If appropriate, we indicate the dependence on the incident direction by writing $u^s(x, \theta)$

For the factorization method we have to consider this problem for more general source terms $f \in L^2(D)$: Determine $v \in H^1_{loc}(\mathbb{R}^3)$ such that

$$\Delta v + k^2 (1 + q) v = -k^2 q f \quad \text{in } \mathbb{R}^3. \tag{4.15}$$

For general $q \in L^\infty(D)$ and $f \in L^2(D)$ this equation has to be understood in the variational sense, i.e., $v \in H^1_{loc}(\mathbb{R}^3)$ has to satisfy

$$\iint_{\mathbb{R}^3} \left[\nabla v \cdot \nabla \phi - k^2 (1 + q) v \phi \right] dx = k^2 \iint_D q f \phi \, dx \tag{4.16}$$

for all functions $\phi \in H^1(\mathbb{R}^3)$ with compact support. A well-known regularity result for elliptic differential equations (see [144]) yields that $v \in H^2_{loc}(\mathbb{R}^3)$ and v is even analytic in $\mathbb{R}^3 \setminus \overline{D}$. In particular, the radiation condition (4.13) makes sense.

An equivalent way to interpret the radiating solution of (4.15) is by the Lippmann–Schwinger integral equation. We denote again by Φ the fundamental solution of the Helmholtz equation as defined in (4.1). Then it can be shown[2] that if $v \in H^1_{loc}(\mathbb{R}^3)$ is a radiating solution of (4.15) (in the sense of (4.16)) then the restriction $v|_D$ solves

$$v(x) = k^2 \iint_D q(y) v(y) \, \Phi(x, y) \, dy + k^2 \iint_D q(y) f(y) \, \Phi(x, y) \, dy, \quad x \in D. \tag{4.17}$$

Furthermore, if $v \in L^2(D)$ solves (4.17) then v can be extended by the right-hand side of (4.17) to a function $v \in H^1_{loc}(\mathbb{R}^3)$ which is a radiating variational solution of (4.15).

Finally, the Lippmann-Schwinger integral equation (4.17) – and therefore also the variational equation (4.16) – have unique solutions for every $f \in L^2(D)$ and $q \in L^\infty(D)$ with $\text{Im } q \geq 0$.

For the particular case $f = u^i$ we observe from (4.17) that u^s satisfies

$$u^s(x, \theta) = k^2 \iint_D q(y) u(y, \theta) \, \Phi(x, y) \, dy, \quad x \in \mathbb{R}^3,$$

from which we derive the explicit form of the far field pattern

$$u^\infty(\hat{x}, \theta) = k^2 \iint_D q(y) u(y, \theta) e^{-ik\hat{x} \cdot y} \, dy, \quad \hat{x} \in S^2. \tag{4.18}$$

We note that in addition u has to solve the variational equation (4.16) for $f = u^i$ or, equivalently, the Lippmann–Schwinger integral equation (4.17) for $f = u^i$, i.e.,

$$u(x) = u^i(x) + k^2 \iint_D q(y) u(y) \, \Phi(x, y) \, dy, \quad x \in D. \tag{4.19}$$

---

[2] We were not able to find a precise reference of this fact. In [43] this is shown for classical solutions, i.e., for smooth functions q. The arguments can easily be extended to the case $q \in L^\infty(D)$.

In the *inverse scattering problem* one wants to determine q or at least its support D from the data $u^\infty(\hat{x}, \theta)$ for all $\hat{x}, \theta \in S^2$.

The following uniqueness theorem is well known and goes back to Sylvester, Uhlmann [176], Ramm [164]), and Novikov [151].

Theorem 4.3 *Under the assumptions on D and q for fixed wavenumber $k > 0$ the far field patterns $u^\infty(\hat{x}, \theta)$ for all $\hat{x}, \theta \in S^2$ uniquely determine the contrast function q, i.e., if there are two contrasts q_1 and q_2 with corresponding far field patterns $u_1^\infty(\hat{x}, \theta)$ and $u_2^\infty(\hat{x}, \theta)$, respectively, then $u_1^\infty(\hat{x}, \theta) = u_2^\infty(\hat{x}, \theta)$ for all $\hat{x}, \theta \in S^2$ implies that $q_1 = q_2$.*

For a **proof** we refer to, e.g., [113].

As in Chapter 1 we collect the data in the far field operator F from $L^2(S^2)$ into itself, defined again as

$$(F\psi)(\hat{x}) := \int_{S^2} u^\infty(\hat{x}, \theta)\, \psi(\theta)\, ds(\theta)\,, \quad \hat{x} \in S^2\,. \tag{4.20}$$

The Lippmann–Schwinger integral equation (4.19) is a fixed point equation for u. For sufficiently small values of $k^2 \max_{x \in \overline{D}} \iint_D |q(y)||\Phi(x, y)|dy$ it can be solved iteratively by the Neumann series. The first iterate is called the *Born approximation* and is given by

$$u_b(x) = u^i(x) + k^2 \iint_D q(y)\, u^i(y, \theta)\, \Phi(x, y)\, dy\,, \quad x \in \mathbb{R}^3\,,$$

i.e., we replaced the total field in the integrand by the incident field. We indicate this approximation by the subscript "b" in the following. Then u_b^∞ is explicitly given by

$$u_b^\infty(\hat{x}, \theta) = k^2 \iint_D q(y)\, u^i(y; \theta)\, e^{-ik\hat{x}\cdot y}\, dy = k^2 \iint_D q(y)\, e^{ik\, y\cdot(\theta - \hat{x})}\, dy\,, \tag{4.21}$$

for $\hat{x}, \theta \in S^2$. We observe that $u_b^\infty(\hat{x}, \theta)$ is just the Fourier transform of $k^2 q$ evaluated at $k(\theta - \hat{x})$. From this observation we note that the knowledge of $u_b^\infty(\hat{x}, \theta)$ for all $\hat{x}, \theta \in S^2$ determines the Fourier transform $\hat{q}$ in the ball $B(0, 2k)$ with radius $2k$ and center 0. The function $\hat{q}$ is analytic since q has compact support. Therefore, the knowledge of $\hat{q}$ on $B(0, 2k)$ determines $\hat{q}$ in $\mathbb{R}^3$ and thus also q. This uniqueness result corresponds to Theorem 4.3 for the Born approximation.

As in the general case we define the corresponding far field operator F_b from $L^2(S^2)$ into itself by

$$(F_b\psi)(\hat{x}) := \int_{S^2} u_b^\infty(\hat{x}, \theta)\, \psi(\theta)\, ds(\theta)\,, \quad \hat{x} \in S^2\,. \tag{4.22}$$

From the explicit form (4.21) of u_b^∞ we observe that u_b^∞ satisfies always the reciprocity relation (4.23) below. Also it follows that for real-valued q the operator F_b is self-adjoint. In the general case of multiple scattering (i.e., where the Born approximation fails to be an appropriate model) the operator F is no longer self-adjoint but only normal. We collect some important properties of F in the following theorem which corresponds to Theorem 1.8.

Theorem 4.4 *Let u^∞ and $F : L^2(S^2) \to L^2(S^2)$ be the far field pattern (4.18) and the far field operator (4.20), respectively.*

(a) The following reciprocity relation holds:

$$u^\infty(-\hat{x}, \theta) = u^\infty(-\theta, \hat{x}) \quad \text{for all } \hat{x}, \theta \in S^2. \tag{4.23}$$

(b) The far field operator F satisfies $F - F^* = \frac{ik}{8\pi^2} F^*F + 2iR$ where F^* denotes again the L^2-adjoint of F and $R : L^2(S^2) \to L^2(S^2)$ is some compact and self-adjoint non-negative operator which vanishes for real-valued contrasts q.

(c) The scattering operator $S := I + \frac{ik}{8\pi^2} F$ is sub-unitary, i.e.,

$$S^*S = I - \frac{k}{4\pi^2} R.$$

In the case where q is real-valued the scattering operator S is unitary, and the far field operator F is normal.

(d) Assume that k^2 is not an eigenvalue of the following interior transmission problem in D, i.e., the only solution $(v, w) \in H^2(D) \times H^2(D)$ of

$$\Delta v + k^2(1 + q)v = 0 \text{ in } D, \qquad \Delta w + k^2 w = 0 \text{ in } D, \tag{4.24}$$

$$v = w \text{ on } \partial D \quad \text{and} \quad \frac{\partial v}{\partial \nu} = \frac{\partial w}{\partial \nu} \quad \text{on} \partial D, \tag{4.25}$$

is the trivial one $v = w = 0$. Then F is one-to-one and its range $\mathcal{R}(F)$ is dense in $L^2(S^2)$.

Proof: (a) The proof of (1.28) depends only on the exterior of D and can literally copied for the present case. Using the second Green's theorem for the functions $u(\cdot, \theta)$ and $u(\cdot, \hat{x})$ in D yields that

$$\int_{\partial D} \left[u(y, \theta) \frac{\partial}{\partial \nu} u(y, \hat{x}) - u(y, \hat{x}) \frac{\partial}{\partial \nu} u(y, \theta) \right] ds(y) = 0$$

because both functions satisfy the same Helmholtz equation in D.

(b) Again, we follow the lines of the corresponding proof for the Dirichlet boundary condition of Theorem 1.8, part (a). Equation (1.34) has to be replaced by

$$2ik^2 \iint_D (\text{Im } q) \, v \overline{w} \, dx = \int_{D_R} \left[v \, \Delta \overline{w} - \overline{w} \, \Delta v \right] dx = \int_{|x|=R} \left[v \frac{\partial \overline{w}}{\partial \nu} - \overline{w} \frac{\partial v}{\partial \nu} \right] ds. \tag{4.26}$$

We introduce the operator $L : L^2(S^2) \to L^2(D)$ which maps $g \in L^2(S^2)$ into $v|_D$ where v is the solution of the scattering problem with incident field $v^i(x) = \int_{S^2} \exp(ikx \cdot \theta) g(\theta) \, ds(\theta)$, $x \in \mathbb{R}^3$. Then the left-hand side of (4.26) can be written in the form $2ik^2 \big((\mathrm{Im}\, q)\, Lg, Lh\big)_{L^2(D)}$. Equation (1.35) takes now the form

$$2ik^2 \big(L^*\big((\mathrm{Im}\, q)\, Lg\big), h\big)_{L^2(S^2)} = -\frac{ik}{8\pi^2} \big(Fg, Fh\big)_{L^2(S^2)} - \big(g, Fh\big)_{L^2(S^2)} + \big(Fg, h\big)_{L^2(S^2)}$$

which proves part (b) with $Rg = k^2 L^*\big((\mathrm{Im}\, q)\, Lg\big)$.

(c) This is proved with the help of Part (b) above exactly as in Theorem 1.8, part (b).

(d) Again, as in the proof of Theorem 1.8, part (d), let $g \in L^2(S^2)$ be such that $Fg = 0$ on S^2. From the definition of the far field operator we note that $Fg = w^\infty$ where w^∞ is the far field pattern which corresponds to the incident field $w^i(x) = \int_{S^2} \exp(ikx \cdot \theta) g(\theta) \, ds(\theta)$, $x \in \mathbb{R}^3$. Rellich's Lemma 1.2 and the analytic continuation argument implies that the scattered field w^s vanishes outside of D.[3] Then $w = w^i + w^s$ and $v = -w^i$ satisfy (4.24) and (4.25). By the assumption v and w have to vanish which is only possible for $g = 0$. The proof of denseness of the range of F follows exactly the lines of the proof of Theorem 1.8, part (d). $\square$

The interior transmission problem (4.24), (4.25) is an unconvential eigenvalue problem and corresponds to the interior Dirichlet, Neumann, or impedance eigenvalue problem in the obstacle scattering case. It has been studied in, e.g., [41, 169] (see also [43, 113]). In particular, it has been shown that, in general, for real-valued contrasts q which satisfy a certain decay condition at the boundary ∂D there exists at most a countably set of eigenvalues k^2. For the spherically stratisfied case they indeed exist. On the other hand, if $\mathrm{Im}\, q > 0$ on some set $U \subset D$ then no eigenvalues exist. We refer to Section 4.5 below for a detailed investigation of this transmission eigenvalue problem.

4.3 Factorization of the far field operators

We begin by considering the far field operator F_b for the Born approximation. This integral operator F_b has an obvious factorization just as in the case of the multistatic response matrix $\mathbf{F}$ of Section 4.1. Indeed, we define the operator $H : L^2(S^2) \to L^2(D)$ with adjoint $H^* : L^2(D) \to L^2(S^2)$ by

$$(H\psi)(x) = \sqrt{|q(x)|} \int_{S^2} \psi(\theta)\, e^{ik\, x \cdot \theta} \, ds(\theta), \quad x \in D, \tag{4.27}$$

$$(H^*\varphi)(\hat{x}) = \iint_D \varphi(y)\, e^{-ik\hat{x}\cdot y} \sqrt{|q(y)|} \, dy, \quad \hat{x} \in S^2. \tag{4.28}$$

---

[3] Here we make use of the assumption that the exterior of $\overline{D}$ is connected.

Substituting (4.21) into (4.22) and changing the orders of integration yields

$$F_b \psi = k^2 H^* \left(\frac{q}{|q|} H \psi \right), \quad \text{i.e.,} \quad F_b = H^* T_b H \tag{4.29}$$

with the operator $T_b f := k^2 (\operatorname{sign} q) f$ for $f \in L^2(D)$. Here, and in the following, we denote again by $\operatorname{sign} z = z/|z|$ the (complex) sign of $z \in \mathbb{C}$. Formula (4.29) corresponds directly to the factorization (4.2).

Now we consider the general case (4.12), (4.13), i.e., we allow multiple scattering. We make the Assumption 4.2 from the previous section.

Theorem 4.5 *Let $F : L^2(S^2) \to L^2(S^2)$ be defined by (4.20). Then*

$$F = H^* T H \tag{4.30}$$

with H and H^* from (4.27), (4.28), respectively. The operator $T : L^2(D) \to L^2(D)$ is defined by $Tf = k^2 (\operatorname{sign} q) [f + \sqrt{|q|}\, v|_D]$ where again $\operatorname{sign} q := q/|q|$ and $v \in H^1_{loc}(\mathbb{R}^3)$ is the radiating solution of

$$\Delta v + k^2(1 + q)v = -k^2 \frac{q}{\sqrt{|q|}} f \quad in \ \mathbb{R}^3 . \tag{4.31}$$

Proof: Analogously to the case of an inpenetrable obstacle we define the data-to-pattern operator $G : L^2(D) \to L^2(S^2)$ by $f \mapsto v^\infty$ where $v \in H^1_{loc}(\mathbb{R}^3)$ solves (4.31). The superposition principle and equation (4.14) yield $F = GH$. From (4.28) we observe that $H^*\varphi$ is the far field pattern w^∞ of the volume potential

$$w(x) = \iint\limits_D \varphi(y)\, \Phi(x,y)\, \sqrt{|q(y)|}\, dy, \quad x \in \mathbb{R}^3 .$$

For Hölder continuous densities $\varphi\sqrt{|q|}$ it is well known (cf. [70]) that w is twice differentiable in $\mathbb{R}^3 \setminus \partial D$ and satisfies

$$\Delta w + k^2 w = -\varphi \sqrt{|q|} \ in \ \mathbb{R}^3 \setminus \partial D . \tag{4.32}$$

By a density argument the identity $H^*\varphi = w^\infty$ holds for all $q \in L^\infty(D)$ and all $\varphi \in L^2(D)$ where $w \in H^1_{loc}(\mathbb{R}^3)$ denotes the (radiating) variational solution of (4.32). Now we set $\varphi = Tf = k^2 (\operatorname{sign} q) [f + \sqrt{|q|}\, v|_D]$ where v solves (4.31). Then $H^* Tf = w^\infty$ where w is the radiating solution of

$$\Delta w + k^2 w = -\sqrt{|q|}\, Tf = -k^2 \frac{q}{\sqrt{|q|}} f - k^2 q\, v = \Delta v + k^2 v \quad in \ \mathbb{R}^3 .$$

Since also v solves the radiation condition we conclude that w and v coincide. This implies $H^* Tf = v^\infty = Gf$. Substituting $H^* T = G$ into $F = GH$ yields the assertion. $\square$

Remarks: This factorization (4.30) has the same form as (4.29) and differs only in the form of T. While T_b is a multiplication operator and thus local we observe that T is non-local because it contains the solution v of (4.31). We formally derive (4.29) from (4.30) by setting v to zero. From the definition of v one observes that the operators T and

T_b differ of order k^4 for small values of k. Also one observes that for real and positive values of q the operator T_b is the identity (times k^2). We will see in Theorem 4.8 below that T is a compact perturbation of T_b. An essential difference is also that for real-valued contrasts q the operator F_b is self-adjoint – as seen also from the factorization – while F is only normal by Theorem 4.4.

4.4 Localization of the support of the contrast

Again, we make the Assumption 4.2 of Section 4.2. First, we will prove the following characterization of D by the ranges of the operator H^*, characterized by (4.28). This result corresponds to Theorem 1.12 in the case of Dirichlet boundary conditions and to Theorem 4.1 in the MUSIC algorithm.

Theorem 4.6 *Let Assumption 4.2 hold. For $z \in \mathbb{R}^3$ define again $\phi_z \in L^2(S^2)$ by (1.41), i.e.,*

$$\phi_z(\hat{x}) = e^{-ik\hat{x}\cdot z}, \quad \hat{x} \in S^2. \tag{4.33}$$

Then it holds that

$$z \in D \quad \text{if, and only if,} \quad \phi_z \in \mathcal{R}(H^*). \tag{4.34}$$

Proof: Assume first that $z \in D$. We have to show that $\phi_z \in \mathcal{R}(H^*)$. Let $B[z, \varepsilon] \subset D$ be some closed ball with center z and radius $\varepsilon > 0$ which is completely contained in D. Choose a function $\psi \in C^\infty(\mathbb{R})$ with $\psi(t) = 1$ for $|t| \geq \varepsilon$ and $\psi(t) = 0$ for $|t| \leq \varepsilon/2$ and define $v \in C^\infty(\mathbb{R}^3)$ by $v(x) = \psi(|x-z|)\,\Phi(x,z)$ in $\mathbb{R}^3$. Then v satisfies $v = \Phi(\cdot, z)$ on ∂D and $\partial v/\partial \nu = \partial \Phi(\cdot, z)/\partial \nu$ on ∂D and $\Delta v + k^2 v = 0$ for $|x - z| \geq \varepsilon$. Therefore, from the representation theorem (see Theorem 2.1 of [43]) we have for $x \in D$:

$$v(x) = \int\limits_{\partial D} \left\{ \Phi(x,y) \frac{\partial v(y)}{\partial \nu} - v(y) \frac{\partial \Phi(x,y)}{\partial \nu(y)} \right\} ds(y) - \iint\limits_{D} \left\{ \Delta v(y) + k^2 v(y) \right\} \Phi(x,y)\,dy$$

$$= \int\limits_{\partial D} \left\{ \Phi(x,y) \frac{\partial \Phi(y,z)}{\partial \nu(y)} - \Phi(y,z) \frac{\partial \Phi(x,y)}{\partial \nu(y)} \right\} ds(y)$$

$$- \iint\limits_{|y-z|<\varepsilon} \left\{ \Delta v(y) + k^2 v(y) \right\} \Phi(x,y)\,dy$$

$$= - \iint\limits_{|y-z|<\varepsilon} \left\{ \Delta v(y) + k^2 v(y) \right\} \Phi(x,y)\,dy$$

since the first integral vanishes by Green's formula ($\Phi(\cdot, x)$ and $\Phi(\cdot, z)$ both satisfy the radiation condition). Since $\Phi(\cdot, z)$ and v coincide outside of $\overline{D}$ we conclude that

$$\phi_z(\hat{x}) = v^\infty(\hat{x}) = - \iint\limits_{|y-z|<\varepsilon} \left\{ \Delta v(y) + k^2 v(y) \right\} e^{-ik\hat{x}\cdot y}\,dy \quad \text{for } \hat{x} \in S^2.$$

Now we set

$$\varphi = \begin{cases} -[\Delta v + k^2 v]/\sqrt{|q|} & \text{in } B[z, \varepsilon]\,, \\ 0\,, & \text{in } D \setminus B[z, \varepsilon]\,. \end{cases}$$

Then $\varphi \in L^2(D)$ since $|q|$ is bounded below on $B[z, \varepsilon]$ and $\phi_z = H^*\varphi$ which proves that $\phi_z \in \mathcal{R}(H^*)$.

Let now $z \notin D$ and assume, on the contrary, that there exists $\varphi \in L^2(D)$ with $H^*\varphi = \phi_z$ on S^2. Then, by Rellich's Lemma 1.2 and unique continuation,

$$\iint\limits_{D} \varphi(y)\,\Phi(x, y)\,\sqrt{|q(y)|}\,dy = \Phi(x, z) \qquad \text{for all } x \text{ in the exterior of } D \cup \{z\}\,.$$

The left-hand side of this equation defines a C^1-function in $\mathbb{R}^3$ (see Lemma 4.1 of [70]) and is a solution of the Helmholtz equation in the exterior of D. The right-hand side, however, has a singularity at $z \notin D$ which is a contradiction. This proves the theorem. $\square$

In what follows we will consider two cases of contrasts. First, we will treat non-absorbing media and will derive a characterization of D by applying Theorem 1.23 since in this case the far field operator F is normal and the scattering operator unitary. Second, we will treat general possibly absorbing media (i.e., q can be complex-valued). In this case we will apply the abstract Theorem 2.15 from Chapter 2. We will not consider a third case where q is complex-valued everywhere on D and Im q is bounded below by a positive constant. This case would lead to the particular simple characterization of D by the (self-adjoint) imaginary part of F alone. Instead, we refer to Chapter 5 where we consider this case for Maxwell's equations.

For both cases we need properties of the operator T in the factorization (4.30) which we collect in Theorem 4.8 below. As in the case of Dirichlet boundary conditions (compare Corollary 1.18) we have to exclude the case that k^2 is the eigenvalue of an eigenvalue problem – which in this case is the interior transmission eigenvalue problem. However, we need it in a weak formulation. First, we define the weighted space $L^2(D, |q|dx)$ as the completion of $L^2(D)$ with respect to the norm corresponding to the inner product

$$(\phi, \psi)_{L^2(D,|q|dx)} = \iint\limits_{D} \phi\,\overline{\psi}\,|q|\,dx\,.$$

We note that for eigenpairs (v, w) of (4.24) and (4.25) the Cauchy data of the difference $u = v - w$ vanish on ∂D. Furthermore, the $u = v - w$ satisfies the differential equation

$$\Delta u + k^2(1 + q)u = -k^2 q\,w \quad \text{in } D\,.$$

This motivates the following definition.

Definition 4.7 k^2 *is called an* interior transmission eigenvalue *if there exists* $(u, w) \in H_0^1(D) \times L^2(D, |q|dx)$ *with* $(u, w) \neq (0, 0)$ *and a sequence* $\{w_j\}$ *in* $H^2(D)$ *with* $w_j \to w$

in $L^2(D, |q|dx)$ and $\Delta w_j + k^2 w_j = 0$ in D and

$$\iint_D \left[\nabla u \cdot \nabla \psi - k^2(1+q)\, u\, \psi \right] dx = k^2 \iint_D q\, w\, \psi\, dx \quad \text{for all } \psi \in H^1(D). \quad (4.35)$$

We note that the boundary condition $\partial u/\partial \nu = 0$ on ∂D is implicitly included in (4.35) by requiring that this equation holds for all $\psi \in H^1(D)$. We can reformulate the approximation of w by w_j by requiring that w is in the closure of $\{\phi \in H^2(D) : \Delta\phi + k^2\phi = 0$ in $D\}$ with respect to the norm $\|\cdot\|_{L^2(D,|q|dx)}$. We will study this eigenvalue problem in Section 4.5 below.

Theorem 4.8 *Let Assumption 4.2 hold. Then we have:*

(a) The operator T from Theorem 4.5 can be written in the form $T = T_0 + K$ where T_0 has the form $T_0 f = k^2 (\text{sign } q) f$ for $f \in L^2(D)$ and $K : L^2(D) \to L^2(D)$ is compact. If there exist $t \in [0, \pi]$ and $c > 0$ such that

$$\text{Re}\left[e^{-it} q(x) \right] \geq c\,|q(x)| \quad \text{almost everywhere on } D, \quad (4.36)$$

then the operator $\text{Re}[\exp(-it)T_0]$ is coercive, in particular,

$$\text{Re}\left[e^{-it} \, (T_0 f, f)_{L^2(D)} \right] \geq k^2 c \,\|f\|_{L^2(D)}^2, \quad f \in L^2(D). \quad (4.37)$$

(b) For all $f \in L^2(D)$ we have

$$\text{Im}(Tf, f)_{L^2(D)} \geq 0.$$

(c) Assume that k^2 is not an eigenvalue of the interior transmission problem in the sense of Definition 4.7. Then

$$\text{Im}(Tf, f)_{L^2(D)} > 0 \quad \text{for all } f \in \overline{\mathcal{R}(H)},\, f \neq 0.$$

We note that (4.36) is always satisfied for real-valued q (with $c = 1$ and $t = 0$ if q is positive or $t = \pi$ if q is negative). For complex-valued q with $\text{Re } q \geq 0$ and $\text{Im } q \geq 0$ this condition is satisfied with $t = \pi/2$ and $c = 1/\sqrt{2}$ since in this case $\text{Re } q + \text{Im } q \geq |q|$.

Proof: (a) The form $T = T_0 + K$ is obvious where K maps $f \in L^2(D)$ into $k^2 q/\sqrt{|q|}\, v|_D$ and v is the radiating solution of (4.31). The operator K is compact because of the compact imbedding of $H^1(D)$ into $L^2(D)$. Furthermore,

$$\text{Re}\left[e^{-it}(T_0 f, f)_{L^2(D)} \right] = k^2 \iint_D \frac{\text{Re}[\exp(-it)\, q]}{|q|} \, |f|^2\, dx \geq k^2 c \iint_D |f|^2\, dx.$$

(b) For $f \in L^2(D)$ we set $w = f + \sqrt{|q|}\, v$ and have

$$(Tf, f)_{L^2(D)} = k^2 \iint_D (\text{sign } q)\, w\overline{f}\, dx = k^2 \iint_D (\text{sign } q)\, w \left[\overline{w} - \sqrt{|q|}\,\overline{v} \right] dx$$

$$= k^2 \iint_D (\text{sign } q)\, |w|^2\, dx - k^2 \iint_D w\, \frac{q}{\sqrt{|q|}}\, \overline{v}\, dx.$$

Now we observe that v satisfies the differential equation

$$\Delta v + k^2 v = -k^2 \frac{q}{\sqrt{|q|}} f - k^2 q \, v = -k^2 \frac{q}{\sqrt{|q|}} \, w \,,$$

i.e.,

$$(Tf,f)_{L^2(D)} = k^2 \iint\limits_{D} (\text{sign } q) \, |w|^2 \, dx + \iint\limits_{D} \bar{v} \left[\Delta v + k^2 v \right] dx \,.$$

By two applications of Green's theorem (in D and in $\{x \notin D : |x| < R\}$, respectively) we arrive at

$$(Tf,f)_{L^2(D)} = k^2 \iint\limits_{D} (\text{sign } q) \, |w|^2 \, dx + \iint\limits_{D} \left[k^2 |v|^2 - |\nabla v|^2 \right] dx + \int\limits_{\partial D} \bar{v} \frac{\partial v}{\partial \nu} \, ds$$

$$= k^2 \iint\limits_{D} (\text{sign } q) \, |w|^2 \, dx + \iint\limits_{|x|<R} \left[k^2 |v|^2 - |\nabla v|^2 \right] dx + \int\limits_{|x|=R} \bar{v} \frac{\partial v}{\partial \nu} \, ds \,.$$

For $R \to \infty$ the last term on the right hand side converges to $ik/(4\pi)^2 \int_{S^2} |v^\infty|^2 \, ds$ since $\partial v(x)/\partial \nu = ikv(x) + \mathcal{O}(1/|x|^2)$ by the radiation condition. Therefore, also the second term on the right hand side converges and

$$(Tf,f)_{L^2(D)} = k^2 \iint\limits_{D} (\text{sign } q) \, |w|^2 \, dx + \iint\limits_{\mathbb{R}^3} \left[k^2 |v|^2 - |\nabla v|^2 \right] dx + \frac{ik}{(4\pi)^2} \int\limits_{S^2} |v^\infty|^2 \, ds.$$

Taking the imaginary part yields

$$\text{Im}(Tf,f)_{L^2(D)} = k^2 \iint\limits_{D} \frac{\text{Im } q}{|q|} \, |w|^2 \, dx + \frac{k}{(4\pi)^2} \int\limits_{S^2} |v^\infty|^2 \, ds \geq 0 \,.$$

(c) Let now $f \in \overline{\mathcal{R}(H)}$ with $\text{Im}(Tf,f)_{L^2(D)} = 0$. Then $v^\infty \equiv 0$ and thus, by Rellich's Lemma, $v \equiv 0$ in $\mathbb{R}^3 \setminus D$. Since $v \in H^2_{loc}(\mathbb{R}^3)$ we conclude that $v|_D \in H^1_0(D)$ with $\partial v/\partial \nu = 0$ on ∂D. Furthermore, we recall that v satisfies the differential equation

$$\Delta v + k^2(1+q)v = -k^2 \frac{q}{\sqrt{|q|}} f \quad \text{in } \mathbb{R}^3 \,.$$

Since $f \in \overline{\mathcal{R}(H)}$ there exists $f_j = H \psi_j \in \mathcal{R}(H)$ with $f_j \to f$ in $L^2(D)$. From the form of H we note that f_j has the form $f_j = \sqrt{|q|} \, w_j$ and w_j is a solution of the Helmholtz equation $\Delta w_j + k^2 w_j = 0$ in $\mathbb{R}^3$. With $\tilde{w} = f/\sqrt{|q|}$ this shows that $(v, \tilde{w})$ satisfies the condition of being an eigenfunction in the sense of Definition 4.7. From the assumption we conclude that $(v, \tilde{w})$ has to vanish and therefore also f. The combination of part (b) yields the assertion. $\square$

We recall the factorization of F in the form (4.30) of Theorem 4.5, i.e.,

$$F = H^* T H$$

which holds without any additional assumptions on k or q. As mentioned above, we consider the real-valued and the general complex-valued cases separately.

First, we consider the case of *real-valued* contrasts q. In this case, by Theorem 4.4 the far field operator F is normal and the scattering operator $S = I + \frac{ik}{8\pi^2} F$ is unitary. In order to apply the characterization of Theorem 1.23 we set $X = L^2(D)$ and have to investigate the operator $A = T$ from $L^2(D)$ onto itself. From the definition of T (or the previous theorem, part (a)) we know that $T = T_0 + K$ with some compact operator K, and T_0 has the form $T_0 f = \pm k^2 I$ (depending on the sign of q). The second main assumption for the application of Theorem 1.23 is the positivity of $\mathrm{Im}\, T$ on $\overline{\mathcal{R}(H)}$. This has been just proved in part (c) of the previous theorem provided that k^2 is not an interior transmission eigenvalue.

Therefore, application of Theorem 1.23 and combination with Theorem 4.6 yields the first main theorem of this section which is the analog to Theorems 1.25 and 1.27.

Theorem 4.9 *In addition to Assumption 4.2 let q be real-valued and assume that k^2 is not an interior transmission eigenvalue in the sense of Definition 4.7.*
For any $z \in \mathbb{R}^3$ define again $\phi_z \in L^2(S^2)$ by (4.33), i.e.,

$$\phi_z(\hat{x}) := e^{-ik\hat{x}\cdot z}, \quad \hat{x} \in S^2.$$

Then

$$z \in D \iff \phi_z \in \mathcal{R}\big((F^*F)^{1/4}\big) \tag{4.38}$$

$$\iff W(z) := \left[\sum_j \frac{\big|(\phi_z, \psi_j)_{L^2(S^2)}\big|^2}{|\lambda_j|}\right]^{-1} > 0. \tag{4.39}$$

Here, $\lambda_j \in \mathbb{C}$ are the eigenvalues of the normal operator F with corresponding eigenfunctions $\psi_j \in L^2(S^2)$.
Therefore, $\chi_D(z) = \mathrm{sgn}\, W(z)$ is the characteristic function of D.

Next, we turn to the problem where q can be *complex-valued*. Under the assumption (4.36) of Theorem 4.8 will apply Theorem 2.15. Again, X is given by $L^2(D)$. The operator $\mathrm{Re}\big[\exp(-it)T\big]$ can be written in the form $\mathrm{Re}\big[\exp(-it)T\big] = C + \mathrm{Re}\big[\exp(-it)K\big]$ with compact K and where $C = \mathrm{Re}\big[\exp(-it)T_0\big]$ is coercive by Theorem 4.8, part (a). Furthermore, $\mathrm{Im}(Tf, f)_{L^2(D)} > 0$ for all $f \in \overline{\mathcal{R}(H)}$ with $f \neq 0$. Therefore, all of the assumptions of Theorem 2.15 are satisfied. Combination with Theorem 4.6 yields:

Theorem 4.10 *In addition to Assumption 4.2 assume that k^2 is not an interior transmission eigenvalue in the sense of Definition 4.7. Furthermore, assume that (4.36) holds, i.e., that there exists $t \in [0, \pi]$ and $c > 0$ such that*

$$\mathrm{Re}\big[e^{-it} q(x)\big] \geq c|q(x)| \quad \text{almost everywhere on } D.$$

For any $z \in \mathbb{R}^3$ define again $\phi_z \in L^2(S^2)$ by (4.33), i.e.,

$$\phi_z(\hat{x}) := e^{-ik\hat{x}\cdot z}, \quad \hat{x} \in S^2.$$

Then

$$z \in D \iff \phi_z \in \mathcal{R}\big(F_\#^{1/2}\big) \tag{4.40}$$

$$\iff W(z) := \left[\sum_j \frac{\big|(\phi_z, \psi_j)_{L^2(S^2)}\big|^2}{\lambda_j} \right]^{-1} > 0 \tag{4.41}$$

where $F_\# = \big|\mathrm{Re}\big[\exp(-it)\,F\big]\big| + \mathrm{Im}\,F$, *and* $\lambda_j \in \mathbb{R}_{>0}$ *are the eigenvalues of the self-adjoint and positive operator* $F_\#$ *with corresponding eigenfunctions* $\psi_j \in L^2(S^2)$.

Therefore, $\chi_D(z) = \mathrm{sign}\,W(z)$ *is the characteristic function of* D.

We note that the implementation of this result requires the knowledge of the parameter $t \in [0, \pi]$. As we have seen in the remark following Theorem 4.8 that for real-valued q we can choose $t = 0$ or $t = \pi$, depending on the sign of q. If q is complex-valued and $\mathrm{Re}\,q \geq 0$ we can set $t = \pi/2$. In the following lemma we give a sufficient condition for the case that $\mathrm{Re}\,q \leq 0$,

Lemma 4.11 *Let there exist* $\delta > 0$ *with* $\mathrm{Re}\,q(x) + \delta\,\mathrm{Im}\,q(x) \geq 0$ *almost everywhere on* D. *Then assumption (4.36) of Theorem 4.8 is satisfied.*

Proof: Choose $\lambda \in \mathbb{R}$ with $1/\sqrt{2} < \lambda < 1$ and set for abbreviation $r = \mathrm{Re}\,q(x)$ and $j = \mathrm{Im}\,q(x)$. Then, because $j \geq 0$ and $r + \delta j \geq 0$:

$$(r + 2\delta j)^2 = \big[(r + \delta j) + \delta j\big]^2 \geq (r + \delta j)^2 + \delta^2 j^2$$

$$= (1 - \lambda^2)\,r^2 + \Big[\lambda^2 r^2 + 2\,r\,\delta j + \frac{\delta^2 j^2}{\lambda^2}\Big] + \Big(2 - \frac{1}{\lambda^2}\Big)\delta^2 j^2$$

$$\geq (1 - \lambda^2)\,r^2 + \Big(2 - \frac{1}{\lambda^2}\Big)\delta^2 j^2$$

$$\geq c\,(r^2 + j^2)$$

with $c = \min\big\{(1 - \lambda^2), \big(2 - \frac{1}{\lambda^2}\big)\delta^2\big\} > 0$. Since $r + 2\delta j = (r + \delta j) + \delta j \geq 0$ this proves

$$\mathrm{Re}\big[(1 - 2\delta i) \cdot q(x)\big] \geq \sqrt{c}\,\big|q(x)\big|.$$

Writing $1 + 2\delta i$ in polar coordinates as $1 + 2\delta i = \sqrt{1 + 4\delta^2}\,\exp(it)$ yields the desired estimate (4.36) (note that $t \in (0, \pi/2)$). $\square$

4.5   The interior transmission eigenvalue problem

This section is devoted to the investigation of the interior transmission eigenvalue problem of Definition 4.7. Throughout this section we strengthen Assumption 4.2 and assume the following:

Assumption 4.12 *Let* k *be real and positive and* $D \subset \mathbb{R}^3$ *be open and bounded such that the complement* $\mathbb{R}^3 \setminus \overline{D}$ *is connected. Let* $q \in L^\infty(D)$ *be real-valued and either* $q > 0$

on D or $q < 0$ on D. In the latter case we assume the existence of some $c_0 > 0$ such that $1 + q \geq c_0$ on D. Furthermore, we assume one of the following properties (A) or (B):

(A) There exists $\gamma_0 > 0$ such that $|q(x)| \geq \gamma_0$ for almost all $x \in D$ i.e. $|q|$ is bounded away from zero.

(B) $q \in C(\overline{D})$ and

$$\gamma(\delta) := \sup_{x \in \partial D} \int_0^\delta \frac{dt}{|q(x - t\nu(x))|} \longrightarrow 0 \quad \text{as } \delta \to 0. \tag{4.42}$$

We extend q by zero outside of D.

Assumption 4.12(B) allows q to vanish at the boundary of D. However, the decay of $|q|$ along the normal vector must exceed t, i.e., $|q(x - t\nu(x))| \geq c\,t^p$ for some $c > 0$ and $p \in [0, 1)$.

In the first part of this section we will prove a regularity result for solutions of the eigenvalue problem. We define the spaces X and W by

$$X = \left\{ u \in H_0^1(D) : u \in L^2(D, dx/|q|) \right\},$$

$$W = \left\{ u \in H_0^1(D) : u, \Delta u \in L^2(D, dx/|q|), \quad \partial u/\partial \nu = 0 \text{ on } \partial D \right\}.$$

Here, $\partial u/\partial \nu$ is well defined (in the sense of the trace theorem) since $\Delta u \in L^2(D)$ (see, e.g., [144]). The condition $\partial u/\partial \nu = 0$ on ∂D can be formulated as

$$\iint_D \left[\nabla u \cdot \nabla \psi + \psi \, \Delta u \right] dx = 0 \quad \text{for all } \psi \in H^1(D).$$

Analogously to $L^2(D, |q|dx)$ the weighted space $L^2(D, dx/|q|)$ is defined as the completion of $C_0^\infty(\overline{D})$ with respect to the norm

$$\|u\|_{L^2(D, dx/|q|)} = \sqrt{\iint_D |u|^2 \frac{dx}{|q|}}.$$

$L^2(D, dx/|q|$ is obviously a subspace of $L^2(D)$. We equip X and W with the inner products

$$(u, v)_X = (u, v)_{L^2(D, dx/|q|)} + (\nabla u, \nabla v)_{L^2(D)} = \iint_D \left[\frac{u \overline{v}}{|q|} + \nabla u \cdot \nabla \overline{v} \right] dx,$$

$$(u, v)_W = (u, v)_{L^2(D, dx/|q|)} + (\Delta u, \Delta v)_{L^2(D, dx/|q|)} = \iint_D \left[u \overline{v} + \Delta u \, \Delta \overline{v} \right] \frac{dx}{|q|}.$$

X and W are normed spaces which are boundedly imbedded in $H_0^1(D)$. For X this is obvious. For W this follows from Green's theorem since

$$\|\nabla u\|_{L^2(D)}^2 = -\iint_D u \, \Delta \overline{u} \, dx \leq \|u\|_{L^2(D)} \|\Delta u\|_{L^2(D)} \leq \|q\|_\infty \|u\|_W^2. \tag{4.43}$$

We note that under Assumption 4.12(A) the spaces X and W coincide with $H_0^1(D)$ and $\left\{u \in H_0^1(D) \cap H^2(D) : \partial u/\partial \nu = 0 \text{ on } \partial D\right\}$, respectively.

Lemma 4.13 *Let Assumption 4.12 be satisfied. The spaces X and W are Hilbert spaces and compactly imbedded in $L^2(D, dx/|q|)$. An equivalent norm on W is given by $\|\Delta\phi\|_{L^2(D,dx/|q|)}$. In particular, there exists $c > 0$ with*

$$\|\Delta u\|_{L^2(D,dx/|q|)} \geq c \, \|u\|_{L^2(D,dx/|q|)} \qquad \textit{for all } u \in W . \tag{4.44}$$

Proof: The completeness of X is obvious. For W this follows from (4.43) as it can easily be verified. Also, the space W is boundedly imbedded in X. Therefore, it suffices to prove that X is compactly imbedded in $L^2(D, dx/|q|)$. We have to prove this only under Assumption 4.12, (B) since under (A) this follows from the classical Rellich imbedding theorem. First, we show the existence of some $c > 0$ such that

$$\left\| \frac{u}{\sqrt{|q|}} \right\|_{L^2(U_\delta)}^2 \leq c\,\gamma(\delta) \left[\left\| \frac{u}{\sqrt{|q|}} \right\|_{L^2(U_{\delta_0})}^2 + \|\nabla u\|_{L^2(U_{\delta_0})}^2 \right] \tag{4.45}$$

for all $\delta \leq \delta_0$ and all $u \in X$. Here, $\gamma(\delta)$ is given by (4.42) and $U_\delta = \left\{x \in D : d(x, \partial D) < \delta\right\}$ and $\delta_0 > 0$ is chosen such that every $z \in U_{\delta_0}$ can uniquely be written as $z = x - t\nu(x)$ for some $x \in \partial D$ and some $t \in (0, \delta_0)$ (cf. [43]). It is sufficient to prove (4.45) for smooth functions $u \in X$. For fixed $x \in \partial D$ and smooth functions $u \in X$ we write (by a slight abuse of notation) $u(t)$ and $q(t)$ instead of $u(x - t\nu(x))$ and $q(x - t\nu(x))$, respectively. Then, for $\tau, t \in (0, \delta_0)$,

$$|u(t)| \leq |u(\tau)| + \left| \int_\tau^t \left|\nabla u(x - s\nu(x))\right| ds \right| \leq |u(\tau)| + \sqrt{\delta_0} \sqrt{\int_0^{\delta_0} \left|\nabla u(x - s\nu(x))\right|^2 ds}$$

and thus by squaring

$$|u(t)|^2 \leq 2\,|u(\tau)|^2 + 2\delta_0 \int_0^{\delta_0} \left|\nabla u(x - s\nu(x))\right|^2 ds$$

$$\leq \frac{2\,\|q\|_\infty |u(\tau)|^2}{|q(\tau)|} + 2\delta_0 \int_0^{\delta_0} \left|\nabla u(x - s\nu(x))\right|^2 ds .$$

Integration with respect to τ from 0 to δ_0 yields

$$\delta_0 |u(t)|^2 \leq 2\,\|q\|_\infty \int_0^{\delta_0} \frac{|u(\tau)|^2}{|q(\tau)|}\, d\tau + 2\delta_0^2 \int_0^{\delta_0} \left|\nabla u(x - s\nu(x))\right|^2 ds .$$

Division by $\delta_0|q(t)|$, integration with respect to t from 0 to δ, and integration with respect to $x \in \partial D$ yields (4.45).

Now we use standard arguments to prove the compact imbedding property. Indeed, let $\{u_j\}$ be a sequence in X with $\|u_j\|_X \leq 1$. Then $\{u_j\}$ is also bounded in $H_0^1(D)$ and thus contains a weakly convergent subsequence which we denote also by $\{u_j\}$. Let u be the weak limit. Then $\|u_j - u\|_{L^2(D)} \to 0$ since $H_0^1(D)$ is compactly imbedded in $L^2(D)$. It remains to show that $\|u_j - u\|_{L^2(D,dx/|q|)} \to 0$. Let $\varepsilon > 0$ be arbitrary. Choose $\delta > 0$ with $4\,c\,\gamma(\delta) < \varepsilon/2$ where c is the constant in (4.45). On $D \setminus U_\delta$ the function $|q|$ is bounded below by some $q_0 > 0$. Using (4.45) we estimate

$$
\begin{aligned}
\|u_j - u\|_{L^2(D,dx/|q|)}^2 &= \left\| \frac{u_j - u}{\sqrt{|q|}} \right\|_{L^2(U_\delta)}^2 + \left\| \frac{u_j - u}{\sqrt{|q|}} \right\|_{L^2(D\setminus U_\delta)}^2 \\
&\leq c\,\gamma(\delta) \left[\left\| \frac{u_j - u}{\sqrt{|q|}} \right\|_{L^2(U_{\delta_0})}^2 + \|\nabla(u_j - u)\|_{L^2(U_{\delta_0})}^2 \right] \\
&\quad + \frac{1}{q_0} \|u_j - u\|_{L^2(D\setminus U_\delta)}^2 \\
&\leq c\,\gamma(\delta) \underbrace{\|u_j - u\|_X^2}_{\leq 4} + \frac{1}{q_0} \|u_j - u\|_{L^2(D)}^2 \\
&\leq \frac{\varepsilon}{2} + \frac{\varepsilon}{2} = \varepsilon
\end{aligned}
$$

for sufficiently large j. This proves that X is compactly imbedded in $L^2(D, dx/|q|)$. Finally, assume that there exists no constant $c > 0$ with (4.44). Then there exists a sequence $u_j \in W$ with $\|u_j\|_{L^2(D,dx/|q|)} = 1$ and $\Delta u_j \to 0$ in $L^2(D, dx/|q|)$. Since the sequence $\{u_j\}$ is bounded in W there exists a weakly convergent subsequence $u_j \rightharpoonup u$ in W. From $\Delta u_j \to 0$ we conclude that $\Delta u = 0$ and thus $u = 0$. On the other hand, since W is compactly imbedded in $L^2(D, dx/|q|)$ the sequence $\{u_j\}$ has to converge in norm to u which contradicts $\|u_j\|_{L^2(D,dx/|q|)} = 1$ and $u = 0$. $\square$

After these preparations we are able to prove the main result of this section.

Theorem 4.14 *Let Assumption 4.12 be satisfied.*

(a) If k^2 is an interior transmission eigenvalue in the sense of Definition 4.7 with corresponding eigenpair $(u, w) \in H_0^1(D) \times L^2(D, |q|dx)$ then $u \in W$ and u satisfies

$$
\iint\limits_D \left[\Delta u + k^2(1+q)u \right] \left[\Delta\psi + k^2\psi \right] \frac{dx}{q} = 0 \quad \text{for all } \psi \in W. \tag{4.46}
$$

(b) The set of interior transmission eigenvalues is discrete. If there exist infinitely many eigenvalues then $+\infty$ and $-\infty$ are the only possible accumulation points.

Proof: (a) From well-known regularity results (see [70], Theorem 8.12) we conclude that $u \in H^2(D)$ and u satisfies

$$
\Delta u + k^2(1+q)\,u = -k^2 q\,w \quad \text{in } D \tag{4.47}
$$

and that u and $\partial u/\partial \nu$ vanish on ∂D. Since the right-hand side is in $L^2(D, dx/|q|)$ it suffices for u being in W to show that $u \in X$. First, we choose $t \geq 0$ such that the variational equation

$$\iint_D \left[\nabla v \cdot \nabla \psi - k^2(1 + tq)\, v\psi\right] dx = 0 \qquad \text{for all } \psi \in H_0^1(D)$$

admits only the trivial solution $v = 0$ in $H_0^1(D)$. (We leave the simple proof that such a parameter $t \geq 0$ exists to the reader.) Then we observe from (4.35) that $u \in H_0^1(D)$ satisfies

$$\iint_D \left[\nabla u \cdot \nabla \overline{\psi} - k^2(1 + tq)\, u\overline{\psi}\right] dx = \ell(\psi), \quad \psi \in H^1(D),$$

where

$$\ell(\psi) = k^2 \iint_D \left[(1 - t)u + w\right] q\, \overline{\psi}\, dx, \quad \psi \in H^1(D).$$

We study this equation in X. More precisely, with the given (and fixed) eigenpair $(u, w) \in H_0^1(D) \times L^2(D, |q|dx)$ we rewrite this equation and consider

$$(v, \psi)_X - a(v, \psi) = \ell(\psi), \quad \psi \in X, \tag{4.48}$$

for $v \in X$ where a is given by

$$a(v, \psi) = \iint_D \left[1 + k^2|q|(1 + tq)\right] v\, \overline{\psi}\, \frac{dx}{|q|}, \quad v, \psi \in X.$$

The choice of t assures that (4.48) has at most one solution $v \in H_0^1(D)$ and thus in X. We show that (4.48) has a solution $v \in X$ which would imply that $u = v \in X$.

Since a is bounded in $X \times X$ and ℓ is bounded in X there exists a bounded operator K from X into itself and $r \in X$ with $a(v, \psi) = (Kv, \psi)_X$ and $\ell(\psi) = (r, \psi)_X$ for all $v, \psi \in X$. Therefore, (4.48) can be written as $v - Kv = r$ in X. Furthermore, the operator K is compact in X. Indeed, let $\{v_j\}$ converge weakly to zero in X. Then, by Lemma 4.13, $\{v_j\}$ converges to zero in the norm of $L^2(D, dx/|q|)$. Therefore, by the form of a,

$$\|Kv_j\|_X^2 = a(v_j, Kv_j) \leq c\, \|v_j\|_{L^2(D,dx/|q|)} \|Kv_j\|_{L^2(D,dx/|q|)} \leq c\, \|v_j\|_{L^2(D,dx/|q|)} \|Kv_j\|_X$$

and thus $\|Kv_j\|_X \leq c\, \|v_j\|_{L^2(D,dx/|q|)} \to 0$ as j tends to infinity. This proves compactness of K in X. The Riesz–Fredholm theory (cf. [42]) implies existence of some $v \in X$ of (4.48). Therefore, we have shown that $u \in W$.

It remains to prove (4.46). Recall that there exists a sequence $w_j \in H^2(D)$ with $w_j \to w$ in $L^2(D, |q|dx)$ and $\Delta w_j + k^2 w_j = 0$ in D. Green's theorem yields $\int_D w_j \left[\Delta\psi + k^2\psi\right] dx = 0$ for $\psi \in W$. Therefore, also $\left|\int_D w \left[\Delta\psi + k^2\psi\right] dx\right| = \left|\int_D (w - w_j) \left[\Delta\psi + k^2\psi\right] dx\right| \leq c\|w - w_j\|_{L^2(D,|q|dx)} \|\psi\|_X \to 0$ as j tends to infinity and thus $\int_D w \left[\Delta\psi + k^2\psi\right] dx = 0$. Now we divide (4.47) by q, multiply it by $\Delta\psi + k^2\psi$, and integrate over D, and arrive at (4.46).

(b) We write (4.46) in the form

$$a_0(u, \phi) + k^2\, a_1(u, \phi) + k^4\, a_2(u, \phi) = 0 \quad \text{for all } \phi \in W$$

where the sequi-linear forms $a_j : W \times W \to \mathbb{C}$ are defined by

$$a_0(v, \phi) = \iint_D \Delta v\, \Delta\overline{\phi}\, \frac{dx}{q}\,,$$

$$a_1(v, \phi) = \iint_D \left[(1 + q)v\,\Delta\overline{\phi} + \Delta v\,\overline{\phi}\right] \frac{dx}{q}\,,$$

$$a_2(v, \phi) = \iint_D (1 + q)\, v\,\overline{\phi}\, \frac{dx}{q}\,.$$

First we observe that all forms are bounded in $W \times W$. Therefore, by a well-known theorem of Riesz there exist linear and bounded operators $A_j, j = 0, 1, 2$, from X into itself with

$$a_j(v, \phi) = \big(A_j v, \phi\big)_W\,, \quad v, \phi \in W\,, \ \ j = 0, 1, 2\,.$$

Furthermore, the form $+a_0$ or $-a_0$ is coercive (depending on whether $q > 0$ or $q < 0$). To see this we set $\sigma = \text{sign}\, q$ and estimate

$$\sigma\, a_0(\phi, \phi) = \sigma \iint_D |\Delta\phi|^2\, \frac{dx}{q} = \iint_D |\Delta\phi|^2\, \frac{dx}{|q|} = \|\Delta\phi\|^2_{L^2(D, dx/|q|)}$$

$$\geq \frac{1}{2}\,\|\Delta\phi\|^2_{L^2(D, dx/|q|)} + \frac{c^2}{2}\,\|\phi\|^2_{L^2(D, dx/|q|)} \geq \frac{1}{2}\,\min\{1, c^2\}\,\|\phi\|^2_W$$

where we used the estimate (4.44). Therefore, the operator σA_0 is coercive, i.e.,

$$\sigma\,\big(A_0\phi, \phi\big)_W = \sigma\, a_0(\phi, \phi) \geq \tilde{c}\,\|\phi\|^2_W \quad \text{for all } \phi \in W\,.$$

Finally, we show that A_1 and A_2 are compact in W. Let ϕ_n converge weakly to zero in W. Then the special form of a_1 allows the estimate:

$$\|A_1\phi_n\|^2_W = a_1(\phi_n, A_1\phi_n) \leq \|1 + q\|_\infty \|\phi_n\|_{L^2(D, dx/|q|)}\, \|A_1\phi_n\|_W$$

$$+ \|\phi_n\|_W\, \|A_1\phi_n\|_{L^2(D, dx/|q|)}\,.$$

By the compact imbedding property we note that $\|\phi_n\|_{L^2(D, dx/|q|)} \to 0$ and also $\|A_1\phi_n\|_{L^2(D, dx/|q|)} \to 0$. This and the boundedness of the sequences $\|\phi_n\|_W$ and $\|A_1\phi_n\|_W$ imply norm convergence $A_1\phi_n \to 0$ in W. The proof of compactness of A_2 follows by analogous arguments.

Therefore, the variational equation (4.46) is equivalent to the equation

$$A_0 u + k^2\, A_1 u + k^4\, A_2 u = 0\,, \tag{4.49}$$

i.e., $u \in W$ is an eigenfunction of this nonlinear eigenvalue problem corresponding to the eigenvalue k^2. Since σA_0 is self-adjoint and coercive there exists a self-adjoint and coercive square root R of σA_0. Furthermore, σA_2 is self-adjoint and positive. Therefore, there exists B_2 from W into itself with $B_2^2 = \sigma R^{-1} A_2 R^{-1}$. Defining $u_1, u_2 \in W$ by

$$u_1 = Ru, \quad u_2 = k^2 B_2 u_1 = k^2 B_2 Ru$$

we observe that (4.49) is equivalent to the system

$$\frac{1}{k^2} \begin{pmatrix} u_1 \\ u_2 \end{pmatrix} = \begin{pmatrix} -\sigma R^{-1} A_1 R^{-1} & -B_2 \\ B_2 & 0 \end{pmatrix} \begin{pmatrix} u_1 \\ u_2 \end{pmatrix}. \tag{4.50}$$

From this the assertion follows since the matrix operator is compact in $W \times W$. $\square$

Remarks: In Chapter 1 we studied the obstacle scattering case while in chapter we treated the inhomogeneous medium case. In most applications a combination of both situations occur: An obstacle is located inside an inhomogeneous medium. The inverse scattering problem is to recover the shape of the obstacle from the far field data when the properties of the inhomogeneous medium (i.e., the index of refraction is known). The question of uniqueness of this inverse problem has been proved in [128]. In a recent paper [148] Nachman, Päivärinta, and Teirilä applied the Factorization Method to the more general situation where the (known) background contrast q does not need to have compact support but is allowed to be in $L^3_{loc}(\mathbb{R}^3)$ satisfying a short-range condition. In this paper a different presentation of the abstract Factorization Method is presented which avoids the introduction of the inf-condition of Theorem 1.16 and replaces it by the following characterization (in the notation of Theorem 1.16):

$$\phi \in \mathcal{R}(B) \iff \exists c > 0 : \left| \langle \phi, \psi \rangle \right| \leq c \left| \langle \psi, F\psi \rangle \right|^{1/2} \forall \psi \in Y.$$

5

The factorization method for Maxwell's equations

5.1 Maxwell's equations

Electromagnetic wave propagation is described by particular equations relating five vector fields $\mathcal{E}$, $\mathcal{D}$, $\mathcal{H}$, $\mathcal{B}$, and $\mathcal{J}$ and the scalar field ρ, where $\mathcal{E}$ and $\mathcal{D}$ denote the *electric field* and *electric induction*, respectively, while $\mathcal{H}$ and $\mathcal{B}$ denote the *magnetic field* and *magnetic induction*. Likewise, $\mathcal{J}$ and ρ denote the *current* and *charge distribution* of the medium. All fields will be assumed to depend both on the space variable $x \in \mathbb{R}^3$ and on the time variable $t \in \mathbb{R}$.

The actual equations that govern the behavior of the electromagnetic field may be expressed easily in integral form. Such a formulation, which has the advantage of being closely connected to the physical situation, has been used effectively by a number of authors, in particular by Sommerfeld [175] and by Müller [147]. We use the more familiar differential form of Maxwell's equations which can be derived very easily from the integral relations by the use of the integral theorems of Stokes and Gauss. Maxwell's equations in *differential form* formulate Ampère's Law, the Law of Induction and Magnetic Laws, respectively, as:

$$\operatorname{curl} \mathcal{H} = \frac{\partial}{\partial t} \mathcal{D} + \mathcal{J} \tag{5.1}$$

$$\operatorname{curl} \mathcal{E} = -\frac{\partial}{\partial t} \mathcal{B} \tag{5.2}$$

$$\operatorname{div} \mathcal{D} = \rho \tag{5.3}$$

$$\operatorname{div} \mathcal{B} = 0 \tag{5.4}$$

Taking the divergence of (5.1), using (5.3), and noting that $\operatorname{div} \operatorname{curl} = 0$ we derive an equation relating the current and charge densities:

$$\operatorname{div} \mathcal{J} + \frac{\partial}{\partial t} \rho = 0 . \tag{5.5}$$

We may consider (5.5), as analogous to the continuity or conservation equation in fluid dynamics. It expresses the fact that charge is conserved in the neighborhood of any point.

The current density $\mathcal{J}$ commonly consists of two terms: one, $\mathcal{J}_e$, associated with external sources of electromagnetic disturbances and the other, $\mathcal{J}_c$, associated with conduction currents produced as a result of the electric field. We will be considering source free regions for which $\mathcal{J}_e$ vanishes.

We will consider electromagnetic wave propagation in *linear* and *isotropic* media. This means, first, that there exist linear relationships (the *constitutive relations*) between $\mathcal{E}$ and $\mathcal{D}$, $\mathcal{H}$ and $\mathcal{B}$:

$$\mathcal{D} = \epsilon\,\mathcal{E}, \quad \text{and} \quad \mathcal{B} = \mu\,\mathcal{H}. \tag{5.6}$$

In general, the quantities ϵ and μ may be space dependent, but we assume that they are independent of time and of direction and are therefore scalar (as opposed to tensor) quantities. Hence the term *isotropic*.

The *permittivity* or *dielectric constant*, ϵ, is related to the ability of the medium to sustain an electric charge. Its value, ϵ_0, in a vacuum has been experimentally determined and is approximately $8.854 \cdot 10^{-12}\,As/Vm$ while that, say, in fused quartz it is approximately $3.545 \cdot 10^{-11}\,As/Vm$.

The *magnetic permeability* for most substances, μ, is close to its value in vacuo $\mu_0 = 4\pi \cdot 10^{-7}\,Vs/Am$. Those substances for which μ is significantly different from this value are called magnetic, either *paramagnetic* or *diamagnetic* if $\mu > \mu_0$ or $\mu < \mu_0$, respectively.

We assume that ϵ and μ are independent of the field strength.

The quantity $c_0 := 1/\sqrt{\epsilon_0\mu_0}$ has the dimension of velocity. It is a consequence of the field equations that this quantity is the velocity of propagation of the electromagnetic field disturbance through free space. Experimental measurements have shown that, in vacuo, this velocity is the same as that of light and hence $c_0 \approx 2.9979 \cdot 10^8\,m/s$.

Two special cases will be considered in the following: media in which the constitutive parameters vary smoothly, and media in which there are manifolds of discontinuity (interfaces) of these parameters. In a medium where ϵ and μ vary smoothly, Maxwell's equations are equivalent to a system of partial differential equations. In the second case where an interface exists, the behavior of the constitutive parameters determine transmission or boundary conditions for the magnetic and electric fields.

To the linear constitutive relations, we add a third, namely *Ohm's Law*, which relates the quantities $\mathcal{J}_c$ and $\mathcal{E}$ by a linear relation of the form

$$\mathcal{J}_c = \sigma\,\mathcal{E}. \tag{5.7}$$

The scalar function σ is called the *conductivity*. Substances for which σ is not negligibly small are called *conductors*. Metals, for example, are good conductors as is brine.

From now on we assume that all fields vary periodically in time with the same angular frequency $\omega = 2\pi/T$ and period T. This could be insured by assuming periodic time dependence of the applied external currents or fields.

It is very convenient to use the complex representation of the fields in the form

$$\mathcal{E}(x,t) = \mathrm{Re}\left(E(x)\,e^{-i\omega t}\right), \qquad \mathcal{D}(x,t) = \mathrm{Re}\left(D(x)\,e^{-i\omega t}\right),$$

$$\mathcal{H}(x,t) = \mathrm{Re}\left(H(x)\,e^{-i\omega t}\right), \qquad \mathcal{B}(x,t) = \mathrm{Re}\left(B(x)\,e^{-i\omega t}\right).$$

Here, E, D, H, and B are now space-dependent complex vector fields. By using these formulas the derivative with respect to time transforms into multiplication by $-i\omega$. Thus, Maxwell's equations (5.1)–(5.4) in conducting, isotropic, and source free media (i.e., $\mathcal{J}_e = 0$ and $\rho = 0$) read for the space dependent parts

$$\operatorname{curl} H = (-i\omega\epsilon + \sigma)\,E, \tag{5.8}$$

$$\operatorname{curl} E = i\omega\mu\,H, \tag{5.9}$$

$$\operatorname{div}(\epsilon E) = 0, \tag{5.10}$$

$$\operatorname{div}(\mu H) = 0. \tag{5.11}$$

We remark that (5.10) and (5.11) follow directly from (5.8) and (5.9) for homogeneous media, i.e., for which ϵ, μ, and σ are constant.

If we consider a situation in which a surface Γ separates the two media D and $\mathbb{R}^3 \setminus \overline{D}$ from each other, the constitutive parameters ϵ, μ, and σ are continuous in both media but have finite jumps on Γ. While Maxwell's equations (5.8)–(5.11) hold in D and in the exterior of $\overline{D}$, the presence of these jumps implies that the fields E and H satisfy the following *transmission conditions* on the surface Γ.

$$\nu \times E_+ - \nu \times E_- = 0 \quad \text{and} \quad \nu \times H_+ - \nu \times H_- = 0 \quad \text{on } \Gamma. \tag{5.12}$$

Here again, $\nu = \nu(x)$ denotes the exterior unit normal vector at $x \in \Gamma$, and $E_\pm$ and $H_\pm$ denote the limits of E and H, respectively, from the exterior $(+)$ and interior $(-)$.

5.2 The direct scattering problem

In this section we study the direct scattering problem for non-magnetic media. Let $k = \omega\sqrt{\varepsilon_0\mu_0} > 0$ be the wavenumber with frequency ω, electric permittivity ε_0, and magnetic permeability μ_0 in vacuum. An incident electromagnetic field consists of a pair H^i and E^i which satisfy the time harmonic Maxwell system in vacuum, i.e.,

$$\operatorname{curl} E^i - i\omega\mu_0 H^i = 0 \quad \text{in } \mathbb{R}^3, \tag{5.13}$$

$$\operatorname{curl} H^i + i\omega\varepsilon_0 E^i = 0 \quad \text{in } \mathbb{R}^3. \tag{5.14}$$

This incident wave is scattered by a medium with space-dependent electric permittivity $\varepsilon = \varepsilon(x)$ and conductivity $\sigma = \sigma(x)$. We assume that the magnetic permeability μ is constant and equal to the permeability μ_0 of vacuum. Furthermore, we assume that $\varepsilon \equiv \varepsilon_0$ and $\sigma \equiv 0$ outside of some bounded domain. The total fields are superpositions of the incident and scattered fields, i.e., $E = E^i + E^s$ and $H = H^i + H^s$ and satisfy the

Maxwell system

$$\operatorname{curl} E - i\omega\mu_0 H = 0 \quad \text{in } \mathbb{R}^3, \tag{5.15}$$

$$\operatorname{curl} H + i\omega\varepsilon E = \sigma E \quad \text{in } \mathbb{R}^3. \tag{5.16}$$

Furthermore, the tangential components of E and H are continuous on interfaces where σ or ε are discontinuous. Finally, the scattered fields have to satisfy the *Silver–Müller radiation condition*

$$\sqrt{\frac{\mu_0}{\varepsilon_0}}\, H^s(x) \times x - |x| E^s(x) = \mathcal{O}\left(\frac{1}{|x|}\right) \quad \text{as } |x| \to \infty \tag{5.17}$$

uniformly w.r.t. $\hat{x} = x/|x|$.

In this chapter we will always work with the magnetic field H only. This is motivated by the fact that the magnetic field is divergence free as seen from (5.15) and the fact that $\operatorname{div}\operatorname{curl} = 0$. In general, this is not the case for the electric field E.

Eliminating the electric field E from the system (5.15), and (5.16) leads to

$$\operatorname{curl}\left[\frac{1}{\sigma - i\omega\varepsilon}\,\operatorname{curl} H\right] - i\omega\mu_0 H = 0,$$

i.e.,

$$\operatorname{curl}\left[\frac{1}{\varepsilon_r}\,\operatorname{curl} H\right] - k^2 H = 0 \quad \text{in } \mathbb{R}^3 \tag{5.18}$$

where ε_r denotes the (complex-valued) relative permittivity

$$\varepsilon_r(x) = \frac{\varepsilon(x)}{\varepsilon_0} + i\,\frac{\sigma(x)}{\omega\,\varepsilon_0}. \tag{5.19}$$

We note that $\varepsilon_r \equiv 1$ outside of some bounded domain. The incident field H^i satisfies

$$\operatorname{curl}^2 H^i - k^2 H^i = 0 \quad \text{in } \mathbb{R}^3. \tag{5.20}$$

Subtracting both equations yields

$$\operatorname{curl}\left[\frac{1}{\varepsilon_r}\,\operatorname{curl} H^s\right] - k^2 H^s = \operatorname{curl}\left[q\,\operatorname{curl} H^i\right] \quad \text{in } \mathbb{R}^3, \tag{5.21}$$

where the contrast q is defined by $q = 1 - 1/\varepsilon_r$. The Silver–Müller radiation condition turns into

$$\operatorname{curl} H^s(x) \times \hat{x} - ikH^s(x) = \mathcal{O}\left(\frac{1}{|x|^2}\right), \quad |x| \to \infty. \tag{5.22}$$

The continuity of the tangential components of E and H translates into analogous requirements for H^s and $\operatorname{curl} H^s/\varepsilon_r$.

It will be necessary to allow more general source terms on the right-hand side of (5.21). In particular, we will consider the following problem:

$$\operatorname{curl}\left[\frac{1}{\varepsilon_r}\,\operatorname{curl} v\right] - k^2 v = \operatorname{curl} f \quad \text{in } \mathbb{R}^3 \tag{5.23}$$

and v satisfies the Silver–Müller radiation condition (5.22). Here $f \in L^2(\mathbb{R}^3, \mathbb{C}^3)$ is an arbitrary vector function with compact support. The solutions v of (5.23) as well as of (5.18) and (5.21) have to be understood in the variational sense, i.e., are sought in the space

$$H_{loc}(\mathrm{curl}, \mathbb{R}^3) = \left\{ v : \mathbb{R}^3 \to \mathbb{C}^3 : v|_B \in H(\mathrm{curl}, B) \text{ for all balls } B \subset \mathbb{R}^3 \right\}$$

where

$$H(\mathrm{curl}, B) = \left\{ v \in L^2(B, \mathbb{C}^3) : \mathrm{curl}\, v \in L^2(B, \mathbb{C}^3) \right\}.$$

We recall that a vector function $v \in L^2(B, \mathbb{C}^3)$ possesses an L^2-curl if there exists $w \in L^2(B, \mathbb{C}^3)$ (which is denoted by $\mathrm{curl}\, v$) with

$$\iint_B \left[w \cdot \psi - v \cdot \mathrm{curl}\,\psi \right] dx = 0 \quad \text{for all } \psi \in C_0^\infty(B, \mathbb{C}^3).$$

This definition is motivated by the Green's formula in the form

$$\iint_B \left[v \cdot \mathrm{curl}\, w - w \cdot \mathrm{curl}\, v \right] dx = \int_{\partial B} w \cdot (v \times \nu)\, ds \tag{5.24}$$

for all sufficiently smooth vector fields v and w.

$H(\mathrm{curl}, B)$ is a Hilbert space with inner product

$$(v, w)_{H(\mathrm{curl},B)} := (v, w)_{L^2(B)} + (\mathrm{curl}\, v, \mathrm{curl}\, w)_{L^2(B)}.$$

Then $v \in H_{loc}(\mathrm{curl}, \mathbb{R}^3)$ is said to be the variational solution of (5.23) if v satisfies

$$\iint_{\mathbb{R}^3} \left[\frac{1}{\varepsilon_r} \mathrm{curl}\, v \cdot \mathrm{curl}\,\psi - k^2 v \cdot \psi \right] dx = \iint_{\mathbb{R}^3} f \cdot \mathrm{curl}\,\psi\, dx \tag{5.25}$$

for all $\psi \in H(\mathrm{curl}, \mathbb{R}^3)$ with compact support. Outside of the supports of $\varepsilon_r - 1$ and f the solution satisfies $\mathrm{curl}^2 v - k^2 v = 0$. Taking the divergence of this equation and using the identities $\mathrm{div}\,\mathrm{curl} = 0$ and $\mathrm{curl}^2 = -\Delta + \nabla\,\mathrm{div}$ this system is equivalent to the pair of equations

$$\Delta v + k^2 v = 0 \quad \text{and} \quad \mathrm{div}\, v = 0.$$

Classical interior regularity results (cf. [70] combined with [43]) yield that v is analytic outside of the supports of $\varepsilon_r - 1$ and f. In particular, the radiation condition (5.22) is well defined.

There are several approaches for proving existence of a solution of this scattering problem. In [127] it was suggested to transform (5.25) into a variational equation on a bounded domain with non-local boundary conditions involving the exterior Calderon operator. A similar – but different – approach has been studied in [59] for the case of a layered background medium, cf. also [146]. In [43] an equivalent integral equation for the electric field is derived from which existence can be proved by the Riesz–Fredholm theory. However, this approach assumes that σ and ε_r are smooth functions in $\mathbb{R}^3$, and it is not clear how this method can be generalized to non-smooth data. In [122] one of us derived a new integral equation (even for the case where the background medium is layered) which we will present in the following.

Assumption 5.1 *In the following, we assume that $D \subset \mathbb{R}^3$ is open and bounded with $\partial D \in C^2$. Furthermore, let $k > 0$ be the (real and positive) wavenumber and $\varepsilon_r \in L^\infty(\mathbb{R}^3)$ such that $\mathrm{Re}\,\varepsilon_r \geq 0$ and $\mathrm{Im}\,\varepsilon_r \geq 0$ and $\varepsilon_r \equiv 1$ on $\mathbb{R}^3 \setminus D$. Furthermore, we assume that there exists $c_0 > 0$ with $\mathrm{Re}\,\varepsilon_r + \mathrm{Im}\,\varepsilon_r \geq c_0$ on D. Then, in particular, $1/\varepsilon_r \in L^\infty(\mathbb{R}^3)$.*

Before we show equivalence of the variational formulation to an integral equation we prove the following crucial lemma.

Lemma 5.2 *Consider the volume potential*

$$v(x) = \mathrm{curl} \iint_D \Phi(x,y) f(y)\, dy, \quad x \in \mathbb{R}^3, \tag{5.26}$$

where again $\Phi(x,y)$ denotes the fundamental solution of the scalar Helmholtz equation in $\mathbb{R}^3$, i.e.,

$$\Phi(x,y) = \frac{\exp(ik|x-y|)}{4\pi\,|x-y|}, \quad x \neq y.$$

(a) For $f \in C^{1,\alpha}(D, \mathbb{C}^3)$ the volume potential $v \in C^2(\mathbb{R}^3 \setminus \partial D, \mathbb{C}^3)$ is the unique classical solution of the transmission problem

$$\mathrm{curl}^2 v - k^2 v = \mathrm{curl} f \quad in\ \mathbb{R}^3 \setminus \partial D, \tag{5.27}$$

$$\nu \times (v_+ - v_-) = 0, \quad \nu \times (\mathrm{curl}\, v_+ - \mathrm{curl}\, v_-) = f \times \nu \quad on\ \partial D, \tag{5.28}$$

satisfying the radiation condition (5.22). (In (5.27) we have set $f = 0$ in the exterior of $\overline{D}$.)

(b) For $f \in L^2(D, \mathbb{C}^3)$ the volume potential is the unique radiating variational solution of (5.27), and (5.28), i.e., $v \in H_{loc}(\mathrm{curl}, \mathbb{R}^3)$ solves

$$\iint_{\mathbb{R}^3} \left[\mathrm{curl}\, v \cdot \mathrm{curl}\, \psi - k^2 v \cdot \psi\right] dx = \iint_D f \cdot \mathrm{curl}\, \psi\, dx \tag{5.29}$$

for all $\psi \in H(\mathrm{curl}, \mathbb{R}^3)$ with compact support.

(c) The restriction of v to D defines a bounded operator from $L^2(D, \mathbb{C}^3)$ into $H(\mathrm{curl}, D)$.

Proof: (a) Let $f \in C^{1,\alpha}(D, \mathbb{C}^3)$ and $a \in \mathbb{C}^3$ be a fixed vector. For $x \notin \partial D$ we have

$$a \cdot v(x) = -\iint_D a \cdot \left[\nabla_y \Phi(x,y) \times f(y)\right] dy$$

$$= \iint_D \operatorname{curl}_y\left[\Phi(x,y)\, a\right] \cdot f(y)\, dy$$

$$= \iint_D \Phi(x,y)\, a \cdot \operatorname{curl} f(y)\, dy + \int_{\partial D} \Phi(x,y)\, a \cdot \left(f(y) \times \nu(y)\right) ds(y)\,.$$

Since this holds for all $a \in \mathbb{C}^3$ we conclude that

$$v(x) = \iint_D \Phi(x,y)\, \operatorname{curl} f(y)\, dy + \int_{\partial D} \Phi(x,y)\left[f(y) \times \nu(y)\right] ds(y)\,.$$

From the classical jump conditions of the volume and single layer potential with Hölder continuous densities (cf. [43]) we conclude that v is continuous in $\mathbb{R}^3$ and $\nu \times (\operatorname{curl} v_+ - \operatorname{curl} v_-) = f \times \nu$ on ∂D and $\Delta v + k^2 v = -\operatorname{curl} f$ in D. Furthermore, the divergence of v vanishes which proves (5.27). We have therefore shown that the volume potential (5.26) is a classical solution of (5.27) and (5.28). The radiation condition (5.22) can be seen directly (see [43] for a proof). It remains to show uniqueness which we will prove later in Theorem 5.5 for a more general case.

(b), (c) Let now $f \in L^2(D, \mathbb{C}^3)$. Then the volume potential v is well defined and a function in $H_{loc}(\operatorname{curl}, \mathbb{R}^3)$. This latter property is seen from the fact that the volume potential

$$x \mapsto \iint_D \Phi(x,y) f(y)\, dy\,, \quad x \in \mathbb{R}^3,$$

is in $H^2_{loc}(\mathbb{R}^3, \mathbb{C}^3)$ (cf. [144]). Therefore, since for smooth f the potential v solves the variational equation (5.29) a density argument implies this also for $f \in L^2(D, \mathbb{C}^3)$. For uniqueness we refer again to Theorem 5.5 below. $\square$

Now we go back to the transmission problem (5.25) and write it in the form

$$\iint_{\mathbb{R}^3} \left[\operatorname{curl} v \cdot \operatorname{curl} \psi - k^2 v \cdot \psi\right] dx = \iint_D \left[f + q\,\operatorname{curl} v\right] \cdot \operatorname{curl} \psi\, dx$$

for all $\psi \in H(\operatorname{curl}, \mathbb{R}^3)$ with compact support. Again, we recall that $q = 1 - 1/\varepsilon_r$. By Lemma 5.2 this equation is equivalent to

$$v(x) = \operatorname{curl} \iint_D \Phi(x,y)\left[f(y) + q(y)\,\operatorname{curl} v(y)\right] dy\,, \quad x \in \mathbb{R}^3\,. \tag{5.30}$$

Therefore, as a corollary of Lemma 5.2 we have:

Lemma 5.3 *Define the integral operator* $A : L^2(D, \mathbb{C}^3) \to H(\text{curl}, D)$ *by*

$$(Af)(x) = \text{curl} \iint_D \Phi(x, y) f(y) \, dy, \quad x \in D. \tag{5.31}$$

(a) *If* $v \in H_{loc}(\text{curl}, \mathbb{R}^3)$ *is a radiating solution of (5.25) then the restriction* $v|_D \in H(\text{curl}, D)$ *solves the equation*

$$v - A(q \, \text{curl} \, v) = Af. \tag{5.32}$$

(b) *If* $v \in H(\text{curl}, D)$ *solves equation (5.32) then the extension of* v *by the right-hand side of (5.30) solves (5.25) and the radiation conditions (5.22).*

We note that A is well defined by part (c) of Lemma 5.2.

It is the aim of the next lemma to prove that the mapping $v \mapsto v - A(q \, \text{curl} \, v)$ is a Fredholm operator of index 0. We introduce the operator $A_i : L^2(D, \mathbb{C}^3) \to H(\text{curl}, D)$ by

$$(A_i f)(x) = \text{curl} \iint_D \Phi_i(x, y) f(y) \, dy, \quad x \in D, \tag{5.33}$$

where

$$\Phi_i(x, y) = \frac{\exp(-|x - y|)}{4\pi |x - y|}, \quad x \neq y,$$

is the fundamental solution for $k = i$.

Lemma 5.4 (a) *The operator* $A - A_i$ *is compact from* $L^2(D, \mathbb{C}^3)$ *into* $H(\text{curl}, D)$.

(b) *The operator* $v \mapsto v - A_i(q \, \text{curl} \, v)$ *is a bounded isomorphism from* $H(\text{curl}, D)$ *onto itself.*

Proof: Part (a) follows from the smoothness of $\Phi - \Phi_i$.

(b) We consider the equation $v - A_i(q \, \text{curl} \, v) = g$ for any $g \in H(\text{curl}, D)$, i.e., $v - g = A_i(q \, \text{curl} \, v)$. The assertion of Lemma 5.2, part (b), holds also for A_i. Therefore, $v \in H(\text{curl}, D)$ is a solution of $v - g = A_i(q \, \text{curl} \, v)$ if, and only if, there exists an extension $w \in H(\text{curl}, \mathbb{R}^3)$ of $v - g$ into $\mathbb{R}^3$ which solves

$$\iint_{\mathbb{R}^3} \left[\text{curl} \, w \cdot \text{curl} \, \psi + w \cdot \psi \right] dx = \iint_D q \, \text{curl} \, v \cdot \text{curl} \, \psi \, dx$$

$$= \iint_D q \, \text{curl} \, w \cdot \text{curl} \, \psi \, dx + \iint_D q \, \text{curl} \, g \cdot \text{curl} \, \psi \, dx$$

for all $\psi \in H(\mathrm{curl}, \mathbb{R}^3)$ with compact support, i.e.,

$$\iint\limits_{\mathbb{R}^3} \left[\frac{1}{\varepsilon_r} \mathrm{curl}\, w \cdot \mathrm{curl}\, \psi + w \cdot \psi \right] dx = \iint\limits_{D} q\, \mathrm{curl}\, g \cdot \mathrm{curl}\, \psi\, dx \qquad (5.34)$$

for all $\psi \in H(\mathrm{curl}, \mathbb{R}^3)$. Note that, by Lemma 5.2 and the fact that $\Phi_i(x, y)$ decays exponentially for $y \in D$ we conclude that $w \in H(\mathrm{curl}, \mathbb{R}^3)$ and this equation holds for all $\psi \in H(\mathrm{curl}, \mathbb{R}^3)$. The left hand side of (5.34) defines a bilinear form $a(w, \psi)$ on $H(\mathrm{curl}, \mathbb{R}^3) \times H(\mathrm{curl}, \mathbb{R}^3)$ with

$$\mathrm{Re}\big[e^{\pi i/4} a(w, \overline{w}) \big] \geq \frac{1}{\sqrt{2}} \min \left\{ \frac{c_0}{\|\varepsilon_r\|_\infty^2}\, ,\, 1 \right\} \|w\|_{H(\mathrm{curl}, \mathbb{R}^3)}^2$$

for all $w \in H(\mathrm{curl}, \mathbb{R}^3)$. Indeed, by Assumptions 5.1 we have

$$\mathrm{Re}\left[e^{\pi i/4} \frac{1}{\varepsilon_r(x)} \right] = \frac{\mathrm{Re}\, \varepsilon_r(x) + \mathrm{Im}\, \varepsilon_r(x)}{\sqrt{2}\, |\varepsilon_r(x)|^2} \geq \frac{c_0}{\sqrt{2}\, \|\varepsilon_r\|_\infty^2}\, .$$

Therefore, equation (5.34) has a unique solution for all $g \in H(\mathrm{curl}, \mathbb{R}^3)$. From this the assertion of part (b) follows easily. Indeed, for $g = 0$ we conclude that the corresponding w has to vanish in $\mathbb{R}^3$, thus also v since w is an extension of $v - g$. For given $g \in H(\mathrm{curl}, D)$ we determine $w \in H(\mathrm{curl}, \mathbb{R}^3)$ as the solution of (5.34). Then $v := w|_D + g$ solves $v - g = A_i(q\, \mathrm{curl}\, v)$. This ends the proof. $\square$

Combining Lemmas 5.3 and 5.4 yields the first part of the following theorem.

Theorem 5.5 *Let $\varepsilon_r \in L^\infty(\mathbb{R}^3)$ satisfy the above Assumptions 5.1.*

(a) *Then (5.25) satisfies the Fredholm alternative,[1] i.e., there exists a unique radiating solution $v \in H_{loc}(\mathrm{curl}, \mathbb{R}^3)$ of (5.25) for every $f \in L^2(D, \mathbb{C}^3)$ provided uniqueness holds. In this case, for any compact set B containing $\overline{D}$ in it's interior there exists a constant $c > 0$ (depending only on B, ω, μ_0, ε_0, and ε_r) such that*

$$\|v\|_{H(curl, B)} \leq c\|f\|_{L^2(D)} \quad \text{for all } f \in L^2(D, \mathbb{C}^3)\,.$$

The restriction of v to D is the unique solution of the integro-differential equation

$$v(x) = \mathrm{curl} \iint\limits_{D} \Phi(x, y) \big[f(y) + q(y)\, \mathrm{curl}\, v(y) \big]\, dy\,, \quad x \in D\,. \qquad (5.35)$$

(b) *Let, in addition, $\mathrm{Im}\, \varepsilon_r > 0$ a.e. in D or $\varepsilon_r \in C^{1,\alpha}(D)$. Then uniqueness holds, i.e., also existence by part (a).*

Proof: It suffices to prove uniqueness. We assume that v is a solution of the homogeneous problem, i.e., of (5.25) for $f = 0$. We set $w(x) = \overline{v(x)}\phi(x)$ in (5.25) where $\phi \in C^\infty(\mathbb{R}^3)$ is some mollifier with $\phi(x) = 1$ for $|x| \leq R$ and $\phi(x) = 0$ for $|x| \geq 2R$. Then, by Green's

---

[1] Actually, we show only one part of it.

theorem (note that v is smooth for $R < |x| < 2R$) and the boundary conditions $\phi = 1$ for $|x| = R$ and $\phi = 0$ for $|x| = 2R$,

$$
\begin{aligned}
0 &= \iint_{|x|<R} \left[\frac{1}{\varepsilon_r} |\operatorname{curl} v|^2 - k^2 |v|^2 \right] dx + \iint_{R<|x|<2R} \left[\operatorname{curl} v \cdot \operatorname{curl}(\bar{v}\phi) - k^2 |v|^2 \phi \right] dx \\
&= \iint_{|x|<R} \left[\frac{1}{\varepsilon_r} |\operatorname{curl} v|^2 - k^2 |v|^2 \right] dx - \int_{|x|=R} (\operatorname{curl} v \times v) \cdot \bar{v} \, ds .
\end{aligned}
\tag{5.36}
$$

Taking the imaginary part yields

$$
\operatorname{Im} \int_{|x|=R} (\operatorname{curl} v \times v) \cdot \bar{v} \, ds \le 0
$$

since $\operatorname{Im}(1/\varepsilon_r) \le 0$. From this and the binomial theorem we estimate

$$
\begin{aligned}
\int_{|x|=R} & \left| \operatorname{curl} v(x) \times \frac{x}{|x|} - ik\, v(x) \right|^2 ds(x) \\
&= \int_{|x|=R} \left[|\operatorname{curl} v|^2 + k^2 |v|^2 \right] ds - 2k \operatorname{Im} \int_{|x|=R} (\operatorname{curl} v \times v) \cdot \bar{v} \, ds \\
&\ge \int_{|x|=R} \left[|\operatorname{curl} v|^2 + k^2 |v|^2 \right] ds .
\end{aligned}
$$

From this estimate and the radiation condition we conclude that $\int_{|x|=R} |v|^2 \, ds$ and $\int_{|x|=R} |\operatorname{curl} v|^2 \, ds$ tend to zero as R tends to infinity. Since the components v_j satisfy the scalar Helmholtz equation $\Delta v_j + k^2 v_j = 0$ outside of D Rellich's Lemma 1.2 and the analytic continuation principle yield that v vanishes outside of D.

If $\varepsilon_r \in C^{1,\alpha}(D)$ we extend ε_r to a $C^{1,\alpha}$-function into a (exterior) neighborhood of ∂D and apply the unique continuation principle (see [43]) which yields that v also vanishes in D.

If $\operatorname{Im} \varepsilon_r > 0$ on D we go back to (5.36) which takes the form (note that v vanishes outside of D)

$$
\iint_D \left[\frac{1}{\varepsilon_r} |\operatorname{curl} v|^2 - k^2 |v|^2 \right] dx = 0 .
\tag{5.37}
$$

Taking the imaginary part of this equation yields $\operatorname{curl} v = 0$ in D and thus also $v = 0$ in D. $\square$

As we saw in the proof, the question of uniqueness of this scattering problem is closely related to the unique continuation property. To the authors knowledge, the unique continuation property has been proved for smooth contrasts only. We followed [43] and assumed $\varepsilon_r \in C^{1,\alpha}(D)$ by modifying their arguments. The weakest assumptions on ε_r for which the unique continuation property holds seems to be due to Vogelsang, see [181] and [152].

Remark 5.6 If $U \subset \mathbb{R}^3$ is any open and bounded set such that $\overline{U} \cap (\mathrm{supp}\,(\varepsilon_r - 1) \cup \mathrm{supp} f) = \emptyset$ then the mapping $f \mapsto v|_U$ is compact from $L^2(B, \mathbb{C}^3)$ into $C^m(U, \mathbb{C}^3)$ for all $m \in \mathbb{N}$. This follows directly from the extension of (5.35) into $\mathbb{R}^3$.

It is well known (see e.g. [43, 146]) that every radiating solution v of (5.23) has an asymptotic behavior of the form

$$v(x) = \frac{\exp(ik|x|)}{4\pi |x|}\, v^{\infty}(\hat{x}) + \mathcal{O}\left(\frac{1}{|x|^2}\right), \quad |x| \to \infty,$$

uniformly w.r.t. $\hat{x} = x/|x|$. The vector field v^{∞} is called the *far field pattern* of v. It is an analytic function on the unit sphere $S^2 = \left\{x \in \mathbb{R}^3 : |x| = 1\right\}$ with respect to $\hat{x}$ and is a tangential field, i.e., it satisfies $v^{\infty}(\hat{x}) \cdot \hat{x} = 0$ for all $\hat{x} \in S^2$.

We go now back to the scattering problem (5.15)–(5.16) and consider the special case where the incident waves H^i and E^i are given by plane waves of the form

$$H^i(x, \theta; p) = p\, e^{ik\theta \cdot x} \text{ and } E^i(x, \theta; p) = -\frac{1}{i\omega\varepsilon_0}\, \mathrm{curl}\, H^i(x, \theta; p) = -\sqrt{\frac{\mu_0}{\varepsilon_0}}\, (\theta \times p)\, e^{ik\theta \cdot x}.$$

$$\tag{5.38}$$

Here, $\theta \in S^2$ denotes the direction of incidence and $p \in \mathbb{C}^3$ the polarization vector. We have to assume that $p \cdot \theta = \sum_{j=1}^3 p_j \theta_j = 0$ in order to ensure that H^i and E^i are divergence free.

Then the far field patterns H^{∞} and E^{∞} of H^s and E^s, respectively, are defined and depend on θ and p as well. We will indicate this dependence by writing $H^{\infty} = H^{\infty}(\hat{x}, \theta; p)$ and $E^{\infty} = E^{\infty}(\hat{x}, \theta; p)$. Note again that they are tangential vector fields, i.e., $H^{\infty}(\hat{x}, \theta; p) \cdot \hat{x} = 0$ and $E^{\infty}(\hat{x}, \theta; p) \cdot \hat{x} = 0$ for all $\hat{x} \in S^2$ and all $\theta \in S^2$ and $p \in \mathbb{C}^3$ with $p \cdot \theta = 0$. Furthermore, $E^{\infty}(\hat{x}, \theta; p) = H^{\infty}(\hat{x}, \theta; p) \times \hat{x}$, see [43]. Therefore, it is sufficient to work only with one of the far field patterns. Our approach suggests to work with H^{∞} rather than with E^{∞}.

The far field patterns depend linearly on p, i.e., we can write $H^{\infty}(\hat{x}, \theta; p) = \hat{H}^{\infty}(\hat{x}, \theta) p$ for all $p \in \mathbb{C}^3$ with $p \cdot \theta = 0$ where $\hat{H}^{\infty}(\hat{x}, \theta) \in \mathbb{C}^{3\times 3}$ is a matrix.

We are now able to define the *inverse scattering problem*. Given $H^{\infty}(\hat{x}, \theta; p)$ for all $\hat{x}, \theta \in S^2$ and $p \in \mathbb{C}^3$ with $p \cdot \theta = 0$, find the support $\overline{D}$ of q. Because of the linear dependence of H^{∞} on p it is sufficient to know H^{∞} only for a basis of three vectors for p. The task of determining only D is rather modest since it is well known that one can even reconstruct q uniquely from this set of data, see [53]. However, the proof of uniqueness is non-constructive while we will give an explicit characterization of the characteristic function of D which can, e.g., be used for numerical purposes. Also, our analysis works also for anisotropic media where it is well known that ε_r can only be determined up to a smooth change of coordinates.

We introduce the subspace $L_t^2(S^2)$ of $L^2(S^2, \mathbb{C}^3)$ consisting of all tangential fields, i.e.,

$$L_t^2(S^2) := \left\{ v \in L^2(S^2, \mathbb{C}^3) : v(\hat{x}) \cdot \hat{x} = 0, \ \hat{x} \in S^2 \right\}.$$

The *far field operator* $F : L_t^2(S^2) \to L_t^2(S^2)$ is defined as

$$(Fp)(\hat{x}) := \int_{S^2} H^\infty\big(\hat{x}, \theta; p(\theta)\big) \, ds(\theta), \quad \hat{x} \in S^2. \tag{5.39}$$

It is linear since H^∞ depends linearly on the polarization p. We note that Fp is the far field pattern of the magnetic field which corresponds to the incident (magnetic) field

$$v_p(x) = \int_{S^2} H^i\big(x, \theta; p(\theta)\big) \, ds(\theta) = \int_{S^2} p(\theta) \, e^{ikx\cdot\theta} \, ds(\theta), \quad x \in \mathbb{R}^3. \tag{5.40}$$

These entire solutions of $\operatorname{curl}^2 v - k^2 v = 0$ are called *Herglotz wave functions* with density p.

Before we turn to the factorization we collect again important properties of F in the following theorem which corresponds exactly to Theorems 1.8 and 4.4.

Theorem 5.7 *Let H^∞ and $F : L_t^2(S^2) \to L_t^2(S^2)$ be the far field pattern and the far field operator (5.39), respectively.*

(a) The following reciprocity relation holds:

$$q \cdot H^\infty\big(-\hat{x}, \theta; p\big) = p \cdot H^\infty\big(-\theta, \hat{x}; q\big) \tag{5.41}$$

for all $\hat{x}, \theta \in S^2$ and $p, q \in \mathbb{C}^3$ with $p \cdot \theta = 0$ and $q \cdot \hat{x} = 0$.

(b) The far field operator F satisfies $F - F^* = \frac{ik}{8\pi^2} F^* F + 2iR$ where F^* denotes again the L^2-adjoint of F and $R : L_t^2(S^2) \to L_t^2(S^2)$ is some compact and self-adjoint non-negative operator which vanishes for real-valued contrasts q.

(c) The scattering operator $S := I + \frac{ik}{8\pi^2} F$ is sub-unitary and has the form

$$S^* S = I - \frac{k}{4\pi^2} R$$

with some non-negative operator R. In the case where q is real-valued the operator R vanishes, the scattering operator S is unitary, and the far field operator F is normal.

Proof: This is proven by very similar arguments as in the proofs of Theorems 1.6, 1.8, and 4.4. We use Green's theorem and Green's formula in the form (cf. [43]):

$$\iint_D [v \cdot \operatorname{curl}^2 w - w \cdot \operatorname{curl}^2 v] dx = \int_{\partial D} [w \times \operatorname{curl} v - v \times \operatorname{curl} w] \cdot v \, ds \tag{5.42}$$

and

$$v(x) = -\operatorname{curl} \int_D \big(\nu(y) \times v(y)\big)\Phi(x,y)ds(y) + \frac{1}{k^2}\operatorname{curl}^2 \int_D \big(\nu(y) \times \operatorname{curl} v(y)\big)\Phi(x,y)ds(y)$$

(5.43)

for $x \in D$. The asymptotic form of the fundamental solution $\Phi(x,y)$ for $|x| \to \infty$ yields the representation of the far field pattern in the form

$$v^\infty(\hat{x}) = \int_{\partial D} \Big[ik\,\hat{x} \times \big(\nu(y) \times v(y)\big) + \hat{x} \times \big(\nu(y) \times \operatorname{curl} v(y)\big) \times \hat{x} \Big] e^{-ik\,\hat{x}\cdot y}\, ds(y) \quad (5.44)$$

for $\hat{x} \in S^2$.

(a) Green's formula (5.42) applied to H^i in D and to H^s in $\mathbb{R}^3 \setminus D$, respectively, yields

$$0 = \int_{\partial D} \Big[H^i(y,\theta,p) \times \operatorname{curl} H^i(y,\hat{x},q) - H^i(y,\hat{x},q) \times \operatorname{curl} H^i(y,\theta,p) \Big] \cdot \nu(y)\, ds(y)\,,$$

$$0 = \int_{\partial D} \Big[H^s(y,\theta,p) \times \operatorname{curl} H^s(y,\hat{x},q) - H^s(y,\hat{x},q) \times \operatorname{curl} H^s(y,\theta,p) \Big] \cdot \nu(y)\, ds(y)\,,$$

and, since $\operatorname{curl} H^i(x,\theta,p) = ik\,(\theta \times p)\exp(ik\,\theta \cdot x)$,

$$q \cdot H^\infty(-\hat{x},\theta,p)$$
$$= \int_{\partial D} \Big[H^s(y,\theta,p) \times \operatorname{curl} H^i(y,\hat{x},q) - H^i(y,\hat{x},q) \times \operatorname{curl} H^s(y,\theta,p) \Big] \cdot \nu(y)\, ds(y)\,,$$

$$-p \cdot H^\infty(-\theta,\hat{x},q)$$
$$= \int_{\partial D} \Big[H^i(y,\theta,p) \times \operatorname{curl} H^s(y,\hat{x},q) - H^s(y,\hat{x},q) \times \operatorname{curl} H^i(y,\theta,p) \Big] \cdot \nu(y)\, ds(y)\,.$$

Adding these four equations yields

$$q \cdot H^\infty(-\hat{x},\theta,p) - p \cdot H^\infty(-\theta,\hat{x},q)$$
$$= \int_{\partial D} \Big[H(y,\theta,p) \times \operatorname{curl} H(y,\hat{x},q) - H(y,\hat{x},q) \times \operatorname{curl} H(y,\theta,p) \Big] \cdot \nu(y)\, ds(y)$$
$$= \iint_D \Big[H(y,\hat{x},q) \cdot \operatorname{curl}^2 H(y,\theta,p) - H(y,\theta,p) \cdot \operatorname{curl}^2 H(y,\hat{x},q) \Big] ds(y)$$
$$= 0$$

by Green's theorem applied in D.

(b) Again, this is proven analogously to the corresponding part in Theorem 4.4: Let $g, h \in L_t^2(S^2)$ and v^i and w^i the corresponding Herglotz functions (5.40) with density g

and h, respectively. By v^s, w^s and v, w we denote the corresponding scattered and total fields, respectively. Then

$$-2i\,k^2 \iint_D (\mathrm{Im}\,q)v \cdot \bar{w}dx = \iint_{|x|<R} [v \cdot \mathrm{curl}^2\,\overline{w} - \overline{w} \cdot \mathrm{curl}^2\,v]dx \tag{5.45}$$

$$= \int_{|x|=R} [\overline{w} \times \mathrm{curl}\,v - v \times \mathrm{curl}\,\overline{w}] \cdot \nu\,ds. \tag{5.46}$$

The integral on the right-hand side is split into four parts by decomposing $v = v^i + v^s$ and $w = w^i + w^s$. The integral involving v^i and w^i vanishes by Green's theorem. We note that by the radiation condition

$$\left[\overline{w}^s \times \mathrm{curl}\,v^s - v^s \times \mathrm{curl}\,\overline{w}^s\right] \cdot \hat{x} = \frac{2ik}{(4\pi\,|x|)^2}\,v^\infty(\hat{x}) \cdot \overline{w^\infty(\hat{x})} + \mathcal{O}(1/r^3)$$

and thus

$$\int_{|x|=R} \left[\overline{w}^s \times \mathrm{curl}\,v^s - v^s \times \mathrm{curl}\,\overline{w}^s\right] \cdot \nu\,ds \longrightarrow \frac{ik}{8\pi^2} \int_{S^2} v^\infty \cdot \overline{w^\infty}ds = \frac{ik}{8\pi^2}(Fg, Fh)_{L^2(S^2)}.$$

The terms involving v^s, w^i and v^i, w^s are treated in the same way (see proof of Theorem 1.8). Finally, we introduce the operator $L : L^2_t(S^2) \rightarrow L^2(D, \mathbb{C}^3)$ which maps $g \in L^2_t(S^2)$ into $v|_D$ where v is the solution of the scattering problem with incident field v^i. Then the left-hand side of (5.45) can be written in the form $-2ik^2\,((\mathrm{Im}\,q)Lg, Lh)_{L^2(D)}$. Therefore, we arrive at

$$-2ik^2 \left(L^*\big((\mathrm{Im}\,q)\,Lg\big), h\right)_{L^2(S^2)} = -\frac{ik}{8\pi^2} \left(Fg, Fh\right)_{L^2(S^2)} - \left(g, Fh\right)_{L^2(S^2)} + \left(Fg, h\right)_{L^2(S^2)}$$

which proves part (b) with $Rg = k^2 L^*\big((\mathrm{Im}\,q)\,Lg\big)$.

(c) This is proven with the help of (b) exactly as in Theorem 1.8, part (b). $\square$

Injectivity of F can be shown analogously to the scalar case of the previous chapter (see Theorem 4.4) provided k^2 is not an eigenvalue of some interior transmission problem in the support $\overline{D}$ of q. Classically, this is formulated as

$$\mathrm{curl}\left[\frac{1}{\varepsilon_r}\,\mathrm{curl}\,v\right] - k^2 v = 0 \text{ in } D, \qquad \mathrm{curl}^2\,w - k^2 w = 0 \text{ in } D, \tag{5.47}$$

$$\nu \times v = \nu \times w \text{ on } \partial D \quad \text{and} \quad \frac{1}{\varepsilon_r}\,\nu \times \mathrm{curl}\,v = \nu \times \mathrm{curl}\,w \text{ on } \partial D, \tag{5.48}$$

or, taking the difference $u = v - w$,

$$\mathrm{curl}\left[\frac{1}{\varepsilon_r}\,\mathrm{curl}\,u\right] - k^2 u = \mathrm{curl}\big[q\,\mathrm{curl}\,w\big] \text{ in } D, \qquad \mathrm{curl}^2\,w - k^2 w = 0 \text{ in } D,$$

$$\nu \times u = 0 \text{ on } \partial D \quad \text{and} \quad \frac{1}{\varepsilon_r}\,\nu \times \mathrm{curl}\,u = q\,\nu \times \mathrm{curl}\,w \text{ on } \partial D,$$

where again $q = 1 - 1/\varepsilon_r$. Analogously to Definition 4.7 we have to formulate this problem in a weak form. First, we introduce the space

$$H_0(\mathrm{curl}, D) = \left\{ v \in H(\mathrm{curl}, D) : \nu \times v = 0 \text{ on } \partial D \right\}.$$

Note that the trace exists due to the trace theorem (cf. [146]).

Definition 5.8 *The wavenumber* k^2 *is called an* interior transmission eigenvalue *if there exists a non-vanishing pair* $(u, w) \in H_0(\mathrm{curl}, D) \times L^2(D, \mathbb{C}^3, |q|dx)$ *and a sequence* $\{w_j\}$ *in* $H(\mathrm{curl}, D)$ *with* $w_j \to w$ *in* $L^2(D, \mathbb{C}^3, |q|dx)$ *and*

$$\mathrm{curl}\left[\frac{1}{\varepsilon_r} \mathrm{curl}\, u \right] - k^2 u = \mathrm{curl}\left[q\, w \right] \text{ in } D, \qquad \mathrm{curl}^2 w_j - k^2 w_j = 0 \text{ in } D, \qquad (5.49)$$

and

$$\frac{1}{\varepsilon_r}\, \nu \times \mathrm{curl}\, u = q\, \nu \times w \text{ on } \partial D. \qquad (5.50)$$

Again, we understand (5.49), (5.50) in the variational sense, i.e.,

$$\iint_D \left[\frac{1}{\varepsilon_r} \mathrm{curl}\, u \cdot \mathrm{curl}\, \psi - k^2 u \cdot \psi \right] dx = \iint_D q\, w \cdot \mathrm{curl}\, \psi\, dx \qquad (5.51)$$

for all $\psi \in H(\mathrm{curl}, D)$.

We recall that $L^2(D, \mathbb{C}^3, |q|dx)$ denotes the weighted L^2 – space of vector fields on D. Note that we replaced $\mathrm{curl}\, w$ from the above motivation by w in this definition. We will study this eigenvalue problem in Section 5.5 below.

5.3 Factorization of the far field operator

For this and the following sections we strengthen Assumption 5.1 and assume the following, compare Assumption 4.2:

Assumption 5.9 *Let* $D \subset \mathbb{R}^3$ *be open and bounded such that* $\partial D \in C^2$ *and the complement* $\mathbb{R}^3 \setminus \overline{D}$ *is connected. Let* $k \in \mathbb{R}_{>0}$ *be the wavenumber and let* $\varepsilon_r \in L^\infty(D)$ *satisfy*

(1) $\mathrm{Im}\, \varepsilon_r \geq 0$ *in* D.

(2) *There exists* $c_1 > 0$ *with* $\mathrm{Re}\, \varepsilon_r \geq c_1$ *on* D.

(3) *For all* $f \in L^2(\mathbb{R}^3, \mathbb{C}^3)$ *with compact support there exists a unique radiating solution of (5.23).*

(4) $|\varepsilon_r - 1|$ *is locally bounded below, i.e., for every compact subset* $M \subset D$ *there exists* $c > 0$ *(depending on* M *) with*

$$|\varepsilon_r(x) - 1| \geq c \quad \text{for almost all } x \in M. \qquad (5.52)$$

We extend ε_r *by one outside of* D.

We note that the relative permittivity ε_r of (5.19) always satisfies conditions (1) and (2) provided ε is bounded below on D. Conditions (3) and (4) are, e.g., satisfied for Hölder continuously differentiable parameters ε_r and σ.

We define the operator $\mathcal{H} : L_t^2(S^2) \to L^2(D, \mathbb{C}^3)$ by

$$(\mathcal{H}p)(x) := \sqrt{|q(x)|}\, \mathrm{curl} \int_{S^2} p(\theta)\, e^{ikx\cdot\theta}\, ds(\theta)\,, \quad x \in D\,, \tag{5.53}$$

where again the contrast is defined by $q = 1 - 1/\varepsilon_r$. This operator $\mathcal{H}$ is one-to-one as it is easily seen.

Recalling the definition of Fp from (5.39) and (5.21) we observe that $Fp = G\mathcal{H}p$ where the data-to-pattern operator $G : L^2(D, \mathbb{C}^3) \to L_t^2(S^2)$ is defined by $Gf := v^\infty$ with v^∞ being the far field pattern corresponding to the radiating (variational) solution v of

$$\mathrm{curl}\left[\frac{1}{\varepsilon_r}\,\mathrm{curl}\, v\right] - k^2 v = \mathrm{curl}\left[\frac{q}{\sqrt{|q|}}f\right] \quad \text{in } \mathbb{R}^3\,. \tag{5.54}$$

We will now show that $G = \mathcal{H}^*\mathcal{T}$ for some operator $\mathcal{T}$ which will give the factorization of F in the form $F = \mathcal{H}^*\mathcal{T}\mathcal{H}$.

Theorem 5.10 *Let conditions (1), (2), and (3) of Assumption 5.9 hold and let F and $\mathcal{H}$ be defined by (5.39) and (5.53), respectively. Then*

$$F = \mathcal{H}^*\mathcal{T}\mathcal{H} \tag{5.55}$$

where $\mathcal{H}^* : L^2(D, \mathbb{C}^3) \to L_t^2(S^2)$ denotes the adjoint of $\mathcal{H}$ and $\mathcal{T} : L^2(D, \mathbb{C}^3) \to L^2(D, \mathbb{C}^3)$ is given by $\mathcal{T}f = (\mathrm{sign}\, q)\left[f + \sqrt{|q|}\,\mathrm{curl}\, v\right]$ and $v \in H_{loc}(\mathrm{curl}, \mathbb{R}^3)$ is the radiating solution of (5.54). Again, the contrast is given by $q = 1 - 1/\varepsilon_r$ and $\mathrm{sign}\, q = q/|q|$ denotes the signum of q.

Proof: First we note again that $F = G\mathcal{H}$. The adjoint $\mathcal{H}^* : L^2(D, \mathbb{C}^3) \to L_t^2(S^2)$ of $\mathcal{H}$ is given by

$$(\mathcal{H}^*\varphi)(\hat{x}) = ik\,\hat{x} \times \iint_D \varphi(y)\, e^{-ik\hat{x}\cdot y}\, \sqrt{|q(y)|}\, dy \quad \text{for } \hat{x} \in S^2\,. \tag{5.56}$$

Using the asymptotic behavior of the fundamental solution $\Phi(x, y)$ as $|x|$ tends to infinity it is easily seen that $\mathcal{H}^*\varphi = w^\infty$ where

$$w(x) = \mathrm{curl} \iint_D \varphi(y)\, \Phi(x, y)\, \sqrt{|q(y)|}\, dy\,, \quad x \in \mathbb{R}^3\,.$$

From Lemma 5.2 we conclude that w is the radiating solution of

$$\mathrm{curl}^2 w - k^2 w = \mathrm{curl}\left[\sqrt{|q|}\,\varphi\right] \quad \text{in } \mathbb{R}^3\,.$$

Now we set $\varphi = \mathcal{T}f = (\text{sign } q)\left[f + \sqrt{|q|}\,\text{curl}\,v\right]$ for any $f \in L^2(D, \mathbb{C}^3)$. Then $\mathcal{H}^*\mathcal{T}f = w^\infty$ where w solves

$$\text{curl}^2\,w - k^2 w = \text{curl}\left[\sqrt{|q|}\,\mathcal{T}f\right] = \text{curl}\left[\frac{q}{\sqrt{|q|}}f + q\,\text{curl}\,v\right] = \text{curl}^2\,v - k^2 v \quad \text{in } \mathbb{R}^3,$$

and both, w and v satisfy the radiation condition. Therefore, $w = v$ follows by the uniqueness of the problem and thus $Gf = v^\infty = w^\infty = \mathcal{H}^*\mathcal{T}f$ for all $f \in L^2(D, \mathbb{C}^3)$. Substituting $G = \mathcal{H}^*\mathcal{T}$ into $F = G\mathcal{H}$ yields the assertion. $\square$

We note that the operator $\mathcal{T}$ is one-to-one. Indeed, $\mathcal{T}f = 0$ implies $f + \sqrt{|q|}\,\text{curl}\,v \equiv 0$ in D, i.e., $\text{curl}^2\,v - k^2 v = 0$ in $\mathbb{R}^3$, and v satisfies the radiation condition. The uniqueness assumption yields $v \equiv 0$ in $\mathbb{R}^3$. Therefore, also $f \equiv 0$.

5.4 Localization of the support of the contrast

This section is build up in the same way as the corresponding Section 4.4 of Chapter 4. First, we prove the connection between the domain D and the range $\mathcal{R}(\mathcal{H}^*)$ of $\mathcal{H}^*$.

Theorem 5.11 *Let the conditions of Assumption 5.9 hold. For any $z \in \mathbb{R}^3$ and fixed $p \in \mathbb{C}^3$, $p \neq 0$ we define $\phi_z \in L_t^2(S^2)$ by*

$$\phi_z(\hat{x}) = ik\,(\hat{x} \times p)\,e^{-ik\hat{x}\cdot z}, \quad \hat{x} \in S^2. \tag{5.57}$$

Then $z \in D$ if and only if $\phi_z \in \mathcal{R}(\mathcal{H}^*)$ where the adjoint $\mathcal{H}^* : L^2(D, \mathbb{C}^3) \to L_t^2(S^2)$ of $\mathcal{H}$ is given by (5.56).

Proof: The proof uses Theorem 4.6 of Chapter 4. First, let $z \in D$. By Theorem 4.6 there exists a scalar function $\tilde{\varphi} \in L^2(D)$ such that

$$e^{-ik\,\hat{x}\cdot z} = \iint_D \tilde{\varphi}(y)\,e^{-ik\,\hat{x}\cdot y}\sqrt{|q(y)|}\,dy \quad \text{for } \hat{x} \in S^2.$$

With $\varphi = p\tilde{\varphi} \in L^2(D, \mathbb{C}^3)$ we have that

$$ik\,(\hat{x} \times p)\,e^{-ik\hat{x}\cdot z} = ik\,\hat{x} \times \iint_D \varphi(y)\,e^{-ik\,\hat{x}\cdot y}\sqrt{|q(y)|}\,dy \quad \text{for } \hat{x} \in S^2,$$

i.e., $\phi_z = \mathcal{H}^*\varphi$.

Now we consider the case $z \notin D$ and assume, on the contrary, that $\phi_z = \mathcal{H}^*\varphi \in \mathcal{R}(H^*)$ for some $\varphi \in L^2(D, \mathbb{C}^3)$. We note that ϕ_z and $\mathcal{H}^*\varphi$ are the far field patterns of

$$v_1(x) = \text{curl}_x\big(p\,\Phi(x, z)\big), \quad x \in \mathbb{R}^3 \setminus \{z\},$$

and

$$v_2(x) = \text{curl}\iint_D \varphi(y)\,\Phi(x, y)\,\sqrt{|q(y)|}\,dy, \quad x \in \mathbb{R}^3,$$

respectively. By Rellich's Lemma and analytic continuation we conclude that $v_1 \equiv v_2$ on $\mathbb{R}^3 \setminus (D \cup \{z\})$. This contradicts the fact that v_2 is continuous on $\mathbb{R}^3$ and v_1 is singular at z. $\square$

In this section we consider three cases of contrasts. First, we will consider the simplest case where D is absorbing everywhere with $\operatorname{Im} q > 0$ on D. This will lead to a self-adjoint problem. Second, we consider the non-absorbing case , i.e., q is real-valued and positive everywhere (or negative everywhere). In this case, the far field operator is again normal and the scattering operator unitary and we can apply the abstract Theorem 1.23. Third, we will consider the general case where only parts of D may be absorbing.

The following theorem collects properties of the operator $\mathcal{T}$ needed for the analysis of the factorization methods.

Theorem 5.12 *Let the conditions of Assumption 5.9 hold and let* $\mathcal{T} : L^2(D,\mathbb{C}^3) \to L^2(D,\mathbb{C}^3)$ *be defined as in Theorem 5.10, i.e.,*

$$\mathcal{T}f = (\operatorname{sign} q)\left[f + \sqrt{|q|}\,\operatorname{curl} v\right]$$

and $v \in H_{loc}(\operatorname{curl}, \mathbb{R}^3)$ *solves (5.54), i.e., is a radiating variational solution of*

$$\operatorname{curl}\left[\frac{1}{\varepsilon_r}\,\operatorname{curl} v\right] - k^2 v = \operatorname{curl}\left[\frac{q}{\sqrt{|q|}}f\right] \quad in\ \mathbb{R}^3 . \tag{5.58}$$

Here, again, $q = 1 - 1/\varepsilon_r$ *denotes the contrast. Then we have:*

(a)

$$\operatorname{Im}(\mathcal{T}f, f)_{L^2(D)} \geq 0 \quad for\ all\ f \in L^2(D,\mathbb{C}^3) .$$

(b) Assume that k^2 *is not an eigenvalue of the interior transmission eigenvalue problem of Definition 5.8. Then*

$$\operatorname{Im}(\mathcal{T}f, f)_{L^2(D)} > 0$$

for all $f \in \overline{\mathcal{R}(\mathcal{H})} \subset L^2(D,\mathbb{C}^3)$ *with* $f \neq 0$*. Here, again* $\mathcal{H} : L^2_t(S^2) \to L^2(D,\mathbb{C}^3)$ *is defined in (5.53), and* $\overline{\mathcal{R}(\mathcal{H})}$ *denotes the closure of* $\mathcal{R}(\mathcal{H})$ *in* $L^2(D,\mathbb{C}^3)$*.*

(c) Assume that there exists a constant $\gamma_0 > 0$ *such that* $\operatorname{Im} q \geq \gamma_0|q|$ *almost everywhere in D. Then there exists* $\gamma_1 > 0$ *such that*

$$\operatorname{Im}(\mathcal{T}f, f)_{L^2(D)} \geq \gamma_1 \|f\|^2_{L^2(D)} \tag{5.59}$$

for all $f \in L^2(D,\mathbb{C}^3)$*.*

(d) Define the operator $\mathcal{T}_0$ *from* $L^2(D,\mathbb{C}^3)$ *into itself by* $\mathcal{T}_0 f = (\operatorname{sign} q)f$ *for* $f \in L^2(D,\mathbb{C}^3)$*. Then* $\mathcal{T} - \mathcal{T}_0$ *is compact in* $L^2(D,\mathbb{C}^3)$*.*

Proof: For $f \in L^2(D,\mathbb{C}^3)$ we have $\mathcal{T}f = (\operatorname{sign} q)\left[f + \sqrt{|q|}\,\operatorname{curl} v\right] = (\operatorname{sign} q)\tilde{w}$ with $\tilde{w} := f + \sqrt{|q|}\,\operatorname{curl} v$ and where v solves (5.58). The variational form of (5.58) can be written as

$$\iiint_{\mathbb{R}^3} \left[\operatorname{curl} v \cdot \operatorname{curl} \psi - k^2 v \cdot \psi\right] dx = \iint_D \frac{q}{\sqrt{|q|}}\,\tilde{w} \cdot \operatorname{curl} \psi\, dx \tag{5.60}$$

for all $\psi \in H(\mathrm{curl}, \mathbb{R}^3)$ with compact support. Then, since $f = \tilde{w} - \sqrt{|q|}\,\mathrm{curl}\,v$,

$$(\mathcal{T}f, f)_{L^2(D)} = \iint_D (\mathrm{sign}\,q)\,|\tilde{w}|^2\,dx - \iint_D (\mathrm{sign}\,q)\,\tilde{w} \cdot \mathrm{curl}\,\overline{v}\,\sqrt{|q|}\,dx$$

$$= \iint_D (\mathrm{sign}\,q)\,|\tilde{w}|^2\,dx - \iint_D \frac{q}{\sqrt{|q|}}\,\tilde{w} \cdot \mathrm{curl}\,\overline{v}\,dx .$$

Now we choose a mollifier $\phi \in C^\infty(\mathbb{R}^3)$ with $\phi \equiv 1$ for $|x| \leq R$ and $\phi \equiv 0$ for $|x| \geq 2R$ and set $\psi = \phi\,\overline{v}$ in (5.60). This yields

$$(\mathcal{T}f, f)_{L^2(D)} = \iint_D (\mathrm{sign}\,q)\,|\tilde{w}|^2\,dx - \iint_{|x|<R} \big[|\,\mathrm{curl}\,v|^2 - k^2|v|^2\big]\,dx$$

$$- \iint_{R<|x|<2R} \big[\mathrm{curl}\,v \cdot \mathrm{curl}(\overline{v}\phi) - k^2|v|^2\phi\big]\,dx$$

$$= \iint_D (\mathrm{sign}\,q)\,|w|^2\,dx - \iint_{|x|<R} \big[|\,\mathrm{curl}\,v|^2 - k^2|v|^2\big]\,dx -$$

$$- \int_{|x|=R} (\hat{x} \times \mathrm{curl}\,v) \cdot \overline{v}\,ds . \tag{5.61}$$

Here we have used Green's theorem in the angular region $\{x : R < |x| < 2R\}$ and the fact that v solves $\mathrm{curl}^2\,v - k^2 v = 0$ in this region.

(a) Taking the imaginary part of (5.61) yields

$$\mathrm{Im}(\mathcal{T}f, f)_{L^2(D)} = \iint_D \frac{\mathrm{Im}\,q}{|q|}\,|\tilde{w}|^2\,dx - \mathrm{Im} \int_{|x|=R} (\hat{x} \times \mathrm{curl}\,v) \cdot \overline{v}\,ds .$$

From the radiation condition

$$\lim_{|x|\to\infty} |x|\,\big(\mathrm{curl}\,v(x) \times \hat{x} - ikv(x)\big) = 0$$

we conclude that for $R \to \infty$

$$\mathrm{Im}(\mathcal{T}f, f)_{L^2(D)} = \iint_D \frac{\mathrm{Im}\,q}{|q|}\,|\tilde{w}|^2\,dx + \mathrm{Im}\left[\frac{ik}{(4\pi)^2} \int_{S^2} |v^\infty|^2 ds\right] \geq 0 . \tag{5.62}$$

(b) Assume now that, for some $f \in \overline{\mathcal{R}(\mathcal{H})}$, the term $\mathrm{Im}(\mathcal{T}f, f)_{L^2(D)}$ vanishes. Then v^∞ vanishes on S^2 by (5.62) and thus $v \equiv 0$ outside of D. We recall that $v \in H_{loc}(\mathrm{curl}, \mathbb{R}^3)$ is the radiating solution of (5.58), i.e., in variational form (using the fact that v vanishes outside of D)

$$\iint_D \left[\frac{1}{\varepsilon_r}\,\mathrm{curl}\,v \cdot \mathrm{curl}\,\psi - k^2\,v \cdot \psi\right]dx = \iint_D \frac{q}{\sqrt{|q|}}\,f \cdot \mathrm{curl}\,\psi\,dx \quad \text{for all } \psi \in H(\mathrm{curl}, D) .$$

Furthermore, $v|_D \in H_0(\mathrm{curl}, D)$. Setting $w = f/\sqrt{|q|}$ we observe that (v, w) satisfies the condition (5.51) from Definition 5.8. Since $f \in \overline{\mathcal{R}(\mathcal{H})}$ there exist $\tilde{w}_j \in \mathcal{R}(\mathcal{H})$ with $\tilde{w}_j \to f$ in $L^2(D, \mathbb{C}^3)$. From the form of $\mathcal{H}$ we note that $\tilde{w}_j = \sqrt{|q|}\, w_j$ for some smooth solutions w_j of

$$\mathrm{curl}^2\, w_j - k^2\, w_j = 0 \quad \text{in } D.$$

Therefore, w_j tends to $w = f/\sqrt{|q|}$ in $L^2(D, \mathbb{C}^3, |q|dx)$. We have thus shown that (v, w) satisfy all conditions from Definition 5.8. Since k^2 is not an eigenvalue we conclude that v and $w = f/\sqrt{|q|}$ vanish which proves this part of the theorem.

(c) This follows from (5.62) by standard arguments. Indeed, if there exists no such constant γ_1 we can find a sequence $\{f_j\}$ such that $\|f_j\|_{L^2(D)} = 1$ and $\mathrm{Im}(Tf_j, f_j)_{L^2(D)} \to 0$. From (5.62) and the definition of $\tilde{w}$ we conclude that $f_j + \sqrt{|q|}\,\mathrm{curl}\, v_j \to 0$ in $L^2(D)$ where v_j denotes the solution of (5.58) for f replaced by f_j. Writing this equation (5.58) as

$$\mathrm{curl}^2\, v - k^2 v = \mathrm{curl}\left[\frac{q}{\sqrt{|q|}}\, (f + \sqrt{|q|}\,\mathrm{curl}\, v) \right]$$

we note from Theorem 5.5 that $\{v_j\}$ converges to zero in $H(\mathrm{curl}, D)$ and therefore $f_j \to 0$ in $L^2(D)$. This contradicts $\|f_j\|_{L^2(D)} = 1$.

(d) From the definitions of T and T_0 we note that $Tf - T_0 f = \frac{q}{\sqrt{|q|}}\,\mathrm{curl}\, v$ where $v \in H_{loc}(\mathrm{curl}, \mathbb{R}^3)$ is the radiating solution of (5.58). In particular, v solves

$$\iiint_{\mathbb{R}^3} \left[\frac{1}{\varepsilon_r}\,\mathrm{curl}\, v \cdot \mathrm{curl}\, \psi - k^2 v \cdot \psi \right] dx = \iint_D \frac{q}{\sqrt{|q|}} f \cdot \mathrm{curl}\, \psi\, dx$$

for all $\psi \in H(\mathrm{curl}, \mathbb{R}^3)$ with compact support. By substituting $\psi = \nabla\varphi$ for scalar functions $\varphi \in H^1(\mathbb{R}^3)$ we note that $\iint_{\mathbb{R}^3} v \cdot \nabla\varphi\, dx = 0$ for all $\varphi \in H^1(\mathbb{R}^3)$ with compact support, i.e., $\mathrm{div}\, v = 0$ in $\mathbb{R}^3$.

Let now the sequence $\{f_j\}$ converge weakly to zero $L^2(D, \mathbb{C}^3)$ and denote by $v_j \in H_{loc}(\mathrm{curl}, \mathbb{R}^3)$ the corresponding radiating solutions of (5.58). Let $B \subset \mathbb{R}^3$ be a ball which contains D in its interior. By the boundedness of the solution operator we conclude that $\{v_j\}$ converges weakly to zero in $H(\mathrm{curl}, B)$. Furthermore, from Remark 5.6 we conclude that v_j is smooth outside of $\overline{D}$ and converges uniformly to zero on ∂B. We determine $p_j \in H^1_\diamond(B)$ as the solution of

$$\iint_B \nabla p_j \cdot \nabla\overline{\varphi}\, dx = \int_{\partial B} (v \cdot v_j)\, \overline{\varphi}\, ds \tag{5.63}$$

for all $\varphi \in H^1_\diamond(B)$. Here, the subspace $H^1_\diamond(B)$ of $H^1(B)$ is defined as $H^1_\diamond(B) = \{\varphi \in H^1(B) : \iint_B \varphi\, dx = 0\}$. The solution of (5.63) exists and is unique since the form

$$(p, \varphi) \mapsto \iint_B \nabla p \cdot \nabla\overline{\varphi}\, dx$$

is bounded and coercive on $H^1_\diamond(B)$ by the inequality of Poincaré (cf. [179]). The latter states that there exists a constant $c > 0$ with

$$\iint\limits_B |\nabla\varphi|^2 \, dx \geq c \, \|\varphi\|^2_{H^1(B)} \quad \text{for all } \varphi \in H^1_\diamond(B) \,. \tag{5.64}$$

Problem (5.63) is the variational form of the Neumann boundary value problem

$$\Delta p_j = 0 \text{ in } B\,, \qquad \frac{\partial p_j}{\partial \nu} = \nu \cdot v_j \text{ on } \partial B \,.$$

We observe that $\int_{\partial B} \nu \cdot v_j \, ds$ vanishes. This follows from $\operatorname{div} v_j = 0$ in the variational form $\iiint_{\mathbb{R}^3} v_j \cdot \nabla\varphi \, dx = 0$ if we choose φ with compact support such that $\varphi = 1$ in B. Then, since v_j is smooth outside of B we have that

$$0 = \iint\limits_{\mathbb{R}^3\setminus B} v_j \cdot \nabla\varphi \, dx = - \int\limits_{\partial B} (\nu \cdot v_j) \, ds$$

by Green's first theorem. Therefore, (5.63) holds even for all $\varphi \in H^1(B)$. Substituting $\varphi = p_j$ in (5.63) yields, using (5.64) and the trace theorem,

$$c \, \|p_j\|^2_{H^1(B)} \leq \iint\limits_B |\nabla p_j|^2 \, dx = \int\limits_{\partial B} (\nu \cdot v_j) \, p_j \, ds \leq \tilde{c} \, \|v_j\|_{C(\partial B)} \, \|p_j\|_{H^1(B)} \,,$$

i.e., $\|p_j\|_{H^1(B)} \leq (\tilde{c}/c) \, \|v_j\|_{C(\partial B)}$ which converges to zero.

Therefore, the functions $\tilde{v}_j := v_j - \nabla p_j \in H(\operatorname{curl}, B)$ satisfy:

- $\tilde{v}_j \in H_{\operatorname{div}}(\operatorname{curl}, B) := \{u \in H(\operatorname{curl}, B) : \iint_B \nabla\varphi \cdot u \, dx = 0 \text{ for all } \varphi \in H^1(B)\}$,
- $\tilde{v}_j \rightharpoonup 0$ weakly in $L^2(B, \mathbb{C}^3)$,
- $\operatorname{curl} \tilde{v}_j = \operatorname{curl} v_j \rightharpoonup 0$ weakly in $L^2(B, \mathbb{C}^3)$.

These three conditions assure that $\tilde{v}_j$ converges to zero in the norm of $L^2(B, \mathbb{C}^3)$ since the closed subspace $H_{\operatorname{div}}(\operatorname{curl}, B)$ of $H(\operatorname{curl}, B)$ is compactly imbedded in $L^2(B, \mathbb{C}^3)$. We refer to Weber [182], see also [146], Theorem 4.7. Since also $\|\nabla p_j\|_{L^2(B)} \to 0$ this yields $\|v_j\|_{L^2(B)} \to 0$ as j tends to infinity which ends the proof. $\square$

After these preparations we are able to treat the following three cases of contrasts:

(1) All of D is absorbing, i.e., $\operatorname{Im} \varepsilon_r > 0$ on D.

(2) No part of D is absorbing, i.e., ε_r is real-valued and positive on D.

(3) Parts of D may be absorbing, i.e., we allow quite general values of ε.

We will apply the abstract characterizations of Corollary 1.22, Theorem 1.23, and Theorem 2.15 in combination with Theorem 5.11 which characterizes D by the range of H^*.

We formulate and prove the first main result of this section in which we treat the absorbing medium. We recall the definition of the self-adjoint operator $\operatorname{Im} F := \frac{1}{2i}(F - F^*)$.

Theorem 5.13 *In addition to Assumption 5.9 we assume that there exists $\gamma_0 > 0$ with* $\operatorname{Im}\varepsilon_0(x) \geq \gamma_0$ *for almost all $x \in D$.*[2] *Furthermore, let ϕ_z be defined in (5.57) for $z \in \mathbb{R}^3$ (where again $p \in \mathbb{C}^3$ is kept fixed). Then $z \in D$ if, and only if, $\phi_z \in \mathcal{R}\big((\operatorname{Im}F)^{1/2}\big)$.*

Proof: Analogously to $\operatorname{Im}F$ we define $\operatorname{Im}\mathcal{T}$ as $\operatorname{Im}\mathcal{T} := \frac{1}{2i}(\mathcal{T} - \mathcal{T}^*)$ where $\mathcal{T}^*$ is the adjoint of $\mathcal{T}$ in $L^2(D, \mathbb{C}^3)$. By our assumption on ε_r we conclude that $\operatorname{Im}q = \operatorname{Im}\varepsilon_r/|\varepsilon_r|^2 \geq \gamma_0/\|\varepsilon_r\|_\infty^2$, i.e., the additional assumption of part (c) of Theorem 5.12 is satisfied. Noting that

$$\operatorname{Im}(\mathcal{T}f, f)_{L^2(D)} = ((\operatorname{Im}\mathcal{T})f, f)_{L^2(D)}$$

we conclude that the self-adjoint operator $(\operatorname{Im}\mathcal{T})$ is coercive on $L^2(D, \mathbb{C}^3)$. Furthermore, the factorization (5.55) yields the factorization of $(\operatorname{Im}F)$ in the form

$$\operatorname{Im}F = \mathcal{H}^*(\operatorname{Im}\mathcal{T})\,\mathcal{H}. \tag{5.65}$$

This is exactly the situation where we can apply Corollary 1.22 of Chapter 1 which yields that the ranges of $\mathcal{H}^*$ and $(\operatorname{Im}F)^{1/2}$ coincide. The combination with the characterization of D by Theorem 5.11 yields the assertion. $\square$

Remark: The additional assumption of part (c) of Theorem 5.12 requires the existence of γ_0 such that $\operatorname{Im}q \geq \gamma_0|q|$ on D. Transforming this condition for $q = 1 - 1/\varepsilon_r$ into one for ε_r yields the condition $\varepsilon_r(x) \in C_1 \cup C_2$ for almost all $x \in D$ where the sets $C_1, C_2 \subset \mathbb{C}$ are given by

$$C_1 = \left\{ z \in C_3^- : \left| z - \frac{1}{2}\left(1 - \frac{\sqrt{1 - \gamma_0^2}}{\gamma_0} i \right) \right| \geq \frac{1}{2\gamma_0} \right\},$$

$$C_2 = \left\{ z \in C_3^+ : \left| z - \frac{1}{2}\left(1 + \frac{\sqrt{1 - \gamma_0^2}}{\gamma_0} i \right) \right| \leq \frac{1}{2\gamma_0} \right\},$$

where

$$C_3^{\mp} = \left\{ z \in \mathbb{C} : \left| z - \frac{1}{2} \right| \lessgtr \frac{1}{2} \right\}.$$

The sets C_1 and C_2 are bounded by circles. They are shown in Figure 5.1. The bounds $\operatorname{Re}\varepsilon_r(x) \geq c_1$ and $|\varepsilon_r(x)| \leq \|\varepsilon_r\|_\infty$ are not shown in the figure.

In the second situation we consider *non-absorbing media*.

Theorem 5.14 *In addition to Assumption 5.9 we assume that ε_r is real valued and either $\varepsilon_r > 1$ on D or $\varepsilon_r < 1$ on D (and equal to one on $\mathbb{R}^3 \setminus D$). Furthermore, we assume that*

---

[2] Recall that $\overline{D}$ is the support of q, i.e., $q = 0$ outside of (the open and bounded set) D.

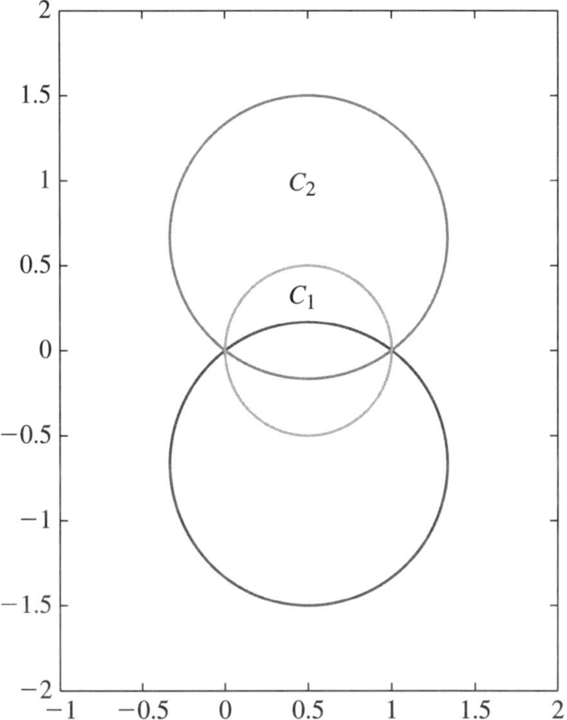

Figure 5.1 The circles which form C_1 and C_2

k^2 is not an interior transmission eigenvalue in the sense of Definition 5.8. Let ϕ_z be again defined in (5.57) for any $z \in \mathbb{R}^3$ (where again $p \in \mathbb{C}^3$ is kept fixed). Then $z \in D$ if, and only if, $\phi_z \in \mathcal{R}\big((F^*F)^{1/4}\big)$.

Proof: This time, we have to check the assumptions of Theorem 1.23 of Chapter 1. By Theorem 5.7 F is normal and the scattering operator $\mathcal{S} := I + \frac{ik}{8\pi^2} F$ is unitary. Furthermore, the factorization

$$F = \mathcal{H}^* \mathcal{T} \mathcal{H}$$

holds and Im $\mathcal{T}$ is positive on $\overline{\mathcal{R}(\mathcal{H})} \setminus \{0\}$. By Theorem 5.12, part (d), the operator $\mathcal{T}$ has the decomposition into $\mathcal{T} = \mathcal{T}_0 + (\mathcal{T} - \mathcal{T}_0)$ where $\mathcal{T}_0 f = (\text{sign } q) f$ for $f \in L^2(D, \mathbb{C}^3)$ and $(\mathcal{T} - \mathcal{T}_0)$ is compact. Since ε_r is real valued so is q and thus either sign $q \equiv 1$ or sign $q \equiv -1$ on D. Therefore, Theorem 1.23 is applicable and yields that the ranges of $\mathcal{H}^*$ and $(F^*F)^{1/4}$ coincide. The combination with Theorem 5.11 yields the assertion. $\square$

In the third situation we consider more **general electric permitivities** ε_r.

Theorem 5.15 *In addition to Assumptions 5.9 we assume that there exists $r > 0$ such that*

$$\left| \varepsilon_r(x) - \frac{1}{2}(1 - r\,i) \right| \geq \frac{\sqrt{1 + r^2}}{2} \tag{5.66}$$

for almost all $x \in D$. Choose $t \in (0, \pi/2)$ such that $\cos t \leq 1/\sqrt{1 + r^2}$.

Furthermore, we assume that k^2 is not an interior transmission eigenvalue in the sense of Definition 5.8. Let ϕ_z be again defined in (5.57) for $z \in \mathbb{R}^3$ (where again $p \in \mathbb{C}^3$ is kept fixed). Then $z \in D$ if, and only if, $\phi_z \in \mathcal{R}\big(F_{\#}^{1/2}\big)$ where $F_{\#} : L_t^2(S^2) \to L_t^2(S^2)$ is defined by

$$F_{\#} = \big| \mathrm{Re}\big(e^{-it} F\big) \big| + \mathrm{Im}\, F \,. \tag{5.67}$$

Remark: The assumption (5.66) describes the exterior of the circle centered at $(1 - ri)/2$ in the lower half plane passing through 0 and 1. For large values of r the part of the circle with $\mathrm{Im}\,\varepsilon \geq 0$ approaches the interval $[0, 1]$ of the real axis. This assumption (5.66) is certainly satisfied for ε_r with $\mathrm{Re}\,\varepsilon_r(x) \geq 1$ and arbitrary $\mathrm{Im}\,\varepsilon_r(x) \geq 0$.

Proof: For this case we verify the assumptions of Theorem 2.15. For (A2) we write $\mathrm{Re}\big[e^{-it}\mathcal{T}\big]$ as

$$\mathrm{Re}\big[e^{-it}\mathcal{T}\big] = \mathrm{Re}\big[e^{-it}\mathcal{T}_0\big] + \mathrm{Re}\big[e^{-it}(\mathcal{T} - \mathcal{T}_0)\big] \,.$$

By Theorem 5.12 the operator $\mathrm{Re}\big[e^{-it}(\mathcal{T} - \mathcal{T}_0)\big]$ is compact. We show that the condition (5.66) implies that the operator $\mathrm{Re}\big[e^{-it}\mathcal{T}_0\big] : L^2(D, \mathbb{C}^3) \to L^2(D, \mathbb{C}^3)$ is coercive. Recalling the definition of the operator $\mathcal{T}_0$ we have to show that there exists $\gamma > 0$ such that

$$(\cos t)\, \mathrm{Re}\, q + (\sin t)\, \mathrm{Im}\, q \geq \gamma\, |q| \quad \text{on } D \,.$$

With $q = 1 - 1/\varepsilon_r = (|\varepsilon_r|^2 - \overline{\varepsilon_r})/|\varepsilon_r|^2$ this estimate is equivalent to

$$(\cos t)\,(z_1^2 + z_2^2 - z_1) + (\sin t)\, z_2 \geq \gamma\, \sqrt{(z_1^2 + z_2^2 - z_1)^2 + z_2^2} \tag{5.68}$$

where we have set $z_1 = \mathrm{Re}\,\varepsilon_r(x)$ and $z_2 = \mathrm{Im}\,\varepsilon_r(x)$. We consider two cases:

First case: $z_1^2 + z_2^2 - z_1 \geq 0$. Then we have since $\cos t \sin t > 0$:

$$\big[(\cos t)\,(z_1^2 + z_2^2 - z_1) + (\sin t)\, z_2\big]^2 \geq (\cos t)^2\,(z_1^2 + z_2^2 - z_1)^2 + (\sin t)^2\, z_2^2$$

$$\geq \gamma^2 \big[(z_1^2 + z_2^2 - z_1)^2 + z_2^2\big]$$

with any $\gamma \leq \min\{\cos t, \sin t\}$. This proves (5.68).

Second case: $z_1^2 + z_2^2 - z_1 < 0$. In terms of z_1 and z_2 the condition (5.66) can be rewritten as

$$-r\, z_2 \leq z_1^2 + z_2^2 - z_1 \leq 0 \,.$$

We define $s \in [\pi/2, \pi]$ by

$$z_1^2 + z_2^2 - z_1 = (\cos s)\sqrt{(z_1^2 + z_2^2 - z_1)^2 + z_2^2} \quad \text{and}$$

$$z_2 = (\sin s)\sqrt{(z_1^2 + z_2^2 - z_1)^2 + z_2^2}\,.$$

Then the previous inequality reads

$$r \sin s \geq -\cos s \geq 0, \tag{5.69}$$

and we have to show the existence of $\gamma > 0$ with

$$\cos t \cos s + \sin t \sin s \geq \gamma\,.$$

First we have that

$$\cos t \cos s + \sin t \sin s \geq \sin s\big[-r \cos t + \sin t\big]\,.$$

Squaring (5.69) we estimate $r^2 \sin^2 s \geq \cos^2 s = 1 - \sin^2 s$, i.e., $\sin s \geq 1/\sqrt{r^2 + 1}$ and thus

$$\cos t \cos s + \sin t \sin s \geq \frac{1}{\sqrt{r^2 + 1}}\big[-r \cos t + \sin t\big]\,.$$

The choice of t guarantees that $-r \cos t + \sin t > 0$. This proves the estimate (5.68) with $\gamma = \big[-r \cos t + \sin t\big]/\sqrt{r^2 + 1}$.

Finally, also assumptions (A3) and (A4) of Theorem 2.15 are satisfied because $\operatorname{Im} \mathcal{T}$ is positive on $\overline{\mathcal{R}(\mathcal{H})} \setminus \{0\}$ by Theorem 5.12, part (b). $\square$

These results characterize the support $\overline{D}$ of $\varepsilon_r - 1$ by the data of the inverse scattering problem which are collected in the operator F. In particular, it gives a new "direct" proof of uniqueness of the inverse scattering problem to determine D from $H^\infty(\hat{x}, \theta; p)$ for all $\hat{x}, \theta \in S^2$ and $p \in \mathbb{C}^3$ with $p \cdot \theta = 0$.

5.5 The interior transmission eigenvalue problem

Analogously to Section 4.5 we will investigate the interior transmission eigenvalue problem of Definition 5.8. It is the aim to prove that the set of eigenvalues is discrete. Analogously to the scalar case we assume the following, compare Assumption 4.12:

Assumption 5.16 *Let $D \subset \mathbb{R}^3$ be open and bounded such that $\partial D \in C^2$ and the complement $\mathbb{R}^3 \setminus \overline{D}$ is connected. Let $k \in \mathbb{R}_{>0}$ be the wavenumber and let $\varepsilon_r \in L^\infty(D)$ be real valued such that either $\varepsilon_r > 1$ on D or $\varepsilon_r < 1$ on D where in the latter case there exists $\gamma_1 > 1/2$ with $\varepsilon_r \geq \gamma_1$ on D. Furthermore, we assume one of the following properties (A) or (B):*

(A) There exists $\gamma_0 > 0$ such that $|\varepsilon_r(x) - 1| \geq \gamma_0$ for almost all $x \in D$, i.e., ε_r is bounded away from one.

(B) $\varepsilon_r \in C(\overline{D})$ and

$$\gamma(\delta) := \sup_{x \in \partial D} \int_0^\delta \frac{dt}{\left|\varepsilon_r(x - tv(x)) - 1\right|} \longrightarrow 0 \quad \text{as } \delta \to 0. \tag{5.70}$$

Finally, we assume that for all $f \in L^2(\mathbb{R}^3, \mathbb{C}^3)$ with compact support there exists a unique radiating solution of (5.23).

We extend ε_r by one outside of D.

We note again that part (B) of Assumption 5.16 is satisfied if $\left|\varepsilon_r(x-tv(x))-1\right| \geq c\,t^p$ for some $c > 0$ and $p \in [0, 1)$.

We recall the definition $q = 1 - 1/\varepsilon_r$ of the contrast and note that for part (A) either $q \geq \gamma_0/(1 + \gamma_0) > 0$ on D or $q \leq -\gamma_0/(1 - \gamma_0) < 0$ on D. Under part (B) of Assumption 5.16 we observe that also

$$\tilde{\gamma}(\delta) := \sup_{x \in \partial D} \int_0^\delta \frac{dt}{\left|q(x - tv(x))\right|} \longrightarrow 0 \tag{5.71}$$

for $\delta \to 0$. We also note that in any case $\|q\|_\infty < 1$ which will be important in the following.

In this section we will work in (subspaces of) $L^2(D, \mathbb{C}^3, |q|dx)$, equipped with the inner product

$$(\psi, \varphi)_{L^2(D, |q|dx)} = \iint_D \psi \cdot \overline{\varphi} \, |q| \, dx. \tag{5.72}$$

We introduce the closed subspaces $L^2_{div}(D, \mathbb{C}^3, |q|dx)$ and V of $L^2(D, \mathbb{C}^3, |q|dx)$ as

$$L^2_{div}(D, \mathbb{C}^3, |q|dx) = \left\{ w \in L^2(D, \mathbb{C}^3, |q|dx) : \iint_D w \cdot \nabla\varphi \, dx = 0 \text{ for all } \varphi \in H_0^1(D) \right\},$$

$$V = \text{closure}\left\{ w \in H^2(D, \mathbb{C}^3) : \text{curl}^2 \, w - k^2 w = 0 \text{ in } D \right\},$$

where the closure is taken with respect to the norm of $L^2(D, \mathbb{C}^3, |q|dx)$. Using the space V we recall that k^2 is an interior transmission eigenvalue if there exists a non-vanishing pair $(u, w) \in H_0(\text{curl}, D) \times V$ such that

$$\iint_D \left[\frac{1}{\varepsilon_r} \text{curl}\, u \cdot \text{curl}\, \psi - k^2 u \cdot \psi \right] dx = \iint_D q\,w \cdot \text{curl}\, \psi \, dx \tag{5.73}$$

for all $\psi \in H(\text{curl}, D)$.

Lemma 5.17 *Let $(u, w) \in H_0(\text{curl}, D) \times V$ solve (5.73) for some real $k > 0$.*

(a) Then $\text{curl}\, u$ and w solve

$$\text{curl}\, u = B_k(w + \text{curl}\, u) \tag{5.74}$$

where the bounded operator B_k from $L^2(D, \mathbb{C}^3, |q|dx)$ into itself is given by $B_k f = \mathrm{curl}^2 \iint_D \Phi(\cdot, y) \, q(y) f(y) \, dy$ for $f \in L^2(D, \mathbb{C}^3, |q|dx)$. Here, we interpret the curl in the distributional sense.

(b) The following orthogonality property holds:

$$(w + \mathrm{curl}\, u, \psi)_{L^2(D, |q|dx)} = \iint_D (w + \mathrm{curl}\, u) \cdot \overline{\psi} \, |q| \, dx = 0 \quad \text{for all } \psi \in V.$$

$$(5.75)$$

Proof: (a) From Lemmas 5.2 and 5.3 we conclude that $u = A(qw + q\, \mathrm{curl}\, u)$ and A is bounded from $L^2(D, \mathbb{C}^3)$ into $H(\mathrm{curl}, D)$. The observation that the multiplication by q is a bounded operator from $L^2(D, \mathbb{C}^3, |q|dx)$ into $L^2(D, \mathbb{C}^3)$ and $L^2(D, \mathbb{C}^3)$ is boundedly imbedded in $L^2(D, \mathbb{C}^3, |q|dx)$ yields assertion (a).

(b) From (5.73) we conclude that

$$\iint_D \left[\mathrm{curl}\, u \cdot \mathrm{curl}\, \overline{\psi} - k^2 u \cdot \overline{\psi} \right] dx = \iint_D q\, (w + \mathrm{curl}\, u) \cdot \mathrm{curl}\, \overline{\psi} \, dx \qquad (5.76)$$

for all $\psi \in H(\mathrm{curl}, D)$. Let now $\psi \in C^2(\overline{D})$ satisfy $\mathrm{curl}^2 \psi - k^2 \psi = 0$ in D. Substitution of $\mathrm{curl}\, \psi$ for ψ in (5.76) yields

$$\iint_D [\mathrm{curl}\, u \cdot \overline{\psi} - u \cdot \mathrm{curl}\, \overline{\psi}] \, dx = \iint_D q\, (w + \mathrm{curl}\, u) \cdot \overline{\psi} \, dx.$$

The left hand side vanishes by Green's theorem since $v \times u = 0$ on ∂D. Formula (5.75) follows since q does not change its sign in D. $\square$

We set $\tilde{u} = w + \mathrm{curl}\, u$ and write (5.74) as

$$\tilde{u} - B_k \tilde{u} = w, \qquad (5.77)$$

and consider this equation in the subspace $L^2_{div}(D, \mathbb{C}^3, |q|dx)$ of $L^2(D, \mathbb{C}^3, |q|dx)$. Similarly to the approach of Colton and Kress in [43] for the scalar case (see also [54] for the anisotropic case) we will project this equation onto the orthogonal complement

$$V^\perp = \left\{ u \in L^2_{div}(D, \mathbb{C}^3, |q|dx) : \iint_D u \cdot \psi \, |q| \, dx = 0 \quad \text{for all } \psi \in V \right\} \qquad (5.78)$$

of V. We note from (5.75) that $\tilde{u} = w + \mathrm{curl}\, u \in V^\perp$. Therefore, if P_k is the orthogonal projector from $L^2_{div}(D, \mathbb{C}^3, |q|dx)$ onto $V^\perp$, and $\tilde{u}$ and $w \in V$ solve (5.77) then $\tilde{u}$ solves

$$\tilde{u} - P_k B_k \tilde{u} = 0. \qquad (5.79)$$

We will now study the projection operator P_k. In particular, we will prove that it can be extended analytically to complex valued k in a neighborhood of the positive real axis.

We have to introduce the following space W which corresponds to the space W of Section 4.5.

$$W = \left\{ \varphi \in H_0(\mathrm{curl}, D) : \nu \times \mathrm{curl}\, \varphi = 0 \text{ on } \partial D, \ \varphi, \mathrm{curl}^2 \varphi \in L^2_{div}(D, \mathbb{C}^3, dx/|q|) \right\} \tag{5.80}$$

equipped with the inner product

$$(\varphi, \psi)_W = \iint\limits_{D} \left[\mathrm{curl}^2 \varphi \cdot \mathrm{curl}^2 \overline{\psi} + \varphi \cdot \overline{\psi} \right] \frac{dx}{|q|}.$$

The trace $\nu \times \mathrm{curl}\, \varphi$ is well defined since $\mathrm{curl}\, \varphi \in H(\mathrm{curl}, D)$.

Lemma 5.18 *The space W is a Hilbert space and compactly imbedded in $L^2(D, \mathbb{C}^3, dx/|q|)$.*

Proof: The completeness of W with respect to the norm $\|\varphi\|_W = (\varphi, \varphi)_W^{1/2}$ follows from Green's formula in the form

$$\| \mathrm{curl}\, \varphi \|^2_{L^2(D)} = \iint\limits_{D} \mathrm{curl}^2 \varphi \cdot \overline{\varphi}\, dx \leq \| \mathrm{curl}^2 \varphi \|_{L^2(D)} \|\varphi\|_{L^2(D)} \leq \|q\|_\infty \|\varphi\|^2_W \tag{5.81}$$

applied to $\varphi_n - \varphi_m$ where $\{\varphi_n\}$ is a Cauchy sequence in W. To show that W is compactly imbedded in $L^2(D, \mathbb{C}^3, dx/|q|)$ we note first that W is boundedly imbedded in $H^1(D, \mathbb{C}^3)$. This follows by a well-known result (see, e.g., Lemma 5.4.3 of [150], Remark 3.48 of [146], or [170]) since for $\varphi \in W$ it holds that φ, $\mathrm{div}\, \varphi$, and $\mathrm{curl}\, \varphi$ are bounded in $L^2(D)$ and $\nu \times \varphi = 0$ on ∂D. (Note that we assumed the boundary of D to be smooth enough.) The boundedness of $\mathrm{curl}\, \varphi$ follows again from estimate (5.81).

Let now $\{\varphi_j\}$ be a sequence in W with $\|\varphi_j\|_W \leq 1$. Then there exists a subsequence which converges weakly in W to some $\varphi \in X$, thus also weakly in $H^1(D, \mathbb{C}^3)$ by the previous remark, and thus in the norm of $L^2(D, \mathbb{C}^3)$ by the compact imbedding of $H^1(D, \mathbb{C}^3)$ in $L^2(D, \mathbb{C}^3)$. Now we use the same arguments as in the proof of Lemma 4.13. In particular, estimate (4.45) of Section 4.5 yields[3]

$$
\begin{aligned}
\|\varphi_j - \varphi\|^2_{L^2(D, dx/|q|)} &\leq \|\varphi_j - \varphi\|^2_{L^2(D \setminus U_\delta, dx/|q|)} \\
&\quad + c\,\tilde{\gamma}(\delta) \left[\|\varphi_j - \varphi\|^2_{L^2(U_{\delta_0}, dx/|q|)} + \|\varphi_j - \varphi\|^2_{H^1(D)} \right] \\
&\leq \|\varphi_j - \varphi\|^2_{L^2(D \setminus U_\delta, dx/|q|)} + \tilde{c}\,\tilde{\gamma}(\delta)\, \|\varphi_j - \varphi\|^2_W,
\end{aligned}
$$

for some $\delta_0 > 0$. Here, $\tilde{\gamma}(\delta)$ is given in (5.71) and U_δ is defined by $U_\delta = \{x \in D : d(x, \partial D) < \delta\}$. Now we proceed as in the proof of Lemma 4.13. Indeed, let $\varepsilon > 0$ be

---

[3] Note that no boundary condition has to hold for this estimate!

arbitrary. Choose $\delta > 0$ such that $4\tilde{c}\,\tilde{\gamma}(\delta) \leq \varepsilon/2$. On $D \setminus U_\delta$ the function $|q|$ is bounded below by some $q_0 > 0$. Therefore,

$$\|\varphi_j - \varphi\|^2_{L^2(D,dx/|q|)} \leq \frac{1}{q_0}\,\|\varphi_j - \varphi\|^2_{L^2(D)} + \frac{\varepsilon}{2} \leq \varepsilon$$

for sufficiently large j. $\square$

For the explicit construction of P_k we have to consider the following variational equation: Given $f \in L^2_{div}(D, \mathbb{C}^3, |q|dx)$, determine $\varphi \in W$ such that

$$\iint_D \left[\mathrm{curl}^2\,\varphi - k^2\varphi\right] \cdot \left[\mathrm{curl}^2\,\overline{\psi} - k^2\overline{\psi}\right] \frac{dx}{q} = \iint_D f \cdot \left[\mathrm{curl}^2\,\overline{\psi} - k^2\overline{\psi}\right] dx \qquad (5.82)$$

for all $\psi \in W$. We note that the right-hand side is well defined.

Theorem 5.19 *Let Assumption 5.16 hold and let $k \geq 0$. Then the operator P_k from $L^2(D, \mathbb{C}^3, |q|dx)$ into itself is given by*

$$P_k f = \frac{1}{q}\left[\mathrm{curl}^2\,\varphi - k^2\varphi\right] \qquad (5.83)$$

where $\varphi \in W$ is the unique solution of the variational equation (5.82). The operator-valued mapping $k \mapsto P_k$ from $\mathbb{R}_{\geq 0}$ to $\mathcal{B}\big(L^2(D, \mathbb{C}^3, |q|dx)\big)$ can be extended analytically in a neighborhood of $\mathbb{R}_{\geq 0}$.

Proof: We define σ to be the signum of q. Then σ is constant and either $+1$ or -1. We multiply (5.82) by σ and write it in the form

$$(\varphi, \psi)_W - k^2 a_1(\varphi, \psi) + (k^4 - 1)\, a_2(\varphi, \psi) = b_1(\psi) - k^2 b_2(\psi) \qquad (5.84)$$

for all $\psi \in W$ where

$$a_1(\varphi, \psi) = \iint_D \left[\mathrm{curl}^2\,\varphi \cdot \overline{\psi} + \varphi \cdot \mathrm{curl}^2\,\overline{\psi}\right] \frac{dx}{|q|}, \qquad (5.85)$$

$$a_2(\varphi, \psi) = \iint_D \varphi \cdot \overline{\psi}\, \frac{dx}{|q|}, \qquad (5.86)$$

$$b_1(\psi) = \sigma \iint_D \left[f \cdot \mathrm{curl}^2\,\overline{\psi}\right] dx, \qquad (5.87)$$

$$b_2(\psi) = \sigma \iint_D \left[f \cdot \overline{\psi}\right] dx. \qquad (5.88)$$

Since all of these forms are bounded there exist bounded operators A_j from W into itself with $a_j(\varphi, \psi) = (A_j\varphi, \psi)_W$ for all $\varphi, \psi \in W$ and $j = 1, 2$. Furthermore, there exists

$r_1, r_2 \in W$ with $b_j(\psi) = (r_j, \psi)_W$ for all $\psi \in W$ and $j = 1, 2$. Equation (5.84) can then be written as

$$\varphi - k^2 A_1 \varphi + (k^4 - 1) A_2 \varphi = r_1 - k^2 r_2. \tag{5.89}$$

The compactness of A_1 and A_2 is shown by the same arguments as in the proof of Theorem 4.14, and we do not repeat the arguments. Therefore, existence of a solution of (5.89) follows provided uniqueness holds. Let $\varphi \in W$ be a solution of (5.89) for real $k \geq 0$ and $r_1 = r_2 = 0$. Then φ solves (5.82) for $f = 0$. Setting $\psi = \varphi$ yields $\text{curl}^2 \varphi - k^2 \varphi = 0$ in D. The boundary conditions $\nu \times \varphi = \nu \times \text{curl}\,\varphi = 0$ on ∂D and the fact that $\text{div}\,\varphi = 0$ imply that φ vanishes in D. Therefore, for any real and non-negative wavenumber k the operator $I - k^2 A_1 + (k^4 - 1) A_2$ is an isomorphism from W onto itself. Therefore, the variational equation (5.82) is uniquely solvable for all $f \in L^2_{div}(D, \mathbb{C}^3, |q| dx)$. Let g be the right-hand side of (5.83). We have to show that $g = P_k f$. From the definition of g we have by Green's formula that

$$\iint\limits_D g \cdot \psi\, q\, dx = \iint\limits_D \left[\text{curl}^2 \varphi - k^2 \varphi\right] \cdot \psi\, dx = 0$$

for all $\psi \in V$, i.e., $g \in V^\perp$ since the signum of q is constant. Substituting the form of g into (5.82) yields

$$\iint\limits_D (g - f) \cdot \left[\text{curl}^2 \overline{\psi} - k^2 \overline{\psi}\right] dx = 0$$

for all $\psi \in W$ which implies that $g - f \in V$. Therefore, f has the decomposition $f = g + (f - g)$ with $g \in V^\perp$ and $f - g \in V$. This proves that $g = P_k f$.

A perturbation argument using the Neumann series applied to $k = k_0$ for fixed non-negative $k_0 \geq 0$ yields that $A_0 - k^2 A_1 + (k^4 - 1) A_2$ are isomorphisms for $k \in U(k_0)$ where $U(k_0)$ is some open set containing k_0. Furthermore, $k \mapsto (A_0 - k^2 A_1 + (k^4 - 1) A_2)^{-1}$ is analytic. This proves that also P_k can be analytically extended to the union $U = \bigcup_{k_0 \geq 0} U(k_0)$. $\square$

We note that for real $k \geq 0$ the projection operator P_k has norm one with respect to $(\cdot, \cdot)_{L^2(D, |q| dx)}$. Next we study the operator B_k from Lemma 5.17.

Lemma 5.20 *Let Assumption 5.16 hold. Denote by B_0 the operator B_k for $k = 0$. Then $\|B_0\| < 1$ in the operator norm of $L^2(D, \mathbb{C}^3, |q| dx)$. The difference $B_k - B_0$ is compact and depends analytically on $k \in U$ with $U \subset \mathbb{C}$ from Theorem 5.19.*

Proof: Consider the operator T defined by

$$(T\psi)(x) = \text{curl}^2 \iint\limits_{\mathbb{R}^3} \frac{1}{4\pi |x - y|}\, \psi(y)\, dy, \quad x \in \mathbb{R}^3, \tag{5.90}$$

where we understand the curl in the distributional sense. T is bounded as an operator from $L^2(\mathbb{R}^3, \mathbb{C}^3)$ into itself with norm equal to one. Indeed, we assume that $\psi \in S$ where

$\mathcal{S}$ denotes the Schwarz space of C^{∞} – functions which, together with their derivatives of any order, decay faster than any polynomial for $|x| \to \infty$. We consider $h(x) = 1/(4\pi |x|)$ as a (regular, tempered) distribution. Then, denoting the right hand side of (5.90) by $g(x)$ we observe that $g = \mathrm{curl}^2(h * \psi)$ where $h * \psi$ denotes the convolution of the distribution h with the test function $\psi \in \mathcal{S}$ and the curl is understood in the distributional sense.

Application of the Fourier transform

$$(\mathcal{F}\psi)(\xi) = \frac{1}{(2\pi)^{3/2}} \iint_{\mathbb{R}^3} e^{-i\xi \cdot x} \psi(x)\, dx, \quad \xi \in \mathbb{R}^3,$$

extended to tempered distributions, shows that

$$(\mathcal{F}g)(\xi) = \mathcal{F}\big(\mathrm{curl}^2(h * \psi)\big)(\xi) = \big[|\xi|^2 I - \xi\,\xi^{\top}\big]\mathcal{F}(h * \psi)(\xi)$$

$$= (2\pi)^{3/2}\big[|\xi|^2 I - \xi\,\xi^{\top}\big](\mathcal{F}h)(\xi)\,(\mathcal{F}\psi)(\xi)\,.$$

The Fourier transform of the distribution h is given by $(\mathcal{F}h)(\xi) = (2\pi)^{-3/2}|\xi|^{-2}$, i.e.,

$$(\mathcal{F}g)(\xi) = \big[I - \hat{\xi}\,\hat{\xi}^{\top}\big](\mathcal{F}\psi)(\xi)$$

where $\hat{\xi} = \xi/|\xi|$. The unitarity of the Fourier transform yields

$$\|g\|_{L^2(\mathbb{R}^3)} = \|\mathcal{F}g\|_{L^2(\mathbb{R}^3)} = \|\mathcal{F}\psi\|_{L^2(\mathbb{R}^3)} = \|\psi\|_{L^2(\mathbb{R}^3)}$$

where we have also used that $f \mapsto \big[I - \hat{\xi}\,\hat{\xi}^{\top}\big]f$ is a projection operator of norm one. We have therefore shown, that the restriction of T to the space $\mathcal{S}$ is bounded with L^2-norm one. Therefore, it has an extension to $L^2(\mathbb{R}^3, \mathbb{C}^3)$ with L^2-norm one. From $B_0 f = T(qf)\big|_D$ we conclude that

$$\|B_0 f\|_{L^2(D)} \leq \|T(qf)\|_{L^2(\mathbb{R}^3)} \leq \|qf\|_{L^2(D)} \leq \sqrt{\|q\|_\infty}\|f\|_{L^2(D,|q|dx)}$$

and thus

$$\|B_0 f\|_{L^2(D,|q|dx)} \leq \|q\|_\infty \|f\|_{L^2(D,|q|dx)}$$

which proves that $\|B_0\| \leq \|q\|_\infty < 1$ by Assumption 5.16. The compactness and analyticity of $B_k - B_0$ follows from the explicit form of B_k. $\square$

We write (5.79) in the form

$$(I - P_k B_0)\tilde{u} - P_k(B_k - B_0)\tilde{u} = 0 \tag{5.91}$$

and consider this equation in $L^2(D, \mathbb{C}^3, |q|dx)$ (with the norm induced by $(\cdot, \cdot)_{L^2(D,|q|dx)}$).

Now we observe that $\|P_k B_0\| < 1$. Therefore, $I - P_k B_0$ is invertible and (5.91) is equivalent to

$$\tilde{u} - (I - P_k B_0)^{-1} P_k(B_k - B_0)\tilde{u} = 0\,. \tag{5.92}$$

Furthermore, the operators $(I - P_k B_0)^{-1} P_k(B_k - B_0)$ are compact and depend analytically on $k \in U$. This is the situation where we can apply the analytic Fredholm theory, see [43], Theorem 8.19. Since $k = 0$ is certainly no eigenvalue we have the following result:

Theorem 5.21 *Let Assumption 5.16 be satisfied. Then the set of interior transmission eigenvalues is discrete.*

Remark: In this chapter we restricted ourselves to non-magnetic and isotropic media. The same technique – for the direct and the inverse scattering problem – can be used to treat the more general case where both, ε and μ, vary with x and are matrix-valued. We refer to [121] and [79]. In the investigation of the Linear Sampling Method (compare Section 7.2 of Chapter 7 for an introduction of the Linear Sampling Method for a scalar scattering problem) the corresponding interior transmission boundary value problem plays an important role. This is the problem where the homogeneous boundary conditions (5.48) have to be replaced by inhomogeneous boundary conditions. Also these problems are treated in [121, 79, 80].

6

The factorization method in impedance tomography

6.1 Derivation of the models

To derive a mathematical model for the problem of electrical impedance tomography we begin again with Maxwell's equations in the frequency domain (compare (5.15) and (5.16)).

$$\operatorname{curl} E = i\omega\mu H, \qquad \operatorname{curl} H = (\sigma - i\omega\varepsilon) E \quad \text{in } B. \tag{6.1}$$

Again, we assume that no interior currents are present. In order to simplify these equations for low frequencies we perform a scaling analysis (cf. [32]). Indeed, let $x = [x]\tilde{x}$ and $H(x) = H([x]\tilde{x}) = [H]\tilde{H}(\tilde{x})$, $E(x) = E([x]\tilde{x}) = [E]\tilde{E}(\tilde{x})$ where $[x], [H]$, $[E] \in \mathbb{R}$ denote the (constant) scaling parameters of x, H, and E, respectively, in which we measure these vectorial quantities. Noting that $\operatorname{curl}_{\tilde{x}} \tilde{E}(\tilde{x}) = \frac{[x]}{[E]} \operatorname{curl}_x E(x)$ and analogous for $\operatorname{curl}_{\tilde{x}} \tilde{H}(\tilde{x})$ we conclude that

$$\operatorname{curl}_{\tilde{x}} \tilde{E} = i\omega\mu \, \frac{[H][x]}{[E]} \, \tilde{H}, \qquad \operatorname{curl}_{\tilde{x}} \tilde{E} = (\sigma - i\omega\varepsilon) \, \frac{[E][x]}{[H]} \, \tilde{E}.$$

As seen from the physical units we note that $\omega\mu \, \frac{[H][x]}{[E]}$ and $(\sigma - i\omega\varepsilon) \, \frac{[E][x]}{[H]}$ are dimensionless. By $\overline{\sigma}$ we denote the mean of σ. If we choose scaling factors for x, H, and E such that $\overline{\sigma} \frac{[E][x]}{[H]} = 1$ we arrive at

$$\operatorname{curl}_{\tilde{x}} \tilde{E} = i\omega\mu \, \overline{\sigma}[x]^2 \tilde{H}, \qquad \operatorname{curl}_{\tilde{x}} \tilde{E} = (\sigma/\overline{\sigma} - i\omega\varepsilon/\overline{\sigma}) \, \tilde{E}. \tag{6.2}$$

In Figure 6.1, taken from [16], we list some electrical properties of tissue. As the magnetic permeability μ of non-ferromagnetic materials is close to the free space permeability $\mu_0 = 4\pi * 10^{-7} \frac{VS}{Am}$ (in SI units) the term $\omega\mu \, \overline{\sigma} \, [x]^2$ is of order $10^{-9} * \omega$ if the extension of the body B is of magnitude $10 \, cm$. Therefore, it is neglible for small frequencies ω. Although small compared to $\sigma/\overline{\sigma}$, the term $\omega\varepsilon/\overline{\sigma}$ is of order $10^{-6} * \omega$ and, therefore, has to be taken into account only for medium range frequencies.

Neglecting the term $\omega\mu \, \overline{\sigma} \, [x]^2$ the first equation of (6.2) yields the existence of a potential u such that $\tilde{E} = \nabla u$. Substituting this into the second equation of (6.2) and

Tissue	σ (in A/Vm)	ε (in $10^{-6}\,As/Vm$)
Lung	0.105	0.22
Muscle	0.132	0.49
Liver	0.146	0.49
Heart	0.167	0.88
Fat	<0.1	0.18

Figure 6.1 Electrical properties of tissue at $\omega = 10\,kHz$

taking the divergence yields the scalar elliptic equation

$$\operatorname{div}\big[\gamma(x)\,\nabla u(x)\big] = 0 \quad \text{in } B. \tag{6.3}$$

Here, the complex-valued admittivity function $\gamma = \gamma(x)$ is defined by

$$\gamma(x) = \sigma/\bar{\sigma} - i\omega\varepsilon/\bar{\sigma}. \tag{6.4}$$

There are different approaches of increasing complexity how to model the electrodes. We refer to [16, 32, 34, 107, 174] for the Gap, Shunt, and Complete Models but restrict ourselves to the simplest model (the "continuum model"). In this model, the normal current density is prescribed at the boundary ∂B of the domain B, i.e.,

$$\gamma\,\frac{\partial u}{\partial v} = f \quad \text{on } \partial B. \tag{6.5}$$

The exact assumptions on γ and f will be formulated below. At this place we only mention that the total flux $\int_{\partial B} \gamma\,\partial u/\partial v\,dx = \int_{\partial B} f\,dx$ has to vanish for any solution u. This follows immediately from (6.3) and the divergence theorem.

One models the anisotropic case by simply letting $\gamma = \gamma(x) \in \mathbb{C}^{3\times 3}$ be a (complex-valued) matrix. Then $\gamma\,\nabla u$ in (6.3) has to be understood as the matrix-vector product of $\gamma(x)$ and $\nabla u(x)$ while the boundary condition (6.5) has to be replaced by

$$\partial_\gamma u := v \cdot (\gamma\,\nabla u) = v^\top \gamma\,\nabla u = f \quad \text{on } \partial B.$$

Here and in the following we use the matrix-vector notation and write $x^\top y$ for $x \cdot y = \sum_{j=1}^{d} x_j y_j$. For complex-valued vectors we also write $x^* y = \bar{x}^\top y = \sum_{j=1}^{d} \bar{x}_j y_j$.

6.2   The Neumann-to-Dirichlet operator and the inverse problem

In this section we will study the following boundary value problem. Let $B \subset \mathbb{R}^d$ where $d = 2$ or $d = 3$ be a bounded and connected domain with sufficiently smooth[1] boundary ∂B. Let $f \in H^{-1/2}(\partial B)$ be given such that $\langle f, \bar{1}\rangle = 0$ where $\bar{1} \in H^{1/2}(\partial B)$ denotes

---

[1] For this and the next section it is sufficient that the boundary is Lipschitz continuous. What we need is the validity of the trace and extension theorems. In Section 6.4 we will use a Green's function and more regularity will be required.

the function with values constant 1 and let $\langle \cdot, \cdot \rangle$ denote again the dual bracket in the dual system[2] $\langle H^{-1/2}(\partial B), H^{1/2}(\partial B) \rangle$. The function f represents the flux applied to the boundary ∂B. We define the closed subspaces $H_{\diamond}^{\pm 1/2}(\partial B)$ and $H_{\diamond}^{1}(B)$ of $H^{\pm 1/2}(\partial B)$ and $H^{1}(B)$, respectively, by

$$H_{\diamond}^{1/2}(\partial B) = \left\{ f \in H^{1/2}(\partial B) : \int_{\partial B} f \, ds = 0 \right\}, \tag{6.6}$$

$$H_{\diamond}^{-1/2}(\partial B) = \left\{ f \in H^{-1/2}(\partial B) : \langle f, \bar{1} \rangle = 0 \right\}, \tag{6.7}$$

$$H_{\diamond}^{1}(B) = \left\{ f \in H^{1}(B) : \int_{\partial B} f(x) \, ds = 0 \right\}. \tag{6.8}$$

Furthermore, let $\gamma \in L^{\infty}\left(B, \mathbb{C}^{d \times d}\right)$ be a matrix-valued complex function such that its imaginary part is negative semi-definite and its real part is uniformly positive-definite, i.e., there exists $c_0 > 0$ such that

$$\mathrm{Im}\left[z^* \gamma(x) z\right] \leq 0 \quad \text{and} \quad \mathrm{Re}\left[z^* \gamma(x) z\right] \geq c_0 \, |z|^2 \tag{6.9}$$

for all $z \in \mathbb{C}^d$ and almost all $x \in B$. We think of γ being of the form $\gamma = \sigma - i\omega\varepsilon$ with conductivity σ, frequency ω, and permittivity ε.

With these data $B \subset \mathbb{R}^d$, $\gamma \in L^{\infty}\left(B, \mathbb{C}^{d \times d}\right)$, and $f \in H_{\diamond}^{-1/2}(\partial B)$ we consider (6.3) and (6.5), i.e.,

$$\mathrm{div}\left[\gamma \, \nabla u\right] = 0 \quad \text{in } B, \qquad \partial_\gamma u = f \quad \text{on } \partial B. \tag{6.10}$$

Again, as in all previous chapters the solution $u \in H_{\diamond}^{1}(B)$ has to be understood in the variational sense. Green's formula for this anisotropic case is formulated as (for sufficiently smooth v, w in some bounded domain D with sufficiently smooth boundary ∂D)

$$\int_{\partial D} v \, \partial_\gamma w \, ds = \int_{\partial D} \left[v \, \gamma \nabla w\right]^\top v \, ds = \iint_{D} \mathrm{div}\left[v \, \gamma \nabla w\right] dx \tag{6.11}$$

$$= \iint_{D} \left[\nabla v^\top \gamma \nabla w + v \, \mathrm{div}(\gamma \nabla w)\right] dx.$$

Therefore, $u \in H_{\diamond}^{1}(B)$ is a variational solution of (6.10) if u satisfies

$$\iint_{B} \nabla \psi(x)^* \gamma(x) \nabla u(x) \, dx = \langle f, \psi \rangle \quad \text{for all } \psi \in H_{\diamond}^{1}(B). \tag{6.12}$$

---

[2] Again, we understand the dual system as a sesqui-linear form. In this sense, it extends the inner product in $L^2(\partial B)$.

On the right hand side $\psi \in H^{1/2}(\partial B)$ denotes the trace of the test function $\psi \in H_\diamond^1(B)$. Although we have used this kind of Neumann boundary value problems already in the proof of Theorem 5.12 we recall the arguments which lead to uniqueness and existence of a solution.

First, we note that the right hand side $\psi \mapsto \langle f, \psi \rangle$ of (6.12) is a bounded conjugate-linear functional on $H_\diamond^1(B)$ because of

$$\left| \langle f, \psi \rangle \right| \leq \|f\|_{H^{-1/2}(\partial B)} \|\psi\|_{H^{1/2}(\partial B)} \leq c \, \|f\|_{H^{-1/2}(\partial B)} \|\psi\|_{H^1(B)}$$

for some $c > 0$ which depends only on B. Furthermore, here, and in the following, $\| \cdot \|_{L^\infty(B)}$ denotes the norm

$$\|\gamma\|_{L^\infty(B)} = \mathrm{essup}_{x \in B} \sqrt{\sum_{j,k=1}^d |\gamma(x)_{jk}|^2}.$$

in $L^\infty(B, \mathbb{C}^{d \times d})$. Then the estimate

$$\left| \iint_B b(x)^* \big[\gamma(x) d(x) \big] \, dx \right| \leq \iint_B |b(x)| \, |\gamma(x) d(x)| \, dx \leq \|\gamma\|_{L^\infty(B)} \|b\|_{L^2(B)} \|d\|_{L^2(B)}$$

holds for all vector fields $b, d \in L^2(B, \mathbb{C}^d)$. Application of this estimate yields boundedness of the Hermitean form

$$a(u, \psi) = \iint_B \nabla \psi(x)^* \gamma(x) \nabla u(x) \, dx, \quad u, \psi \in H_\diamond^1(B),$$

in $H_\diamond^1(B)$ because

$$\left| a(u, \psi) \right| \leq \|\gamma\|_{L^\infty(B)} \|u\|_{H^1(B)} \|\psi\|_{H^1(B)} \quad \text{for all } \psi \in H_\diamond^1(B).$$

Finally, a is coercive because

$$\mathrm{Re} \iint_B \nabla \psi(x)^* \gamma(x) \nabla \psi(x) \, dx \geq c_0 \iint_B \left| \nabla \psi(x) \right|^2 dx \geq \hat{c} \, \|\psi\|_{H^1(B)}^2 \tag{6.13}$$

for all $\psi \in H_\diamond^1(B)$. The latter estimate is known as inequality of Poincaré, compare (5.64). Therefore, by the theorem of Lax and Milgram (see [146]) there exists a unique solution $u \in H_\diamond^1(B)$ of (6.12) for every $f \in H_\diamond^{-1/2}(\partial B)$ and

$$\|u\|_{H^1(B)} \leq \frac{c}{\hat{c}} \, \|\gamma\|_{L^\infty(B)} \, \|f\|_{H^{-1/2}(\partial B)} \tag{6.14}$$

The unique solvability guarantees the existence of the *Neumann-to-Dirichlet operator* $\Lambda : H_\diamond^{-1/2}(\partial B) \longrightarrow H_\diamond^{1/2}(\partial B)$ which assigns to each $f \in H_\diamond^{-1/2}(\partial B)$ the trace $u|_{\partial B} \in H_\diamond^{1/2}(\partial B)$ of the solution $u \in H_\diamond^1(B)$ of (6.12). The following theorem lists the most important properties of Λ.

Theorem 6.1 *Let $B \subset \mathbb{R}^d$ and $\gamma \in L^\infty(B)$ satisfy the conditions formulated at the beginning of the section.*

(a) Then Λ is a bounded isomorphism from $H_\diamond^{-1/2}(\partial B)$ onto $H_\diamond^{1/2}(\partial B)$.

(b) If, in addition, γ is Hermitian (i.e., $\gamma(x)^* = \gamma(x)$) and uniformly positive definite (in the sense of the second estimate of (6.9)) then Λ is self-adjoint and coercive, i.e., there exists $c > 0$ with

$$\langle g, \Lambda f \rangle = \overline{\langle f, \Lambda g \rangle} \quad \text{and} \quad \langle f, \Lambda f \rangle \geq c\|f\|_{H^{-1/2}(\partial B)}^2 \quad \text{for all } f, g \in H_\diamond^{-1/2}(\partial B).$$

(c) For all $C_0 > c_0 > 0$ the mapping $\gamma \mapsto \Lambda_\gamma$ depends Lipschitz-continuously on γ in the set

$$U := \left\{ \gamma \in L^\infty(B, \mathbb{C}^{d \times d}) : \|\gamma\|_{L^\infty(B)} \leq C_0, \ \mathrm{Re}\big[z^* \gamma(x) z\big] \geq c_0|z|^2 \text{ for all } z \text{ a.e. on } B \right\},$$

i.e., there exists $c > 0$ with

$$\big\|(\Lambda_\gamma - \Lambda_{\tilde\gamma})f\big\|_{H^{1/2}(\partial B)} \leq c \, \|\gamma - \tilde\gamma\|_{L^\infty(B)} \|f\|_{H^{-1/2}(\partial B)}$$

$$\text{for all } \gamma, \tilde\gamma \in U, \ f \in H^{-1/2}(\partial B).$$

Proof: (a) It is easy to check that the inverse $\Lambda^{-1} : H_\diamond^{1/2}(\partial B) \longrightarrow H_\diamond^{-1/2}(\partial B)$ is given by $\Lambda^{-1}g = \partial_\gamma v$ where $v \in H_\diamond^1(B)$ denotes the unique solution of the interior Dirichlet boundary value problem with boundary values $v = g$ on ∂B. Of course, $\partial_\gamma v \in H_\diamond^{-1/2}(\partial B)$ has to be understood as the form

$$\langle \partial_\gamma v, \psi \rangle = \iint_B \nabla\psi(x)^* \gamma(x)\, \nabla v(x)\, dx \quad \text{for all } \psi \in H_\diamond^{1/2}(\partial B)$$

where on the right-hand side $\psi \in H_\diamond^1(B)$ is any extension of ψ.

(b) This is seen by Green's theorem from (6.12). Indeed, for $f, g \in H_\diamond^{-1/2}(\partial B)$ let $u, v \in H_\diamond^1(B)$ be the corresponding solutions of the interior Neumann problem (6.10). Then

$$\langle g, \Lambda f \rangle = \langle g, u \rangle = \iint_B \nabla u(x)^* \gamma(x)\, \nabla v(x)\, dx,$$

and, analogously,

$$\langle f, \Lambda g \rangle = \langle f, v \rangle = \iint_B \nabla v(x)^* \gamma(x)\, \nabla u(x)\, dx = \iint_B \nabla u(x)^\top \gamma(x)^\top\, \nabla\overline{v(x)}\, dx.$$

Taking the complex conjugate and noting that $\gamma(x)$ is Hermitian yields the self-adjointness and, furthermore, for $g = f$ by the inequality (6.13) of Poincaré

$$\langle f, \Lambda f \rangle = \iint_B \nabla u(x)^* \gamma(x) \, \nabla u(x) \, dx \geq c_0 \iint_B |\nabla u(x)|^2 dx \geq c_1 \, \|u\|^2_{H^1(B)}.$$

The boundedness of the trace operator and of Λ^{-1} yield

$$\|u\|_{H^1(B)} \geq c_2 \, \|u\|_{H^{1/2}(\partial B)} = c_2 \, \|\Lambda f\|_{H^{1/2}(\partial B)} \geq c_3 \, \|f\|_{H^{-1/2}(\partial B)}$$

which proves the desired estimate.

(c) Let $\gamma, \tilde{\gamma} \in U$. We denote the corresponding solutions by u and $\tilde{u}$, respectively, and the bilinear forms by a and $\tilde{a}$, respectively. Then

$$a(u, \psi) = \langle f, \psi \rangle \quad \text{and} \quad \tilde{a}(\tilde{u}, \psi) = \langle f, \psi \rangle \quad \text{for all } \psi \in H^1_\diamond(B),$$

and thus by taking the difference

$$\iint_B \nabla \psi(x)^* \gamma(x) \, \nabla \big[u(x) - \tilde{u}(x)\big] \, dx = \iint_B \nabla \psi(x)^* \big[\tilde{\gamma}(x) - \gamma(x)\big] \, \nabla \tilde{u}(x) \, dx. \quad (6.15)$$

The right-hand side defines an conjugate-linear functional with respect to ψ and can be estimated by

$$\left| \iint_B \nabla \psi(x)^* \big[\tilde{\gamma}(x) - \gamma(x)\big] \, \nabla \tilde{u}(x) \, dx \right| \leq \|\gamma - \tilde{\gamma}\|_{L^\infty(B)} \|\tilde{u}\|_{H^1(B)} \|\psi\|_{H^1(B)}.$$

By (6.14) we conclude that $\|\tilde{u}\|_{H^1(B)} \leq c \, \|f\|_{H^{-1/2}(\partial B)}$ where $c > 0$ depends only on c_0 through the inequality (6.13) of Poincaré and on C_0. Application of the theorem of Lax–Milgram to (6.15) yields

$$\|u - \tilde{u}\|_{H^1(B)} \leq c \, \|\gamma - \tilde{\gamma}\|_{L^\infty(B)} \|f\|_{H^{-1/2}(\partial B)}$$

for some $c > 0$ which proves the theorem. $\square$

In electrical impedance tomography the admittivity function γ is unknown, and it is to be determined from simultaneous measurements of boundary values u ("boundary voltages") and $\partial_\gamma u$ ("current densities"). In terms of the Neumann-to-Dirichlet operator Λ the *inverse problem* of electrical impedance tomography is to determine the function γ from Λ.

The first question to address is whether or not the data, i.e., the operator Λ, determines the function γ uniquely. This turns out to be an extremely difficult problem. Simple arguments using a change of variables show that a matrix-valued function γ is, in general, not uniquely determined. The best one can hope that it is unique up to a change of variables. So far, a complete and "optimal" answer has been given only very recently in [12, 13] by K. Astala, M. Lassas and L. Päivärinta for the plane problem (i.e., $d = 2$) and real-valued (scalar and matrix-valued) L^∞–functions γ.

The situation is less well understood in $\mathbb{R}^3$. For scalar and sufficiently smooth functions (e.g., $\gamma \in C^2(\overline{B})$) the change of variables $u = \gamma^{1/2}\tilde{u}$ transforms the differential equation into the equation of Schrödinger type

$$\Delta\tilde{u} - \frac{\Delta\gamma^{1/2}}{\gamma^{1/2}}\,\tilde{u} = 0.$$

This is the kind of equation which has been considered in Chapter 4 already. In particular, uniqueness holds in $\mathbb{R}^3$ for scalar functions $\gamma \in C^2(\overline{B})$. We refer to the survey article [16] for further results and references on the question of uniqueness.

The second important problem is stability of the inverse problem. Note that by Theorem 6.1 the direct problem is Lipschitz-continuous. The inverse problem, however, is ill-posed in the sense of Hadamard, i.e., large changes in γ in the interior can result in small changes in Λ (with respect to the operator norm of the space $\mathcal{B}\big(H_\diamond^{-1/2}(\partial B), H_\diamond^{1/2}(\partial B)\big)$ of linear and bounded operators from $H_\diamond^{-1/2}(\partial B)$ into $H_\diamond^{1/2}(\partial B)$. On the boundary, however, the determination of γ from Λ is stable. Indeed, in [177, 180] it is shown that for real-valued, continuous, and strictly positive functions γ_1, γ_2 the following estimate holds:

$$\|\gamma_1 - \gamma_2\|_{L^\infty(\partial B)} \leq c\,\|\Lambda_1 - \Lambda_2\|_{\mathcal{B}(H_\diamond^{-1/2}(\partial B), H_\diamond^{1/2}(\partial B))}.$$

For the interior, Alessandrini has shown in [1] logarithmic stability under some bounds on γ and the smoothness assumption $\gamma \in H^{2+s}(B)$ for some $s > d/2$ and $\delta > 0$:

$$\|\gamma_1 - \gamma_2\|_{L^\infty(B)} \leq c\,\log\big(\|\Lambda_1 - \Lambda_2\|_{\mathcal{B}(H_\diamond^{-1/2}(\partial B), H_\diamond^{1/2}(\partial B))}\big)^{-\delta}.$$

In light of these results we will study a more modest problem which is accessible to the Factorization Method. We assume that the unknown function γ is the perturbation of a known "background" admittivity γ_0 which is real-valued. For simplicity, we assume furthermore that it is scalar and even constant. Therefore, we assume that $\gamma \in L^\infty\big(B, \mathbb{C}^{d\times d}\big)$ has the form

$$\gamma(x) = \begin{cases} I + q(x), & x \in D, \\ I, & x \in B \setminus D, \end{cases} \tag{6.16}$$

for some domain D with $\overline{D} \subset B$ such that $B \setminus \overline{D}$ is connected and some (symmetric) matrix-valued function $q \in L^\infty(D, \mathbb{C}^{d\times d})$ with support $\overline{D}$. Here, and in the following, I denotes the identity matrix. Assumption (6.9) is equivalent to the existence of $c_0 > 0$ such that

$$\mathrm{Im}[z^*q(x)\,z] \leq 0 \quad \text{and} \quad |z|^2 + \mathrm{Re}[z^*q(x)\,z] \geq c_0\,|z|^2 \tag{6.17}$$

for all $z \in \mathbb{C}^d$ and almost all $x \in D$. Further assumptions on q will be added below in Assumption 6.4.

The *inverse problem* which we will treat by the factorization method is to determine the support D of q from the Neumann-to-Dirichlet operator Λ.

6.3   Factorization of the Neumann-to-Dirichlet operator

We recall from the previous section that the Neumann-to-Dirichlet operator $\Lambda :$ $H_\diamond^{-1/2}(\partial B) \to H_\diamond^{1/2}(\partial B)$ is given $\Lambda f = u|_{\partial D}$ where $u \in H_\diamond^1(B)$ solves the boundary value problem (6.10), i.e., the variational equation (6.12) for γ given by (6.16). The particular case that $\gamma \equiv I$ corresponds to the operator Λ_0 which is therefore characterized by $\Lambda_0 f = u_0|_{\partial D}$ where $u_0 \in H_\diamond^1(B)$ solves the boundary value problem

$$\Delta u_0 = 0 \quad \text{in } B, \qquad \frac{\partial u_0}{\partial \nu} = f \quad \text{on } \partial B. \tag{6.18}$$

The basis of the factorization method is the following factorization of the operator $\Lambda - \Lambda_0$. We define the operators

$$G : H_\diamond^{-1/2}(\partial D) \to H_\diamond^{1/2}(\partial B) \qquad \text{and} \quad T : H_\diamond^{1/2}(\partial D) \to H_\diamond^{-1/2}(\partial D)$$

as follows:

For $\psi \in H_\diamond^{-1/2}(\partial D)$ we set $G\psi = v|_{\partial B}$ where $v \in H^1(B \setminus \overline{D})$ with $v|_{\partial B} \in H_\diamond^{1/2}(\partial B)$ solves the following boundary value problem in $B \setminus \overline{D}$:

$$\Delta v = 0 \quad \text{in } B \setminus \overline{D}, \qquad \frac{\partial v}{\partial \nu} = \psi \quad \text{on } \partial D, \qquad \frac{\partial v}{\partial \nu} = 0 \quad \text{on } \partial B. \tag{6.19}$$

For $h \in H_\diamond^{1/2}(\partial D)$ we set $Th = \partial w_+ / \partial \nu$ on ∂D where $w \in H^1(B \setminus \overline{D}) \cap H^1(D)$ with $w|_{\partial B} \in H_\diamond^{1/2}(\partial B)$ solves the boundary value problem of transmission type

$$\text{div}(\gamma \, \nabla w) = 0 \quad \text{in } B \setminus \partial D, \qquad \frac{\partial w}{\partial \nu} = 0 \quad \text{on } \partial B, \tag{6.20}$$

$$\frac{\partial w_+}{\partial \nu} - \partial_\gamma w_- = 0 \quad \text{on } \partial D, \qquad w_+ - w_- = h \quad \text{on } \partial D. \tag{6.21}$$

Here, we denote again by $w_\pm$ the trace of w from the exterior and interior, respectively. The variational form of (6.20) and (6.21) is to find $w \in H^1(B \setminus \overline{D}) \cap H^1(D)$ with $w|_{\partial B} \in H_\diamond^{1/2}(\partial B)$ and $w_+ - w_- = h$ such that

$$\iint\limits_B \nabla \psi^* \, \gamma \, \nabla w \, dx = 0 \quad \text{for all } \psi \in H_\diamond^1(B). \tag{6.22}$$

By T_0 we denote the operator T when γ is replaced by I. We note that T and T_0 are even defined and bounded as operators from $H^{1/2}(\partial D)$ into $H_\diamond^{-1/2}(\partial D)$.

Theorem 6.2 *The following factorization holds:*

$$\Lambda - \Lambda_0 = G \, (T - T_0) \, G^* \tag{6.23}$$

where G and T are defined by (6.19) and (6.20), (6.21), respectively.

Furthermore, the operator T_0 is self-adjoint, and the adjoint $T^* : H_\diamond^{1/2}(\partial D) \to H_\diamond^{-1/2}(\partial D)$ of T is given by $T^*h = \partial w_+ / \partial \nu$ on ∂D where $w \in H^1(B \setminus \overline{D}) \cap H^1(D)$ with

$w|_{\partial B} \in H_\diamond^{1/2}(\partial B)$ *solves the boundary value problem*

$$\mathrm{div}(\gamma^* \nabla w) = 0 \quad \text{in } B \setminus \partial D, \qquad \frac{\partial w}{\partial \nu} = 0 \quad \text{on } \partial B, \tag{6.24}$$

$$\frac{\partial w_+}{\partial \nu} - \partial_{\gamma^*} w_- = 0 \quad \text{on } \partial D, \qquad w_+ - w_- = h \quad \text{on } \partial D. \tag{6.25}$$

In particular, T is self-adjoint if γ is Hermitian.

Proof: First, we prove that T^* has the given form. Indeed, let $Th = \partial w_+/\partial \nu = \partial_\gamma w_-$ and let v be the solution of (6.24) and (6.25) for g instead of h. By Green's formula (6.11) we have for sufficiently smooth g, h:

$$\langle Th, g \rangle - \overline{\langle \partial v_+/\partial \nu, h \rangle} = \int_{\partial D} \left\{ \frac{\partial w_+}{\partial \nu} \left[\overline{v}_+ - \overline{v}_- \right] - [w_+ - w_-] \frac{\partial \overline{v}_+}{\partial \nu} \right\} ds$$

$$= \int_{\partial B} \left\{ \overline{v} \frac{\partial w}{\partial \nu} - w \frac{\partial \overline{v}}{\partial \nu} \right\} ds - \int_{\partial D} \left\{ \overline{v}_- \, \partial_\gamma w_- - w_- \, \partial_{\gamma^\top} \overline{v}_- \right\} ds$$

$$= \iint_{D} \left[\nabla \overline{v} \cdot (\gamma \nabla w) - \nabla w \cdot (\gamma^\top \nabla \overline{v}) \right] dx = 0$$

which proves the desired form of T^*.

By the definitions of $\Lambda - \Lambda_0$ and G we observe that $(\Lambda - \Lambda_0)\varphi = (u - u_0)|_{\partial B} = G\psi$ where $\psi = \partial(u - u_0)_+/\partial \nu$ on ∂D. We introduce the auxiliary operator $L : H_\diamond^{-1/2}(\partial B) \to H_\diamond^{-1/2}(\partial D)$ by $Lf = \partial u_+/\partial \nu$ where u solves (6.10). Analogously, L_0 is defined. Then we observe that $\Lambda - \Lambda_0 = G(L - L_0)$. By two applications of Green's formula as above it is easily shown that the adjoint $L^* : H_\diamond^{1/2}(\partial D) \to H_\diamond^{1/2}(\partial B)$ of L is given by $L^* h = w|_{\partial B}$ where w solves (6.24) and (6.25). Therefore, $L^* h = w|_{\partial B} = G T^* h$. Analogously, $L_0^* = G T_0^*$ and thus $L - L_0 = (T - T_0) G^*$. Substituting this into $\Lambda - \Lambda_0 = G(L - L_0)$ yields the assertion. $\square$

It is well known that the operator $\Lambda - \Lambda_0$ is compact from $H_\diamond^{-1/2}(\partial B)$ into $H_\diamond^{1/2}(\partial B)$. This follows also from the factorization of Theorem 6.2 and the following theorem. Its proof is quite standard and therefore omitted.

Theorem 6.3 *The operator $G : H_\diamond^{-1/2}(\partial D) \to H_\diamond^{1/2}(\partial B)$ is compact and one-to-one with dense range.*

The operator G corresponds to the data-to-pattern operators for the exterior problems in scattering theory. As in the previous chapters the range of G depends in a very specific way on the support D of q. In Theorem 6.6 below we will construct functions ϕ_z, parametrized by points $z \in \mathbb{R}^d$, with the property that z belongs to D if, and only if, ϕ_z belongs to the range of G. Therefore, it is the aim to characterize the range of G by properties which depend on the data $\Lambda - \Lambda_0$ only. From the functional analytic point of view this is the situation studied already in the previous chapters.

6.4 Characterization of the inclusion

The factorization of $\Lambda - \Lambda_0$ in the form (6.23) suggests the application of the factorization method. As we will see below, in this case only the self-adjoint form of Theorem 1.21 is needed which we had formulated in Corollary 1.22. In order to apply this corollary to (6.23) it is again necessary that $T - T_0$ satisfies certain coercivity conditions. For real admittivities γ it is well known that Λ depends monotonically on γ in the sense that

$$z^* \left(\gamma(x) - \tilde{\gamma}(x) \right) z \geq 0 \quad \text{for all } z \in \mathbb{C}^d \quad \text{a.e. on } B$$

implies that $\langle \varphi, \Lambda \varphi \rangle \leq \langle \varphi, \tilde{\Lambda} \varphi \rangle$ for all φ. We prove a similar result for the operators T and T_0 for general admittivities $\gamma \in L^2(B, \mathbb{C}^{d \times d})$ which satisfies the following assumption which is stronger than (6.17).

Assumption 6.4 *Let $B \subset \mathbb{R}^d$ be a connected domain with C^2-boundary ∂B. Let $\gamma \in L^\infty(B, \mathbb{C}^{d \times d})$ has the form*

$$\gamma(x) = \begin{cases} I + q(x), & x \in D, \\ I, & x \in B \setminus D, \end{cases}$$

for some domain D with $\overline{D} \subset B$ and some matrix-valued function $q \in L^\infty(D, \mathbb{C}^{d \times d})$ with support $\overline{D}$. Furthermore, we assume that $B \setminus \overline{D}$ is connected and

(a) there exists $c_0 > 0$ with

$$\mathrm{Im}[z^* q(x) z] \leq 0 \quad \text{and} \quad |z|^2 + \mathrm{Re}[z^* q(x) z] \geq c_0 |z|^2 \tag{6.26}$$

for all $z \in \mathbb{C}^d$ and almost all $x \in B$, and

(b) there exists $c_1 > 0$ such that either

$$- \mathrm{Re}\left[z^* q(x) z\right] - z^* (\mathrm{Im}\, q)(x)\left(I + (\mathrm{Re}\, q)(x)\right)^{-1}(\mathrm{Im}\, q)(x) z \geq c_1 |z|^2 \tag{6.27}$$

for all $z \in \mathbb{C}^d$ and almost all $x \in D$, or

$$\mathrm{Re}\left[z^* q(x) z\right] \geq c_1 |z|^2 \tag{6.28}$$

for all $z \in \mathbb{C}^d$ and almost all $x \in D$.

Here again, $\mathrm{Re}\, q = (q + q^*)/2$ and $\mathrm{Im}\, q = (q - q^*)/(2i)$ for any matrix $q \in \mathbb{C}^{d \times d}$.

We have formulated these assumptions in terms of q. With respect to γ (6.27) and (6.28) have the forms

$$z^*\left[I - \gamma(x)\left(\mathrm{Re}\, \gamma(x)\right)^{-1}\gamma(x)^*\right] z \geq c_1 |z|^2 \tag{6.29}$$

and

$$\mathrm{Re}\left[z^*(\gamma(x) - I) z\right] \geq c_1 |z|^2, \tag{6.30}$$

respectively. Note that for hermitian matrices $q(x)$, (6.27) is equivalent to assume that $-q$ is uniformly positive definite on D.

Theorem 6.5 *Let γ satisfy the assumption (6.26). Then we have:*

(a) $\mathrm{Im}\langle (T - T_0)h, h \rangle \geq 0$ *for all $h \in H_\diamond^{1/2}(\partial D)$.*

(b) Assume that in addition (6.27) holds. Then there exists $c > 0$ with

$$\mathrm{Re}\langle (T - T_0)h, h \rangle \geq c \, \|h\|_{H^{1/2}(\partial D)}^2 \quad \text{for all } h \in H_\diamond^{1/2}(\partial D). \tag{6.31}$$

(c) Assume that (6.28) holds. Then there exists $c > 0$ with

$$\mathrm{Re}\langle (T_0 - T)h, h \rangle \geq c \, \|h\|_{H^{1/2}(\partial D)}^2 \quad \text{for all } h \in H_\diamond^{1/2}(\partial D). \tag{6.32}$$

Proof: First, we compute since $Th = \partial w_+/\partial \nu = \partial_\gamma w_-$:

$$\langle Th, h \rangle = \int\limits_{\partial D} \frac{\partial w_+}{\partial \nu} \left(\overline{w}_+ - \overline{w}_- \right) ds$$

$$= \int\limits_{\partial B} \overline{w} \, \frac{\partial w}{\partial \nu} \, ds \iint\limits_{B \setminus \overline{D}} |\nabla w|^2 \, dx - \iint\limits_{D} \nabla w^* \, \gamma \nabla w \, dx$$

$$= - \iint\limits_{B} \nabla w^* \, \gamma \nabla w \, dx.$$

Analogously,

$$\langle T_0 h, h \rangle = - \iint\limits_{B} |\nabla w_0|^2 \, dx$$

where w_0 solves (6.20) and (6.21) for I instead of γ. Therefore,

$$\mathrm{Im}\langle (T - T_0)h, h \rangle = \mathrm{Im}\langle Th, h \rangle = - \mathrm{Im} \iint\limits_{D} \nabla w^* \, \gamma \nabla w \, dx \geq 0$$

by assumptions (6.26) on γ. In the following we will write γ_0 for I to indicate the analogies.

$$\langle Th, h \rangle - \langle T_0 h, h \rangle = - \iint\limits_{B} \left[\nabla w^* \gamma \nabla w - \nabla w_0^* \, \gamma_0 \nabla w_0 \right] dx$$

$$= -2 \iint\limits_{B} \left(\nabla w - \nabla w_0 \right)^* \gamma \nabla w \, dx$$

$$+ \iint\limits_{B} \left[\nabla w_0^* \, \gamma_0 \nabla w_0 - 2 \nabla w_0^* \, \gamma \nabla w + \nabla w^* \, \gamma \nabla w \right] dx.$$

Since $w - w_0 \in H_\diamond^1(B)$ we note that by (6.22) for $\psi = w - w_0$ the first integral vanishes, thus

$$\langle (T - T_0)h, h \rangle = \iint\limits_{B} \left[\nabla w_0^* \, \gamma_0 \nabla w_0 - 2 \nabla w_0^* \, \gamma \nabla w + \nabla w^* \, \gamma \nabla w \right] dx. \tag{6.33}$$

Interchanging the roles of γ_0 and γ yields

$$\langle (T_0 - T)h, h \rangle = \iint_B \left[\nabla w^* \gamma \nabla w - 2 \nabla w^* \gamma_0 \nabla w_0 + \nabla w_0^* \gamma_0 \nabla w_0 \right] dx. \qquad (6.34)$$

Let now (6.27) (i.e., (6.29)) hold. Since $\mathrm{Re}\,\gamma$ is real, symmetric, and positive-definite there exists the positive-definite square root $(\mathrm{Re}\,\gamma)^{1/2}$. Therefore,

$$\mathrm{Re}\langle (T - T_0)h, h \rangle$$

$$= \iint_B \left[\nabla w^* (\mathrm{Re}\,\gamma) \nabla w - 2 \mathrm{Re}\big(\nabla w_0^* \gamma \nabla w \big) + \nabla w_0^* \gamma_0 \nabla w_0 \right] dx$$

$$= \iint_B \left[\left| (\mathrm{Re}\,\gamma)^{1/2} \nabla w \right|^2 - 2 \, \mathrm{Re}\big\{ \nabla w_0^* \gamma (\mathrm{Re}\,\gamma)^{-1/2} (\mathrm{Re}\,\gamma)^{1/2} \nabla w \big\} + \nabla w_0^* \gamma_0 \nabla w_0 \right] dx$$

$$= \iint_B \left[\left| (\mathrm{Re}\,\gamma)^{1/2} \nabla w - (\mathrm{Re}\,\gamma)^{-1/2} \gamma^* \nabla w_0 \right|^2 + \nabla w_0^* \gamma_0 \nabla w_0 \right.$$

$$\left. - \nabla w_0^* \gamma (\mathrm{Re}\,\gamma)^{-1} \gamma^* \nabla w_0^* \right] dx$$

$$\geq \iint_B \nabla w_0^* \left[\gamma_0 - \gamma (\mathrm{Re}\,\gamma)^{-1} \gamma^* \right] \nabla w_0 \, dx$$

$$\geq c_1 \iint_D |\nabla w_0|^2 \, dx$$

by assumption (6.29). The assertion now follows from standard arguments. Indeed, if $\mathrm{Re}\langle (T - T_0)h, h \rangle = 0$ then $\nabla w_0 \equiv 0$ in D. Therefore, $\partial_{\gamma_0} w_{0+} = \partial_{\gamma_0} w_{0-} = 0$ on ∂D and thus $w_0 \equiv 0$ in $B \setminus \overline{D}$. This implies that $h = w_{0+} - w_{0-} = -w_{0-}$ is constant on ∂D and, therefore, has to vanish since $h \in H_\diamond^{1/2}(\partial D)$. We have thus shown that

$$\mathrm{Re}\langle (T - T_0)h, h \rangle > 0 \quad \text{for all } h \in H_\diamond^{1/2}(\partial D), \ \ h \neq 0.$$

Assume now that there exists a sequence $\{h^{(j)}\}$ in $H_\diamond^{1/2}(\partial D)$ with $\|h^{(j)}\|_{H^{1/2}(\partial D)} = 1$ and $\mathrm{Re}\langle (T - T_0)h^{(j)}, h^{(j)} \rangle \to 0, j \to \infty$. Then $\|\nabla w^{(j)}\|_{L^2(D)} \to 0$. We define the functions $\tilde{w}^{(j)}$ in D by

$$\tilde{w}^{(j)} := w^{(j)} - \alpha_j \chi \quad \text{in } D,$$

where $\alpha_j = \int_{\partial D} w_-^{(j)} ds$ and χ denotes the constant function with value $1 / \int_{\partial D} ds$. Since in $\{ w \in H^1(D) : \int_{\partial D} w \, ds = 0 \}$ the norm $w \mapsto \sqrt{\iint_D |\nabla w|^2 \, dx}$ is equivalent to the standard norm of $H^1(D)$ we conclude that $\tilde{w}^{(j)} \to 0$ in $H^1(D)$. Therefore, also $\partial_{\gamma_0} w_+^{(j)} = \partial_\gamma \tilde{w}_-^{(j)} \to 0$ in $H^{-1/2}(\partial D)$ and thus $w^{(j)} \to 0$ in $H^1(B \setminus \overline{D})$ by the well-posedness of the Neumann problem in $B \setminus \overline{D}$. This yields that

$$h^{(j)} + \alpha_j \chi = w_+^{(j)} - \tilde{w}_-^{(j)} \to 0, \quad \text{in } H^{1/2}(\partial D).$$

Since also $\int_{\partial D} h^{(j)} ds = 0$ this implies $\alpha_j \to 0$ and thus $h^{(j)} \to 0$ in $H^{1/2}(\partial D)$. This contradicts the fact that $\|h^{(j)}\|_{H^{1/2}(\partial D)} = 1$. Therefore, (6.31) is proven.

Let us now consider the case that (6.28) (i.e., (6.30)) holds. Analogously, as above, we write

$$\text{Re}\langle (T_0 - T)h, h \rangle = \iint_B \left[|\nabla w_0|^2 - 2\,\text{Re}\{\nabla w_0)^* \nabla w\} + \text{Re}(\nabla w^* \, \gamma \, \nabla w) \right] dx$$

$$= \iint_B \left[|\nabla w_0 - \nabla w|^2 + \text{Re}\{\nabla w^* \, (\gamma - I)\nabla w\} \right] dx$$

$$\geq c_1 \iint_D |\nabla w|^2 dx.$$

Now we can argue as before. $\square$

The factorization (6.23), combined with the positivity properties of $T - T_0$ allow several forms of the factorization method. By comparing the factorization (6.23) with those of the previous chapters we observe that now Corollary 1.22 is not directly applicable because the operator $\Lambda - \Lambda_0$ – which plays the role of F in these abstract "range identities" – acts between reflexive Banach spaces rather than Hilbert spaces.

We can, however, restrict this operator $\Lambda - \Lambda_0$ to an operator from $L^2_\diamond(\partial B)$ into itself. Then we have to consider the operator G, defined in (6.19), as an operator from $H_\diamond^{-1/2}(\partial B)$ into $L^2_\diamond(\partial B)$. Its adjoint maps $L^2_\diamond(\partial B)$ into $H_\diamond^{1/2}(\partial B)$. A sketch of the factorization is given in Figure 6.2.

The next step in the characterization of D by $\Lambda - \Lambda_0$ is to construct functions ϕ_z in $L^2_\diamond(\partial B)$ which depend on arbitrary $z \in B$ with the property that $\phi_z \in \mathcal{R}(G)$ if, and only if, $z \in D$. From the definition $G\psi = v|_{\partial B}$ where $v \in H^1(B \setminus \overline{D})$ solves (6.19) it is clear

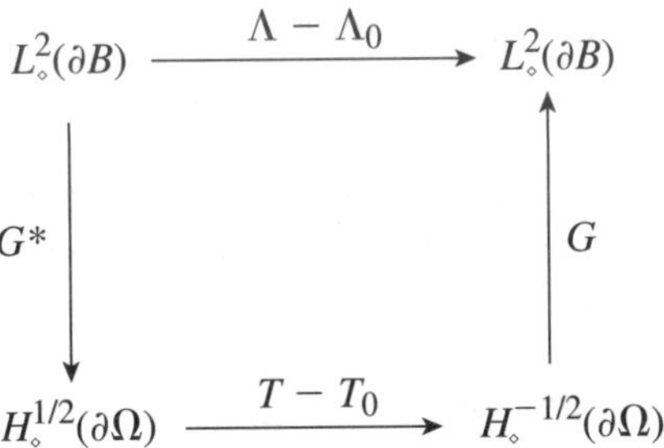

Figure 6.2 The factorization $\Lambda - \Lambda_0 = G(T - T_0)G^*$

that functions ϕ are in the range of G if they are of the form $\phi = v$ and v is harmonic in $B \setminus \overline{D}$ with vanishing Neumann data on ∂B. Therefore, we construct harmonic functions with vanishing Cauchy data which are singular at the given point $z \in B$.

For doing this we have to introduce the Green's function $\Phi_N = \Phi_N(x, z)$ in B with respect to Neumann boundary conditions on ∂B.[3] More precisely, it is well known (see, e.g., [42]) that there exists a function $N \in C^2(B \times B)$ such that for every $z \in B$ we have that $N(\cdot, z) \in C^1(\overline{B})$ satisfies

$$\Delta_x N(x, z) = 0 \quad \text{in } B, \qquad \frac{\partial N(x, z)}{\partial \nu(x)} = -\frac{1}{|\partial B|} - \frac{\partial \Phi_0(x, z)}{\partial \nu(x)} \quad \text{on } \partial B.$$

Here, Φ_0 denotes the fundamental solution of the Laplace equation in $\mathbb{R}^d$, i.e.,

$$\Phi_0(x, z) \; := \; \begin{cases} -\frac{1}{2\pi} \ln |x - z|, & d = 2, \\ 1/(4\pi |x - z|), & d = 3. \end{cases}$$

Then $\Phi_N(x, z) = \Phi_0(x, z) + N(x, z)$ is smooth for $x \neq z$ and satisfies $\Delta_x \Phi_N(\cdot, z) = 0$ in $B \setminus \{z\}$ and $\partial \Phi_N(\cdot, z)/\partial \nu(x) = -1/|\partial B|$ on ∂B. We normalize Φ_N, i.e., N, such that $\int_{\partial B} \Phi_N(x, z)\, ds(x) = 0$ for all $z \in B$.

It is well known (cf. [145]) that Φ_N is symmetric with respect to both variables, i.e., $\Phi_N(x, z) = \Phi_N(z, x)$ for all $x, z \in B$, $x \neq z$. In particular, Φ_N is analytic also with respect to $z \neq x$.

With this Green's function Φ_N we construct the sampling functions $\phi_z \in L^2_\diamond(\partial B)$ as follows: Choose an arbitrary unit vector $\hat{a} \in \mathbb{R}^d$ and set

$$\phi_z(x) \; := \; \hat{a} \cdot \nabla_z \Phi_N(x, z), \quad x \in \partial B. \tag{6.35}$$

Then we have the following characterization of D by the range of G which is completely analogous to the results of Theorems 1.12, 4.6, and 5.11:

Theorem 6.6 *Let $B \subset \mathbb{R}^d$ be a bounded domain with a Lipschitz–continuous boundary ∂B. Furthermore, assume that $D \subset \mathbb{R}^d$ is a domain such that $\overline{D} \subset B$ and $B \setminus \overline{D}$ is connected. For any $z \in B$ define $\phi_z \in L^2_\diamond(\partial B)$ by (6.35). Then $\phi_z \in \mathcal{R}(G)$ if, and only if, $z \in D$.*

Proof: Let first $z \in D$. By differentiating the Laplace equation and the boundary conditions with respect to y we observe that the function $v(x) = \hat{a} \cdot \nabla_z \Phi_N(x, z)$ satisfies $\Delta v = 0$ in $B \setminus \overline{D}$ and $\partial v/\partial \nu = 0$ on ∂B. Furthermore, $v|_{\partial B} = \phi_z$ on ∂B. Therefore, $G(\partial v/\partial \nu) = \phi_z$ which shows that $\phi_z \in \mathcal{R}(G)$.

Second, let $z \notin D$, and assume on the contrary that $\phi_z = G(\partial w/\partial \nu)$ for some $w \in H^1(B \setminus \overline{D})$ with $\Delta w = 0$ in $B \setminus \overline{D}$ and $\partial w/\partial \nu = 0$ on ∂B. As before, define $v(x) = \hat{a} \cdot \nabla_z \Phi_N(x, z)$ for $x \in B \setminus (\overline{D} \cup \{z\})$. Then the Cauchy data of v and w coincide on ∂B. Now we fix $y \in \partial B$ and a ball U centered in y such that $U \cap B \subset B \setminus \overline{D}$ and $z \notin U$ and define the function

$$u = \begin{cases} v - w & \text{in } U \cap B, \\ 0 & \text{in } U \setminus \overline{B}. \end{cases}$$

---

[3] It is only here where we make use of the stronger regularity assumption on ∂B.

Since $v - w$ vanish on ∂B we conclude that $u \in H^1(U)$. Since the Neumann data of v and w vanish on ∂B the variational formulations yields that

$$\iint_U \nabla u \cdot \nabla \psi \, dx = \iint_{U \cap B} \nabla(v - w) \cdot \nabla \psi \, dx = 0$$

for all $\psi \in H_0^1(U)$. Therefore, $u \in H^1(U)$ satisfies $\Delta u = 0$ in U. Classical interior regularity results yield that u is analytic in U. Since u vanishes on $U \setminus \overline{B}$ an analytic continuation argument implies $u \equiv 0$ also in $U \cap B$ and thus, again by analytic continuation, $v \equiv w$ also in $B \setminus (\overline{D} \cup \{z\})$. In particular, this implies that $v \in H^1(B \setminus (\overline{D} \cup \{z\}))$. On the other hand, v has the same type of singularity at $x = z$ as $v_0(x) = \hat{a} \cdot \nabla_z \Phi_0(x, z)$. By Lemma 6.7 below v_0 is not square integrable in a cone of the form $C = \{x = z + t\hat{x} : 0 < t < r_0, \ \hat{x} \cdot \hat{b} > 1 - \varepsilon\}$ where the unit vector $\hat{b} \in \mathbb{R}^d$ is chosen such that $C \subset B \setminus \overline{D}$. This implies, in particular, that v_0 is not in $H^1(B \setminus \overline{D})$, a contradiction. $\square$

Lemma 6.7 *Let* $z, \hat{a}, \hat{b} \in \mathbb{R}^d$ *with* $|\hat{a}| = |\hat{b}| = 1$ *and* $\varepsilon, r_0 > 0$. *Then* $\hat{a} \cdot \nabla_z \Phi_0(\cdot, z) \notin L^2(C)$ *where the cone* C *is defined by*

$$C = \{x = z + t\hat{x} : 0 < t < r_0, \ \hat{x} \cdot \hat{b} > 1 - \varepsilon\}.$$

Proof: For $r \in (0, r_0)$ we set

$$C_r = \{x = z + t\hat{x} : r < t < r_0, \ \hat{x} \cdot \hat{b} > 1 - \varepsilon\}.$$

For the proof we restrict ourselves to the case $d = 3$. (The case $d = 2$ is treated analogously.) Then we have with $x = z + t\hat{x}$ and the transformation formula for multiple integrals (where $dx = t^2 dt \, ds(\hat{x})$)

$$\iint_{C_r} |\hat{a} \cdot \nabla_z \Phi(x, z)|^2 dx = \iint_{C_r} \frac{|\hat{a} \cdot (x - z)|^2}{(4\pi)^2 |x - z|^6} dx = \frac{1}{(4\pi)^2} \int_r^{r_0} \frac{dt}{t^2} \int_{\hat{x} \cdot \hat{b} > 1 - \varepsilon} |\hat{a} \cdot \hat{x}|^2 ds(\hat{x}),$$

and this tends to infinity as r tends to zero since the second integral is certainly positive. $\square$

First, we consider the case that $q(x)$ is hermitian for almost all $x \in D$. From Theorem 6.2 we know that $T - T_0$ is self-adjoint. Since $s(T - T_0)$ for $s = 1$ or $s = -1$ is also coercive by (6.31) or (6.32), respectively, we can apply Corollary 1.22 of Chapter 1 which yields that the ranges of $|\Lambda - \Lambda_0|^{1/2} : L_\diamond^2(\partial B) \to L_\diamond^2(\partial B)$ and $G : H_\diamond^{-1/2}(\partial B) \to L_\diamond^2(\partial B)$ coincide, i.e.,

$$\mathcal{R}\big(|\Lambda - \Lambda_0|^{1/2}\big) = \mathcal{R}(G). \tag{6.36}$$

Combining this equation with Theorem 6.6 yields the following characterization of D.

Theorem 6.8 *Let Assumption 6.4 hold and assume that* $q(x)$ *is hermetian for almost all* $x \in D$. *Then* $z \in D$ *if, and only if,* $\phi_z \in \mathcal{R}\big(|\Lambda - \Lambda_0|^{1/2}\big)$ *where* ϕ_z *is defined by* (6.35).

Now we consider arbitrary functions $q \in L^\infty(D, \mathbb{C}^{d \times d})$ which satisfy Assumption 6.4. From the factorization (6.23) we see that also the real and imaginary parts of $\Lambda - \Lambda_0$ can be factorized. Indeed, if we recall the definitions $\operatorname{Re} B = (B + B^*)/2$ and $\operatorname{Im} B = (B - B^*)/(2i)$ for any bounded operator B, we have

$$\operatorname{Re}(\Lambda - \Lambda_0) = G \operatorname{Re}(T - T_0) G^*, \qquad \operatorname{Im}(\Lambda - \Lambda_0) = G \operatorname{Im}(T - T_0) G^*,$$

and thus

$$\Lambda_\# := \operatorname{Re}(\Lambda - \Lambda_0) + \rho \operatorname{Im}(\Lambda - \Lambda_0) = G A G^*, \tag{6.37}$$

where $\rho \in \mathbb{R}$ is an arbitrary coupling parameter and

$$A := \operatorname{Re}(T - T_0) + \rho \operatorname{Im}(T - T_0).$$

We choose $\rho \geq 0$ if $\operatorname{Re}(T - T_0)$ is coercive and $\rho \leq 0$ if $-\operatorname{Re}(T - T_0)$ is coercive. Then $A : H_\diamond^{1/2}(\partial B) \to H_\diamond^{-1/2}(\partial B)$ is self-adjoint and $+A$ or $-A$ is coercive by Theorem 6.5. Therefore, we can again apply Corollary 1.22 of Chapter 1 which yields that the ranges of $|\Lambda_\#|^{1/2}$ and G coincide. Therefore, combination with Theorem 6.6 yields

Theorem 6.9 *Let Assumption 6.4 hold. Define $F_\#$ and ϕ_z by (6.37) and (6.35), respectively, for any $z \in B$ and $\rho \in \mathbb{R}$ with $\rho \geq 0$ if Assumption (6.27) holds and $\rho \leq 0$ if Assumption (6.28) holds. Then $z \in D$ if, and only if, $\phi_z \in \mathcal{R}\bigl(|\Lambda_\#|^{1/2}\bigr)$.*

Before we present two numerical examples we want to add some remarks.

The Green's function Φ_N can be computed from the Neumann-to-Dirichlet operator Λ_0. Indeed, from the boundary value problem for N in the decomposition $\Phi_N = \Phi_0 + N$ we observe that

$$N(\cdot, z) = -\Lambda_0 \left(\frac{\partial \Phi_0}{\partial \nu(x)}(\cdot, z) + \frac{1}{|\partial B|} \right) + c(z)$$

where $c(z)$ is chosen such that $\int_{\partial B} \Phi_N(x, z)\, ds(x) = 0$.

We have chosen (directional) derivatives of the Green's function Φ_N as sampling functions. If a point $\hat{z}$ inside of the unknown domain D is already known one can also choose

$$\phi_z(x) = \Phi_N(x, z) - \Phi_N(x, \hat{z}), \quad x \in \partial B,$$

as sampling functions instead of ϕ_z from (6.35). By modifying the above arguments it is easily seen that Theorems 6.6 and 6.8 hold also for this choice of ϕ_z.

We can formulate the characterization of Theorem 6.9 also by using an eigensystem $\{\lambda_j, \psi_j\}$ of the self-adjoint and positive operator $\Lambda_\#$ as in the previous chapters. Therefore, a point $z \in B$ belongs to D if, and only if,

$$W(z) := \left[\sum_{j=1}^\infty \frac{|(\phi_z, \psi_j)|^2}{\lambda_j} \right]^{-1} > 0. \tag{6.38}$$

This characterization of D makes use of the full knowledge of the spectral set $\{\lambda_j, \psi_j\}$. In [139] Armin Lechleiter investigated the behavior of the corresponding (regularized) sum W_M for finite-dimensional approximations $\Lambda^{(M)} \in \mathbb{C}^{M \times M}$ of Λ. He shows convergence of $W_M(z)$ to zero for points outside of D while the $W_M(z)$ remains positive for z inside of D.

The treatment of the inverse problem of electrical impedance tomography by the Factorization Method has been generalized to quite general inverse elliptic boundary value problems in [120] and [68]. There are very recent applications of the Factorization Method to related problems in heat conduction (see [66]) and optical tomography (see [89, 91]).

We finish this chapter with two numerical examples in $\mathbb{R}^2$ and plot the finite sum. The disk of radius 1 contains one or two inclusions. In the first example D is an ellipse

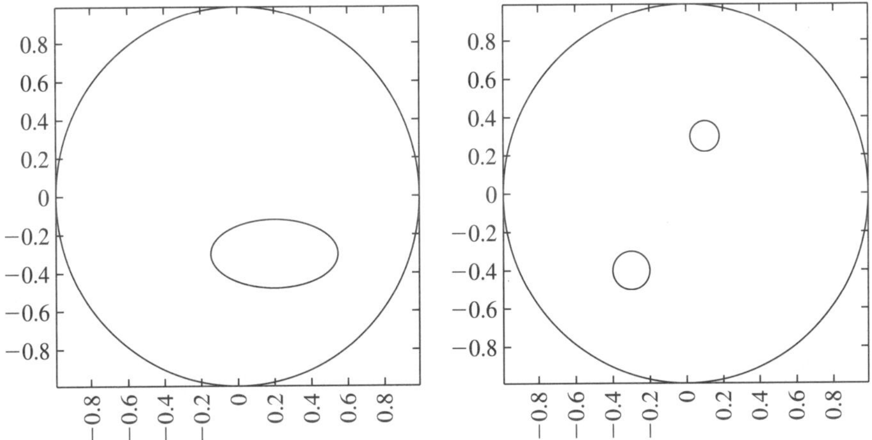

Figure 6.3 The shape of the inclusions for the examples

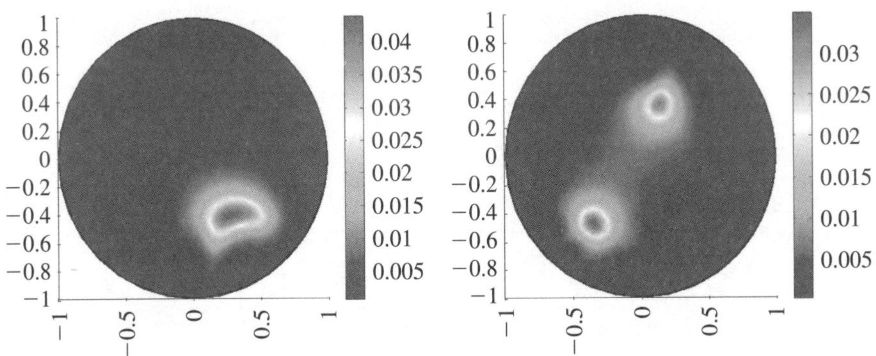

Figure 6.4 Contour plot of W for the examples with 0.1% noise

and in the second example D consists of two circles (see Figure 6.3). The boundary of B is partitioned into 32 equidistant parts which correspond to 32 electrodes. The continuous Neumann-to-Dirichlet operator is then replaced by a 31×31 – matrix $A \in \mathbb{C}^{31 \times 31}$ corresponding to adjacent current patterns. For a precise description of the implementation we refer to [94]. The reconstructions in Figure 6.4 are carried out by Armin Lechleiter.

7
Alternative sampling and probe methods

The Factorization Method is only one of several approaches to solve the inverse scattering problem by avoiding the computation of solutions of direct scattering problems. In this chapter we present some methods which all share the common feature that the region where the unknown domain D is expected is "sampled" by points or curves. In contrast to the factorization method, however, all of them make use of the fact that the fundamental solution $\Phi(\cdot, z)$ – or any other solution of the Helmholtz equation – in any bounded domain G can be approximated by functions from a suitable class of special solutions of the Helmholtz equation – provided the domain G is such that its exterior is connected. For problems in impedance tomography similar approximation results are needed for the Laplace equation. Therefore, the first section collects two basic approximation theorems for the Helmholtz and the Laplace equation.

7.1 Two approximation results

As mentioned above, all of the methods in this chapter make use of approximation theorems for the Laplace equation and the Helmholtz equation. Sometimes they are referred to as "Runges Approximation Theorem,"[1], sometimes they are cited as "Malgrange Approximation Theorem" (cf. [142]). We will need only special cases of these results and will state and prove them for the convenience of the reader.

In the first theorem we show that every solution u of the Helmholtz equation in some region G can be approximated arbitrarily well by solutions v of the Helmholtz equation in a larger domain. This result holds also for $k = 0$, i.e., for the Laplace equation. For $k > 0$ we will prove a stronger result in Theorem 7.3 below. We will make use of the fact that the interior boundary value problem with impedance boundary condition $\partial u / \partial \nu + i u = f$ on ∂G is uniquely solvable for every $f \in H^{-1/2}(\partial G)$. We note that this lemma holds also for the case $k = 0$.

Lemma 7.1 *Let $k \in \mathbb{R}$ with $k \geq 0$ and $G \subset \mathbb{R}^3$ be a bounded Lipschitz domain. For every $f \in H^{-1/2}(\partial G)$ there exists a unique $u \in H^1(G)$ such that $\Delta u + k^2 u = 0$ in G*

---

[1] This notion goes back to Peter Lax [137], cf. also [99, 149], because of the analogy to the problem of approximating holomorphic functions by polynomials.

and $\partial u/\partial v + i\,u = f$ on ∂G. Furthermore, there exists $c > 0$ depending only on k and G such that

$$\|u\|_{H^1(G)} \le c\,\|f\|_{H^{-1/2}(\partial G)}\,.$$

Proof: The solution $u \in H^1(G)$ is again understood in the variational sense, i.e.,

$$\iint_G \left[\nabla u \cdot \nabla \overline{\psi} - k^2 u\,\overline{\psi}\right] dx + i \int_{\partial G} u\,\overline{\psi}\,ds = \langle f, \psi \rangle \quad \text{for all } \psi \in H^1(G)\,. \tag{7.1}$$

To show uniqueness let $f = 0$. Substituting $\psi = \overline{u}$ in (7.1) and taking the imaginary part yields $u = 0$ on ∂G, i.e., $u \in H_0^1(G)$ and

$$\iint_G \left[\nabla u \cdot \nabla \psi - k^2 u\,\psi\right] dx = 0 \quad \text{for all } \psi \in H^1(G)\,.$$

Therefore, extending u by zero outside of G we observe that $u \in H^1(\mathbb{R}^3)$ and $\Delta u + k^2 u = 0$ in $\mathbb{R}^3$ in the variational sense. The unique continuation implies that u vanishes in all of $\mathbb{R}^3$. Existence is then shown by the Riesz-Fredholm theory by the same arguments which have been used already in this monograph. Nevertheless, we sketch the proof and write (7.19) in the form

$$(u, \psi)_{H^1(G)} - a(u, \psi) = \langle f, \psi \rangle \quad \text{for all } \psi \in H^1(G)\,, \tag{7.2}$$

where the sesqui-linear form a on $H^1(G)$ is defined by

$$a(u, \psi) = (k^2 + 1) \iint_G u\,\overline{\psi}\,dx - i \int_{\partial G} u\,\overline{\psi}\,ds\,, \quad u, \psi \in H^1(G)\,.$$

Since a is certainly bounded there exists a linear and bounded operator A such that $a(u, \psi) = (Au, \psi)_{H^1(G)}$ for all $u, \psi \in H^1(G)$. The operator A is even compact as seen as follows. Let $\{u_n\}$ converge weakly to zero in $H^1(G)$. The trace theorem yields that $u_n|_{\partial G}$ converges weakly in $H^{1/2}(\partial G)$. Then, by the compact imbeddings of $H^1(G) \subset L^2(G)$ and $H^{1/2}(\partial G) \subset L^2(\partial G)$ we have that $\|u_n\|_{L^2(G)}$ and $\|u_n\|_{L^2(\partial G)}$ converge to zero. Then we estimate

$$\|Au_n\|_{H^1(G)}^2 = a(u_n, Au_n) \le (k^2 + 1)\|u_n\|_{L^2(G)}\|Au_n\|_{L^2(G)} + \|u_n\|_{L^2(\partial G)}\|Au_n\|_{L^2(\partial G)}$$

$$\le \left[(k^2 + 1)\|u_n\|_{L^2(G)} + \|u_n\|_{L^2(\partial G)}\right]\|Au_n\|_{H^1(G)}$$

from which we see that $\{Au_n\}$ converges to zero in $H^1(G)$. $\quad\square$

Now we can prove the following approximation result which is true for both, the Laplace equation and the Helmholtz equation.

Theorem 7.2 *Let $k \in \mathbb{R}$ with $k \ge 0$ and B, G be bounded Lipschitz domains with $\overline{G} \subset B$ and such that $B \setminus \overline{G}$ is connected and $\partial G \in C^2$. Furthermore, let $u \in H^1(G)$ be a solution* *of the Helmholtz equation $\Delta u + k^2 u = 0$ in G in the variational sense, i.e.,*

$$\iint\limits_{G} \left[\nabla u \cdot \nabla \psi - k^2 u\, \psi \right] dx = 0 \quad \text{for all } \psi \in H_0^1(G)\,.$$

Then, for every $\varepsilon > 0$, there exists a solution $v \in H^1(B)$ of $\Delta v + k^2 v = 0$ in B such that

$$\|v - u\|_{H^1(G)} \leq \varepsilon\,.$$

Furthermore, if U is any open set with $\overline{U} \subset G$ then for every $\varepsilon > 0$ and $m \in \mathbb{N}$ there exists a solution $v \in C^m(\overline{U})$ such that

$$\|v - u\|_{C^m(\overline{U})} \leq \varepsilon\,.$$

Proof: By enlarging B we can assume that also ∂B is smooth and that the exterior of B is connected. We approximate u by solutions which are single layer potentials with L^2-densities on ∂B, i.e., v is of the form

$$v(x) = \int\limits_{\partial B} \Phi(x, y)\, \psi(y)\, ds(y)\,, \quad x \in B\,, \tag{7.3}$$

for $\psi \in L^2(\partial B)$. From [144] we know that $v \in H^1(B)$. By the previous lemma it is sufficient to prove that the traces $\partial v / \partial \nu + i\, v$ on ∂G are dense in $H^{-1/2}(\partial G)$. Therefore, let $\varphi \in H^{1/2}(\partial G)$ such that

$$\int\limits_{\partial G} \varphi \left[\frac{\partial v}{\partial \nu} + i\, v \right] ds = 0\,,$$

for all ψ, i.e.,

$$\int\limits_{\partial G} \varphi(x) \int\limits_{\partial B} \left[\frac{\partial}{\partial \nu(x)} \Phi(x, y)\, \psi(y) + i\, \Phi(x, y)\, \psi(y) \right] ds(y)\, ds(x) = 0 \quad \text{for all } \psi \in L^2(\partial B)\,.$$

Interchanging the orders of integration yields

$$\int\limits_{\partial B} \psi(y) \left\{ \int\limits_{\partial G} \varphi(x) \left[\frac{\partial}{\partial \nu(x)} \Phi(x, y) + i\, \Phi(x, y) \right] ds(x) \right\} ds(y) = 0$$

for all $\psi \in L^2(\partial B)$ and thus

$$w(x) = \int\limits_{\partial G} \varphi(y) \left[\frac{\partial}{\partial \nu(y)} \Phi(x, y) + i\, \Phi(x, y) \right] ds(y) = 0 \quad \text{for all } x \in \partial B\,. \tag{7.4}$$

Therefore, the potential w defined by (7.4) for all $x \notin \partial G$ solves the boundary value problem in the exterior of B with homogeneous Dirichlet boundary conditions on ∂B. Furthermore, in the case $k > 0$ it satisfies the Sommerfeld radiation condition and has to vanish in the exterior of B by Theorem 1.1. In the case $k = 0$ we observe that $w(x) = \mathcal{O}(1/R)$ and $\nabla w(x) = \mathcal{O}(1/R^2)$ for $|x| = R \to \infty$. The uniqueness of the exterior boundary value problem for the Laplace equation (shown again by an application of Greens first theorem) yields that w vanishes in the exterior of B. In any case (i.e., if $k > 0$ or $k = 0$) a unique continuation argument yields that w vanishes in the exterior of G. Now we make use of the jump conditions of the single and double layer potentials for densities in Sobolev spaces, cf. [144], Theorem 6.11. For the traces of w and $\partial w/\partial v$ from the exterior $(+)$ and interior $(-)$, respectively, we have

$$w_+ - w_- = \varphi, \qquad \frac{\partial w_+}{\partial v} - \frac{\partial w_-}{\partial v} = -i\varphi,$$

and thus, because the traces from the exterior vanish,

$$\frac{\partial w_-}{\partial v} + i\,w_- = 0 \quad \text{on } \partial G.$$

The uniqueness result of Lemma 7.1 yields that w vanishes in G and thus also φ. Denseness in the interior domain $\overline{U}$ follows from interior regularity results of the Helmholtz equation. $\square$

By essentially the same arguments we can show:

Theorem 7.3 *Let $k > 0$ and $G \subset \mathbb{R}^3$ be a bounded Lipschitz domain such that $\mathbb{R}^3 \setminus \overline{G}$ is connected and $\partial G \in C^2$. Furthermore, let $u \in H^1(G)$ be a solution of the Helmholtz equation $\Delta u + k^2 u = 0$ in G in the variational sense. Then, for every $\varepsilon > 0$, there exists a Herglotz wave function, i.e., a solution v_ψ of the Helmholtz equation of the form*

$$v_\psi(x) = \int_{S^2} \psi(\theta)\, e^{ik\,x\cdot\theta}\, ds(\theta), \quad x \in \mathbb{R}^3, \tag{7.5}$$

for some $\psi \in L^2(S^2)$, such that

$$\|v_\psi - u\|_{H^1(G)} \le \varepsilon.$$

Furthermore, if U is any open set with $\overline{U} \subset G$ then for every $\varepsilon > 0$ and $m \in \mathbb{N}$ there exists a Herglotz function v_ψ such that

$$\|v_\psi - u\|_{C^m(\overline{U})} \le \varepsilon.$$

Proof: (cf. [58]) Again it suffices to prove that the traces $\partial v_\psi/\partial v + i v_\psi$ on ∂G are dense in $H^{-1/2}(\partial G)$. Let again $\varphi \in H^{1/2}(\partial G)$ such that

$$\int_{\partial G} \varphi \left[\frac{\partial v_\psi}{\partial v} + i\,v_\psi \right] ds = 0 \quad \text{for all } \psi \in L^2(S^2).$$

Substituting the form of v_ψ and interchanging the orders of integration yields

$$\int_{S^2} \psi(\theta) \left\{ \int_{\partial G} \varphi(y) \left[\frac{\partial}{\partial v(y)}\, e^{ik\,y\cdot\theta} + i\,e^{ik\,y\cdot\theta} \right] ds(y) \right\} ds(\theta) = 0$$

for all $\psi \in L^2(S^2)$ and thus

$$\int_{\partial G} \varphi(y) \left[\frac{\partial}{\partial \nu(y)} e^{ik\, y \cdot \theta} + i\, e^{ik\, y \cdot \theta} \right] ds(y) = 0 \quad \text{for all } \theta \in S^2 \,.$$

We note that this term is the far field pattern $w^\infty(-\theta)$ of

$$w(x) = \int_{\partial G} \varphi(y) \left[\frac{\partial}{\partial \nu(y)} \Phi(x,y) + i\, \Phi(x,y) \right] ds(y) \,, \quad x \notin \partial G \,.$$

Rellich's Lemma 1.2 and a unique continuation argument yields that w vanishes in $\mathbb{R}^3 \setminus \overline{G}$. Now we can argue exactly as in the proof of the previous theorem. $\square$

In the following sections we will need corollaries of these basic approximation results. We will formulate and prove them when they appear.

7.2 The dual space method and the linear sampling method

The Linear Sampling Method has been introduced by David Colton and one of the authors in [39] for the scattering of plane harmonic waves by impenetrable obstacles. Actually, the origin of this method traces back to the "Dual Space Method," investigated by David Colton and Peter Monk in a series of papers [45–49]. We will recall this method for the same simple model problem of Chapter 1 and follow the formulations in [43].

Let $D \subset \mathbb{R}^3$ be an open and bounded domain with $C^{1,\alpha}$-boundary ∂D such that the exterior $\mathbb{R}^3 \setminus \overline{D}$ of $\overline{D}$ is connected. Furthermore, let $k > 0$ be the wavenumber and

$$u^i(x, \theta) = e^{ik\, x \cdot \theta} \,, \quad x \in \mathbb{R}^3 \,, \tag{7.6}$$

be the incident plane wave of direction $\theta \in S^2$. Again, $S^2 = \{x \in \mathbb{R}^3 : |x| = 1\}$ denotes the unit sphere in $\mathbb{R}^3$. The scattered field and the total field are again denoted by u^s and u, respectively. u satisfies the Helmholtz equation

$$\Delta u + k^2 u = 0 \quad \text{outside } D \,, \tag{7.7}$$

and the Dirichlet boundary condition

$$u = 0 \quad \text{on } \partial D \,. \tag{7.8}$$

The scattered field u^s satisfies the Sommerfeld radiation condition (1.17) which yields the existence of the far field pattern u^∞ as the first coefficient in the asymptotic expansion

$$u^s(x) = \frac{\exp(ik|x|)}{4\pi |x|} \left[u^\infty(\hat{x}, \theta) + \mathcal{O}(1/r) \right] \quad \text{for } r = |x| \to \infty \tag{7.9}$$

uniformly with respect to $\hat{x} = x/|x|$. Again, we indicate the direction of incidence by writing $u^\infty(\hat{x}, \theta)$.

The *inverse scattering problem* is to find the obstacle D from the knowledge of the far field patterns $u^\infty(\hat{x}, \theta)$ for all $\hat{x}, \theta \in S^2$.

Let again $\Phi = \Phi(x, y)$ denote the fundamental solution of the Helmholtz equation in $\mathbb{R}^3$, i.e.,

$$\Phi(x, y) = \frac{\exp(ik|x - y|)}{4\pi|x - y|}, \quad x \neq y.$$

The idea of the *Dual Space Method* of Colton and Monk,[2] formulated in our notations, is first to approximate the special radiating field $\Phi(\cdot, 0)$ of a source at $z = 0$ – which is a priori assumed to lie inside the scatterer D – by a superposition v of scattered fields $u^s(\cdot, \theta)$ corresponding to plane waves as incident fields. In other words, one tries to determine $g \in L^2(S^2)$ such that

$$v(x) := \int_{S^2} g(\theta)\, u^s(x, \theta)\, ds(\theta) \approx \Phi(x, 0) \quad \text{for } x \notin D. \tag{7.10}$$

The observation that $u^s(x, \theta) = -u^i(x, \theta)$ for $x \in \partial D$ motivates the second step in which the unknown boundary ∂D is found by searching for points $x \in \mathbb{R}^3$ for which the Herglotz wave function

$$v_g(x) := \int_{S^2} g(\theta)\, u^i(x, \theta)\, ds(\theta) = \int_{S^2} g(\theta)\, e^{ik\, x \cdot \theta}\, ds(\theta) \tag{7.11}$$

coincides (approximately) with $-\Phi(x, 0)$. As in [43] we assume that the unknown surface is starlike with respect to the origin and can be parametrized as $x = \rho(\hat{x})\hat{x}$ for $\hat{x} \in S^2$ for some positive $\rho \in C^{1,\alpha}(S^2)$. Therefore, for some convex and compact subset $\mathcal{U}$ of $\{\rho \in C^{1,\alpha}(S^2) : \rho > 0 \text{ on } S^2\}$ we try to determine $\rho \in \mathcal{U}$ such that $v_g(\rho(\hat{x})\hat{x}) \approx -\Phi(\rho(\hat{x})\hat{x}, 0)$ for $\hat{x} \in S^2$ or, in terms of the surface $\Gamma_\rho = \{x = \rho(\hat{x})\hat{x} : \hat{x} \in S^2\}$, such that $v_g|_{\Gamma_\rho} = -\Phi(\cdot, 0)|_{\Gamma_\rho}$.

Since only far field patterns $u^\infty(\cdot, \theta)$ are available the function $g \in L^2(S^2)$ can not be determined by (7.10). Instead, one tries to find g such that the far field patterns of v and $\Phi(\cdot, 0)$ coincide, i.e. one solves the integral equation of the first kind

$$\int_{S^2} g(\theta)\, u^\infty(\hat{x}, \theta)\, ds(\theta) = 1, \quad \hat{x} \in S^2, \tag{7.12}$$

approximately, i.e., $Fg = \bar{1}$ by using the far field operator F from (1.30). Here again, $\bar{1}$ denotes the constant function with value one.

Since (7.12) is an integral equation of the first kind which has, in general, no solution in [43] a combined regularization technique is proposed. For example, one can minimize

---

[2] We will follow the presentation in [43] where this method is called *method of superposition of incident fields* in contrast to the method of superposition of scattered fields.

the functional

$$
J_\varepsilon(g,\rho) = \varepsilon \|v_g\|^2_{L^2(\Gamma)} + \left\| \int_{S^2} g(\theta)\, u^\infty(\cdot,\theta)\, ds(\theta) - \bar{1} \right\|^2_{L^2(S^2)} + \|v_g + \Phi(\cdot,0)\|^2_{C^1(\Gamma_\rho)}
$$

(7.13)

with respect to the pair $(g,\rho) \in L^2(S^2) \times \mathcal{U}$. Here, Γ is some given surface which contains the unknown surface ∂D in its interior, v_g is given by (7.11), and $\varepsilon > 0$ is a regularization parameter. In [43] the following convergence result has been proven:

Theorem 7.4 *Assume that the origin is inside of D and that k^2 is not a Dirichlet eigenvalue of $-\Delta$ in D.*

(a) For every $\varepsilon > 0$ there exists an optimal surface Γ_{ρ_0}, i.e., $\rho_0 \in \mathcal{U}$ satisfies

$$
\inf_{g \in L^2(S^2)} J_\varepsilon(g,\rho_0) = \inf_{(g,\rho) \in L^2(S^2) \times \mathcal{U}} J_\varepsilon(g,\rho)\,.
$$

(b) Let $\{\varepsilon_n\}$ be a null sequence and let $\rho_n \in \mathcal{U}$ be a corresponding sequence of parametrizations of optimal surfaces for the regularization parameter ε_n. Then there exist convergent subsequences of $\{\rho_n\}$. Assume that $\partial D = \Gamma_{\hat{\rho}}$ for some $\hat{\rho} \in \mathcal{U}$. Assume further that the solution of the interior Dirichlet problem

$$
\Delta v + k^2 v = 0 \text{ in } D\,, \quad v = -\Phi(\cdot,0) \text{ on } \partial D\,, \tag{7.14}
$$

can be extended analytically across the boundary ∂D into the interior of Γ with continuous boundary values on Γ. Then every limit point ρ of $\{\rho_n\}$ represents a surface Γ_ρ with $v = -\Phi(\cdot,0)$ on Γ.

Since one cannot prove uniqueness of the solution of the optimization problem we can not assure the existence of only one accumulation point. In particular, it is not clear whether or not the computed surface Γ_ρ coincides with ∂D. Also, a convergence result for the general situation without the restricting assumption that v can be extended across ∂D has not been proved.

The assumption that the origin lies in D is crucial. It is used in the proof of Theorem 7.4 in the way that if v satisfies (7.14) then the solution v^s of the exterior Dirichlet boundary value problem with boundary data v on ∂D coincides with $-\Phi(\cdot,0)$ in the exterior of D.

The origin 0 can, of course, be replaced by the any other known point z inside D. The constant function $\bar{1}$ on the right hand side of (7.12) or in the definition (7.13) of J_ε has then to be replaced by the far field pattern $\exp(-ik\, z \cdot \hat{x})$ of $\Phi(\cdot,z)$. In particular, equation (7.12) has to be replaced by

$$
\int_{S^2} g(\theta)\, u^\infty(\hat{x},\theta)\, ds(\theta) = e^{-ik\, z \cdot \hat{x}}\,, \quad \hat{x} \in S^2\,. \tag{7.15}
$$

In 1996 it was also numerically observed that for a point z outside of D any regularized solution of (7.15) is of much larger L^2-norm than for points z inside of D. This observation led to the idea of the *Linear Sampling Method,* first presented in [39] and [56] for obstacle scattering problems. For survey papers discussing this method we refer the reader to [36] and [38].

The following function $\phi_z \in L^2(S^2)$ has been used already in the Factorization Method.

$$\phi_z(\hat{x}) = e^{-ik\,\hat{x}\cdot z}, \quad \hat{x} \in S^2, \tag{7.16}$$

for any $z \in \mathbb{R}^3$.

Comparing the Linear Sampling Method with the Factorization Method where one checks solvability of the equation $(F^*F)^{1/4}g = \phi_z$ we observe that the Linear Sampling Method computes approximate solutions of the equation $Fg = \phi_z$.

The following theorem is a slight modification of the formulation in, e.g., [25] and provides the mathematical basis of the Linear Sampling Method.

Theorem 7.5 *Let $u^\infty = u^\infty(\hat{x}, \theta)$ be the far field pattern corresponding to the scattering problem (7.6) – (7.9) with associated far field operator F, and assume that k^2 is not a Dirichlet eigenvalue of $-\Delta$ in D.*

(1)   For every $z \in D$ and $\varepsilon > 0$ there exists $g_{z,\varepsilon} \in L^2(S^2)$ such that

$$\left\| Fg_{z,\varepsilon} - \phi_z \right\|_{L^2(S^2)} \leq \varepsilon \tag{7.17}$$

and, for every $z^* \in \partial D$ and every choice of $g_{z,\varepsilon} \in L^2(S^2)$ with (7.17),

$$\lim_{z \to z^*} \left\| g_{z,\varepsilon} \right\|_{L^2(S^2)} = \infty \quad and \quad \lim_{z \to z^*} \left\| v_{g_{z,\varepsilon}} \right\|_{H^1(D)} = \infty \tag{7.18}$$

where v_g is the Herglotz wave function (7.11).

(2)   For every $z \notin D$ and $\varepsilon > 0$ there exists $g_{z,\varepsilon} \in L^2(S^2)$ such that

$$\lim_{\varepsilon \to 0} \left\| Fg_{z,\varepsilon} - \phi_z \right\|_{L^2(S^2)} = 0 \tag{7.19}$$

and

$$\lim_{\varepsilon \to 0} \left\| g_{z,\varepsilon} \right\|_{L^2(S^2)} = \infty \quad and \quad \lim_{\varepsilon \to 0} \left\| v_{g_{z,\varepsilon}} \right\|_{H^1(D)} = \infty. \tag{7.20}$$

In this case $g_{z,\varepsilon}$ can be chosen as the Tikhonov regularization of the equation $Fg = \phi_z$, i.e., as the unique solution of

$$(\varepsilon I + F^*F)g = F^*\phi_z. \tag{7.21}$$

Proof: By the trace theorem there exists $c_1 > 0$ with $\left\| w|_{\partial D} \right\|_{H^{1/2}(\partial D)} \leq c_1 \|w\|_{H^1(D)}$ for all $w \in H^1(D)$. Furthermore, by the well-posedness of the exterior Dirichlet problem there exists $c_2 > 0$ with $\|w^\infty\|_{L^2(S^2)} \leq c_2 \left\| w|_{\partial D} \right\|_{H^{1/2}(\partial D)}$ for all radiating solutions w of the Helmholtz equation $\Delta w + k^2 w = 0$ in $\mathbb{R}^3 \setminus \overline{D}$.

(1)   Let $z \in D$ and $\varepsilon > 0$. From the assumption on k^2 there exists a unique solution $w_z \in H^1(D)$ of $\Delta w + k^2 w = 0$ in D with $w_z = \Phi(\cdot, z)$ on ∂D. By Runge's Approximation Theorem 7.3 there exists $g_{z,\varepsilon} \in L^2(S^2)$ with

$$\left\| v_{g_{z,\varepsilon}} + w_z \right\|_{H^1(D)} \leq \frac{1}{c_1 c_2}\,\varepsilon$$

and thus

$$\left\| v_{g_{z,\varepsilon}} + \Phi(\cdot, z) \right\|_{H^{1/2}(\partial D)} \leq \frac{1}{c_2}\, \varepsilon \,. \tag{7.22}$$

Since $F g_{z,\varepsilon}$ and ϕ_z are the far fields of the scattered fields with boundary data $-v_{g_{z,\varepsilon}}$ and $\Phi(\cdot, z)$ on ∂D, respectively, we conclude that

$$\left\| F g_{z,\varepsilon} - \phi_z \right\|_{L^2(S^2)} \leq \varepsilon$$

which proves (7.17). Assume now, on the contrary, that there exists $z^* \in \partial D$ and $M > 0$ and a sequence $z_n \in D$ with $z_n \to z^*$ and $\|v_n\|_{H^1(D)} \leq M$ where we have set $v_n = v_{g_{z_n,\varepsilon}}$ for abbreviation. Therefore, by (7.22),

$$\|\Phi(\cdot, z_n)\|_{H^{1/2}(\partial D)} \leq \|\Phi(\cdot, z_n) + v_n\|_{H^{1/2}(\partial D)} + \|v_n\|_{H^{1/2}(\partial D)}$$

$$\leq \frac{1}{c_2}\, \varepsilon + c_1 M \quad \text{for all } n\,.$$

This contradics the fact that $\lim_{n\to\infty} \|\Phi(\cdot, z_n)\|_{H^1(G\setminus\overline{D})} \to \infty$ for any bounded region $G \subset \mathbb{R}^3$ which contains $\overline{D}$ in it's interior. Therefore, the second statement of (7.18) is proven. The first statement follows now directly from the boundedness of the mapping $g \mapsto v_g|_D$ from $L^2(S^2)$ into $H^1(D)$.

(2) For every $z \notin D$ and $\varepsilon > 0$ define $g_{z,\varepsilon} \in L^2(S^2)$ by the solution of (7.21).[3] We introduce the Tikhonov functional $J_{z,\varepsilon} : L^2(S^2) \to \mathbb{R}$ by

$$J_{z,\varepsilon}(g) = \|Fg - \phi_z\|^2_{L^2(S^2)} + \varepsilon \, \|g\|^2_{L^2(S^2)}, \quad g \in L^2(S^2)\,.$$

From the binomial equation in the form

$$J_{z,\varepsilon}(g) - J_{z,\varepsilon}(g_{z,\varepsilon}) = 2\,\mathrm{Re}\big[(Fg_{z,\varepsilon} - \phi_z, Fg - Fg_{z,\varepsilon})_{L^2(S^2)} + \varepsilon\,(g_{z,\varepsilon}, g - g_{z,\varepsilon})_{L^2(S^2)} \big]$$

$$+ \|F(g - g_{z,\varepsilon})\|^2_{L^2(S^2)} + \varepsilon\,\|g - g_{z,\varepsilon}\|^2_{L^2(S^2)}$$

$$= 2\,\mathrm{Re}\big[(F^*F g_{z,\varepsilon} - F^*\phi_z + \varepsilon\, g_{z,\varepsilon}, g - g_{z,\varepsilon})_{L^2(S^2)} \big]$$

$$+ \|F(g - g_{z,\varepsilon})\|^2_{L^2(S^2)} + \varepsilon\,\|g - g_{z,\varepsilon}\|^2_{L^2(S^2)}$$

$$= \|F(g - g_{z,\varepsilon})\|^2_{L^2(S^2)} + \varepsilon\,\|g - g_{z,\varepsilon}\|^2_{L^2(S^2)}$$

for all $g \in L^2(S^2)$ we conclude that $g_{z,\varepsilon}$ is the unique minimizer of $J_{z,\varepsilon}$ on $L^2(S^2)$. Furthermore, from

$$\|Fg_{z,\varepsilon} - \phi_z\|^2_{L^2(S2)} \leq J_{z,\varepsilon}(g_{z,\varepsilon}) \leq J_{z,\varepsilon}(g) = \|Fg - \phi_z\|^2_{L^2(S2)} + \varepsilon\,\|g\|^2_{L^2(S2)}$$

and the denseness of the range of F by Theorem 1.8 we derive (7.19).

---

[3] Since the operator is obviously a Fredholm operator of index zero existence follows from uniqueness which is easily seen after multiplication of the homogeneous equation by g.

As in part (1) it remains to prove the second statement of (7.20). Assume on the contrary that $\left\| v_{g_{z,\varepsilon_n}} \right\|_{H^1(D)} \leq M$ for some $M > 0$ and some sequence $\varepsilon_n \to 0$. Then there exists a subsequence of $v_n = v_{g_{z,\varepsilon_n}}$ which converges weakly to some $v \in H^1(D)$. Let v^s be the radiating solution of the Helmholtz equation in the exterior of D with boundary data $v|_{\partial D}$ on ∂D and v^∞ be the corresponding far field pattern. Since Fg_{z,ε_n} is the far field pattern of the scattered field with boundary data $-v_n$ we conclude that $Fg_{z,\varepsilon_n} \to -v^\infty$ and thus $v^\infty = -\phi_z$ on S^2. Rellich's Lemma and unique continuation yield that $v^s = -\Phi(\cdot, z)$ in $\mathbb{R}^3 \setminus (D \cup \{z\})$. As in the proof of Theorem 1.12 this leads to a contradiction since v^s is in $H^1(U \setminus \overline{D})$ for some neighborhood U of z while this is not the case for $\Phi(\cdot, z)$. This proves the second statement of (7.20). $\square$

As we can imagine from the tools used in the proof of this theorem this kind of justification of the Linear Sampling Method can be done for a wide variety of inverse scattering problems. We refer to [24, 23, 26, 28, 35, 37, 38, 40, 51, 52, 80,65]. The drawback of Theorem 7.5 – and all other results which try to explain the fact that, for a given regularization strategy, the regularized solution of (7.15) is much better for z inside of D than for z outside of D – is that there is no guarantee that for $z \in D$ the regularization technique for solving (7.15) will actually pick the density $g_{z,\varepsilon}$. However, numerically the method has proven to be very effective for a large class of obstacle scattering problems.

In [8, 11] Tilo Arens and Armin Lechleiter gave a rigorous mathematical justification of the Linear Sampling Method for those inverse scattering problems for which the characterization by the $(F^*F)^{1/4}$ – method of Subsection 1.4.3 holds. We restrict ourselves to the case where $g = g_{z,\varepsilon}$ is determined as the Tikhonov regularization of the equation (7.15), i.e., the solution of (7.21).

Theorem 7.6 *Let the assumptions of the previous Theorem 7.5 hold. Furthermore, for every $z \in D$ let $g_z \in L^2(S^2)$ denote the solution of $(F^*F)^{1/4}g_z = \phi_z$, i.e. the solution obtained by the Factorization Method, and for every $z \in \mathbb{R}^3$ and $\varepsilon > 0$ let $g = g_{z,\varepsilon} \in L^2(S^2)$ be the Tikhonov-approximation of (7.15), i.e. the unique solution of $\varepsilon g + F^*Fg = F^*\phi_z$ which is computed by the Linear Sampling Method. $v_{g_{z,\varepsilon}}$ denotes the corresponding Herglotz wave function.*

(a) For every $z \in D$ the limit

$$\lim_{\varepsilon \to 0} v_{g_{z,\varepsilon}}(z)$$

exists. Furthermore, there exists $c > 0$, depending on F only, such that for all $z \in D$ the following estimates hold:

$$c \left\| g_z \right\|_{L^2(S^2)}^2 \leq \lim_{\varepsilon \to 0} \left| v_{g_{z,\varepsilon}}(z) \right| \leq \left\| g_z \right\|_{L^2(S^2)}^2. \tag{7.23}$$

(b) For $z \notin D$ the absolute values $\left| v_{g_{z,\varepsilon}}(z) \right|$ tend to infinity as ε tends to zero.

Proof: Since F is one-to-one and normal (Theorem 1.8) there exists an orthonormal system $\{\psi_j : j \in \mathbb{N}\}$ of eigenfunctions ψ_j corresponding to eigenvalues $\lambda_j \in \mathbb{C}$ of F.

From

$$Fg = \sum_{j=1}^{\infty} \lambda_j \, (g, \psi_j)_{L^2(S^2)} \, \psi_j \,, \quad g \in L^2(S^2) \,, \quad \text{and}$$

$$F^*g = \sum_{j=1}^{\infty} \overline{\lambda_j} \, (g, \psi_j)_{L^2(S^2)} \, \psi_j \,, \quad g \in L^2(S^2) \,,$$

one computes the Tikhonov approximation $g_{z,\varepsilon}$ as

$$g_{z,\varepsilon} = \sum_{j=1}^{\infty} \frac{\overline{\lambda_j}}{|\lambda_j|^2 + \varepsilon} \, (\phi_z, \psi_j)_{L^2(S^2)} \, \psi_j \,. \tag{7.24}$$

From $v_g(z) = (g, \phi_z)_{L^2(S^2)}$ for any $g \in L^2(S^2)$ we conclude that

$$v_{g_{z,\varepsilon}}(z) = \sum_{j=1}^{\infty} \frac{\overline{\lambda_j}}{|\lambda_j|^2 + \varepsilon} \, \left| (\phi_z, \psi_j)_{L^2(S^2)} \right|^2 \,. \tag{7.25}$$

(a) Let now $z \in D$. Then $(F^*F)^{1/4} g_z = \phi_z$ is solvable in $L^2(S^2)$ by Theorem 1.25 and thus $(\phi_z, \psi_j)_{L^2(S^2)} = ((F^*F)^{1/4} g_z, \psi_j)_{L^2(S^2)} = (g_z, (F^*F)^{1/4} \psi_j)_{L^2(S^2)} = \sqrt{|\lambda_j|} (g_z, \psi_j)_{L^2(S^2)}$. Therefore, we can express $v_{g_{z,\varepsilon}}(z)$ as

$$v_{g_{z,\varepsilon}}(z) = \sum_{j=1}^{\infty} \frac{\overline{\lambda_j} \, |\lambda_j|}{|\lambda_j|^2 + \varepsilon} \, \left| (g_z, \psi_j)_{L^2(S^2)} \right|^2 = \|g_z\|^2_{L^2(S^2)} \sum_{j=1}^{\infty} \rho_j \frac{\overline{\lambda_j} \, |\lambda_j|}{|\lambda_j|^2 + \varepsilon} \,, \tag{7.26}$$

where $\rho_j = \left| (g_z, \psi_j)_{L^2(S^2)} \right|^2 / \|g_z\|^2_{L^2(S^2)}$ is non-negative with $\sum_j \rho_j = 1$. An elementary argument (theorem of dominated convergence) yields convergence

$$\sum_{j=1}^{\infty} \rho_j \frac{\overline{\lambda_j} \, |\lambda_j|}{|\lambda_j|^2 + \varepsilon} \longrightarrow \sum_{j=1}^{\infty} \rho_j \frac{\overline{\lambda_j}}{|\lambda_j|} = \sum_{j=1}^{\infty} \rho_j \overline{s_j}$$

as ε tends to zero where we denote again by $s_j = \lambda_j / |\lambda_j|$ the signum of λ_j. The properties of ρ_j imply that the limit belongs to the closure M of the convex hull of the complex numbers $\{s_j : j \in \mathbb{N}\}$. Now we use the fact that s_j lie on the upper half circle in $\mathbb{C}$ with center 0 and radius 1 and converge to 1 (cf. Figure 1.5). Therefore, M has a positive distance c from the origin (see Figure 1.6) and thus

$$\left| \sum_{j=1}^{\infty} \rho_j \overline{s_j} \right| \geq c$$

which proves the first estimate of (7.23). The second estimate is seen directly from (7.26).

(b) Let now $z \notin D$ and assume on the contrary that there exists a sequence $\{\varepsilon_n\}$ which tends to zero and such that $|v_n(z)|$ is bounded. Here we have set $v_n = v_{g_{z,\varepsilon_n}}$ for abbreviation. Since s_j converge to 1 there exists $j_0 \in \mathbb{N}$ with $\operatorname{Re} \lambda_j > 0$ for $j \geq j_0$. From (7.25) for $\varepsilon = \varepsilon_n$ we get

$$v_n(z) = \sum_{j=1}^{j_0-1} \frac{\overline{\lambda_j}}{|\lambda_j|^2 + \varepsilon_n} \left|(\phi_z, \psi_j)_{L^2(S^2)}\right|^2 + \sum_{j=j_0}^{\infty} \frac{\overline{\lambda_j}}{|\lambda_j|^2 + \varepsilon_n} \left|(\phi_z, \psi_j)_{L^2(S^2)}\right|^2.$$

Since the finite sum is certainly bounded for $n \in \mathbb{N}$ there exists $c_1 > 0$ such that

$$\left| \sum_{j=j_0}^{\infty} \frac{\lambda_j}{|\lambda_j|^2 + \varepsilon_n} \left|(\phi_z, \psi_j)_{L^2(S^2)}\right|^2 \right| \leq c_1 \quad \text{for all } n \in \mathbb{N}.$$

Observing that for any complex number $w \in \mathbb{C}$ with $\operatorname{Re} w \geq 0$ and $\operatorname{Im} w \geq 0$ we have that $\operatorname{Re} w + \operatorname{Im} w \geq |w|$ we conclude (note that also $\operatorname{Im} \lambda_j > 0$)

$$2 c_1 \geq 2 \left| \sum_{j=j_0}^{\infty} \frac{\lambda_j}{|\lambda_j|^2 + \varepsilon_n} \left|(\phi_z, \psi_j)_{L^2(S^2)}\right|^2 \right|$$

$$\geq \sum_{j=j_0}^{\infty} \frac{\operatorname{Re} \lambda_j + \operatorname{Im} \lambda_j}{|\lambda_j|^2 + \varepsilon_n} \left|(\phi_z, \psi_j)_{L^2(S^2)}\right|^2$$

$$\geq \sum_{j=j_0}^{\infty} \frac{|\lambda_j|}{|\lambda_j|^2 + \varepsilon_n} \left|(\phi_z, \psi_j)_{L^2(S^2)}\right|^2$$

$$\geq \sum_{j=j_0}^{J} \frac{|\lambda_j|}{|\lambda_j|^2 + \varepsilon_n} \left|(\phi_z, \psi_j)_{L^2(S^2)}\right|^2$$

for all $n \in \mathbb{N}$ and all $J \geq j_0$. Letting n tend to infinity yields convergence of the series

$$\sum_{j=j_0}^{\infty} \frac{1}{|\lambda_j|} \left|(\phi_z, \psi_j)_{L^2(S^2)}\right|^2.$$

Therefore, the function

$$g_z = \sum_{j=1}^{\infty} \frac{1}{\sqrt{|\lambda_j|}} (\phi_z, \psi_j)_{L^2(S^2)} \, \psi_j$$

is well defined in $L^2(S^2)$ and a solution of $(F^*F)g_z = \phi_z$. This contradicts the fact that $z \notin D$ by Theorem 1.25 and ends the proof. $\square$

Remark: It can be shown that $v_{g_{z,\varepsilon}}$ converges in $H^1(D)$ to the solution v of the interior boundary value problem for the Helmholtz equation with Dirichlet boundary values $\Phi(\cdot, z)$ (cf. [11]).

From the Factorization Method we know that $\|g_z\|_{L^2(S^2)}$ tends to infinity as z approaches the boundary Γ. Therefore, also the limit $\lim_{\varepsilon \to 0} |v_{g_{z,\varepsilon}}(z)|$ tends to infinity as z approaches the boundary Γ. This is exactly what is observed in the Linear Sampling Method (see [11]).

7.3 The singular sources method

As in the previous section (and in Chapter 1) we consider the simple inverse scattering problem where the underlying model is the Dirichlet problem for the Helmholtz equation in $\mathbb{R}^3$. Therefore, let $u^s = u^s(x, \theta)$ be the scattered field corresponding to the wavenumber $k > 0$, domain $D \subset \mathbb{R}^3$, and incident field $u^i(x, \theta) = \exp(ikx \cdot \theta)$ for $x \in \mathbb{R}^3$. We assume that D is bounded such that $\mathbb{R}^3 \setminus \overline{D}$ is connected and $\partial D \in C^2$. The scattered field u^s satisfies the exterior boundary value problem (1.18), (1.19), (1.20) for $f = -u^i$, i.e.,

$$\Delta u^s + k^2 u^s = 0 \quad \text{outside } D, \tag{7.27}$$

$$u^s = -u^i \quad \text{on } \partial D, \tag{7.28}$$

and

$$\frac{\partial u^s}{\partial r} - ik\, u^s = \mathcal{O}\!\left(r^{-2}\right) \quad \text{for } r = |x| \to \infty \tag{7.29}$$

uniformly with respect to $\hat{x} = x/|x| \in S^2$.

The far field pattern of u^s is again denoted by $u^\infty = u^\infty(\hat{x}, \theta)$ for $\hat{x}, \theta \in S^2$. The *inverse scattering problem* is to determine D from $u^\infty(\hat{x}, \theta)$ for $\hat{x}, \theta \in S^2$. As before we denote by $F : L^2(S^2) \to L^2(S^2)$ the far field operator, defined by

$$(Fg)(\hat{x}) = \int_{S^2} u^\infty(\hat{x}, \theta)\, g(\theta)\, ds(\theta), \quad \hat{x} \in S^2. \tag{7.30}$$

The basic tool in the *Singular Sources Method* is to consider the scattered field $v^s = v^s(x, z)$ which corresponds to the incident field $v^i(x) = \Phi(x, z)$ where $z \notin \overline{D}$ is a given source point. First, we give a new proof of the following result (cf. [160], Theorem 2.1.15).

Theorem 7.7 *There exists $c > 0$ (depending on D and k only) such that*

$$\left| v^s(z, z) \right| \geq \frac{c}{d(z, \partial D)} \quad \text{for all } z \notin \overline{D}. \tag{7.31}$$

Here, $d(z, \partial D) = \inf\{|z - y| : y \in \partial D\}$ denotes the distance of z to the boundary of D.

Proof: It is sufficient to prove the assertion locally, i.e., in a strip $D_h = \{y + t\nu(y) : y \in \partial D,\ 0 < t \leq h\}$ around ∂D. As shown in [42], Chapter 2, we can choose h small enough such that every $z \in D_h$ has a unique representation as $z = y + tv(y)$ for some $y \in \partial D$ and some $t \in (0, h]$. Furthermore, the smoothness of ∂D assures the existence of a constant $c_1 > 0$ with $\left| (y - x) \cdot v(x) \right| \leq c_1 |x - y|^2$ for all $x, y \in \partial D$ and thus

$$|x - y \pm tv(y)| = \sqrt{|x - y|^2 + t^2 \pm 2t\,(x - y) \cdot v(y)} \geq \sqrt{(1 - 2tc_1)|x - y|^2 + t^2}$$

$$\geq \sqrt{(1 - 2hc_1)|x - y|^2 + t^2} \geq \frac{1}{2}\sqrt{|x - y|^2 + t^2}$$

for $h \leq 3/(8c_1)$. In particular, this yields $d(z, \partial D) \geq t/2$.

In the first part of the proof we show that there exists a constant $c_2 > 0$ such that

$$\left| \Phi(x, y + tv(y)) - \Phi(x, y - tv(y)) \right| \leq c_2 \quad \text{for all } x, y \in \partial D, \ t \in (0, h]. \tag{7.32}$$

Indeed, since $\exp(ik|x - z|)/(4\pi|x - z|) - 1/(4\pi|x - z|)$ is bounded it is sufficient to prove the existence of $c > 0$ such that

$$A(x, y, t) := \left| \frac{1}{|x - y - tv|} - \frac{1}{|x - y + tv|} \right| \leq c \quad \text{for all } x, y \in \partial D, \ t \in (0, h].$$

Here, we have written v for $v(y)$. We estimate

$$A(x, y, t) = \frac{\left| |x - y + tv| - |x - y - tv| \right|}{|x - y - tv||x - y + tv|}$$

$$= \frac{\left| |x - y + tv|^2 - |x - y - tv|^2 \right|}{|x - y - tv||x - y + tv|\big[|x - y + tv| + |x - y - tv| \big]}$$

$$\leq \frac{4t\left| (x - y) \cdot v \right|}{\frac{1}{4}\left(|x - y|^2 + t^2 \right)\sqrt{|x - y|^2 + t^2}} \leq 16c_1 \frac{t\,|x - y|^2}{\left(|x - y|^2 + t^2 \right)^{3/2}}$$

It is easy to see that the function $\psi(t) = t\,a^2/(a^2 + t^2)^{3/2}$ is bounded by $\left(\frac{2}{3} \right)^{3/2}/\sqrt{2}$ for all $t \geq 0$ and $a \in \mathbb{R}$. This proves (7.33).

For any $z = y + tv(y) \in D_h$ we define $h_z \in C(\partial D)$ by

$$h_z(x) = \Phi(x, y - tv(y)), \quad x \in \partial D.$$

From (7.33) we conclude that $\|h_z - \Phi(\cdot, z)\|_{C(\partial D)} \leq c_2$ for all $z \in D_h$. Since $y - tv(y) \in D$ we conclude that $\Phi(\cdot, y - tv(y))$ is the solution of the exterior boundary value problem with boundary data h_z on ∂D. The continuous dependence of solutions of the exterior Dirichlet boundary value problem yields the existence of a constant c_3 (only dependent on D and k) such that

$$\left\| v^s(\cdot, z) + \Phi(\cdot, y - tv(y)) \right\|_{C(\mathbb{R}^3 \setminus D)} \leq c_3 \| - \Phi(\cdot, z) + h_z \|_{C(\partial D)} \leq c_3 c_2.$$

Therefore, by the triangle inequality,

$$
\begin{aligned}
\left| v^s(z,z) \right| &= \left| \left[v^s(z,z) + \Phi\big(z, y - tv(y)\big) \right] - \Phi\big(y + tv(y), y - tv(y)\big) \right| \\
&\geq \left| \Phi\big(y + tv(y), y - tv(y)\big) \right| - \left\| v^s(\cdot, z) + \Phi\big(\cdot, y - tv(y)\big) \right\|_{C(\mathbb{R}^3 \setminus D)} \\
&\geq \frac{1}{8\pi\, t} - c_3 c_2 \geq \frac{1}{8\pi\, t} - \frac{1}{16\pi\, h} \geq \frac{1}{16\pi\, t} \geq \frac{c}{d(z, \partial D)}
\end{aligned}
$$

for $h \leq 1/(16\pi c_2 c_3)$. This ends the proof. $\square$

Now we proceed with the derivation of the Singular Sources Method. We fix $z \notin \overline{D}$ and $\varepsilon > 0$ and a bounded domain $G_z \subset \mathbb{R}^3$ such that its exterior is connected and $z \notin \overline{G_z}$ and $\overline{D} \subset G_z$. Applying Theorem 7.3 to $u = \Phi(\cdot, z)$ in some open and bounded domain G such that its exterior is connected and $\overline{G_z} \subset G$ and $z \notin \overline{G}$ yields the existence of $g \in L^2(S^2)$ depending on z, G_z, and ε such that

$$
\| v_g - \Phi(\cdot, z) \|_{C(\overline{G_z})} \leq \varepsilon \, . \tag{7.33}
$$

In the following we indicate the dependence on ε by writing g_ε. The dependence on z, and G_z is not indicated. The previous formula implies, in particular, that

$$
\| v_{g_\varepsilon} - \Phi(\cdot, z) \|_{C(\partial D)} \leq \varepsilon \, .
$$

By the superposition principle, the scattered wave corresponding to the incident field v_{g_ε} is given by

$$
\int_{S^2} u^s(\cdot, \theta)\, g_\varepsilon(\theta)\, ds(\theta) \, .
$$

From the continuous dependence property of solutions of the exterior Dirichlet problem there exists $c_1 > 0$ such that

$$
\left| \int_{S^2} u^s(x, \theta)\, g_\varepsilon(\theta)\, ds(\theta) - v^s(x, z) \right| \leq c_1 \| v_{g_\varepsilon} - \Phi(\cdot, z) \|_{C(\partial D)} \leq c_1 \varepsilon
$$

for all $x \notin D$. In particular,

$$
\left| \int_{S^2} u^s(z, \theta)\, g_\varepsilon(\theta)\, ds(\theta) - v^s(z, z) \right| \leq c_1 \varepsilon \, .
$$

Furthermore, the continuous dependence result yields

$$
\left| \int_{S^2} u^\infty(\hat{x}, \theta)\, g_\varepsilon(\theta)\, ds(\theta) - v^\infty(\hat{x}, z) \right| \leq c_2 \varepsilon \quad \text{for all } \hat{x} \in S^2 ,
$$

for some $c_2 > 0$, i.e.,

$$\| v^\infty(\cdot, z) - F g_\varepsilon \|_{C(S^2)} \le c_2 \, \varepsilon \,.$$

Now we use the mixed reciprocity principle of Theorem 1.7 which yields that $u^s(z, \theta) = v^\infty(-\theta, z)$. Combining these estimates for different g_δ and g_ε yields:

$$\left| v^s(z, z) - \int_{S^2} (F g_\varepsilon)(-\theta) \, g_\delta(\theta) \, ds(\theta) \right| \le \left| v^s(z, z) - \int_{S^2} u^s(z, \theta) \, g_\delta(\theta) \, ds(\theta) \right|$$

$$+ \int_{S^2} \left| v^\infty(-\theta, z) - (F g_\varepsilon)(-\theta) \right| \left| g_\delta(\theta) \right| ds(\theta)$$

$$\le c_1 \, \delta + c_2 \, \varepsilon \int_{S^2} \left| g_\delta(\theta) \right| ds(\theta) \,.$$

Therefore, we have shown the following convergence result for the singular sources method:

Theorem 7.8 *Let $u^\infty = u^\infty(\hat{x}, \theta)$, $\hat{x}, \theta \in S^2$, be the far field pattern of problem (7.27), (7.28), (7.29) and denote the far field operator (7.30) by F. Fix $z \notin \overline{D}$ and a bounded domain $G_z \subset \mathbb{R}^3$ such that its exterior is connected and $z \notin \overline{G_z}$ and $\overline{D} \subset G_z$. For any $\varepsilon > 0$ choose $g = g_\varepsilon \in L^2(S^2)$ with (7.33). Then*

$$\lim_{\delta \to 0} \lim_{\varepsilon \to 0} \int_{S^2} (F g_\varepsilon)(-\theta) \, g_\delta(\theta) \, ds(\theta) = v^s(z, z) \,,$$

i.e., by substituting the form of F,

$$\lim_{\delta \to 0} \lim_{\varepsilon \to 0} \int_{S^2} \int_{S^2} u^\infty(-\theta, \eta) \, g_\varepsilon(\eta) \, g_\delta(\theta) \, ds(\eta) \, ds(\theta) = v^s(z, z) \,.$$

Combining this with (7.31) yields

$$\lim_{\delta \to 0} \lim_{\varepsilon \to 0} \left| \int_{S^2} \int_{S^2} u^\infty(-\theta, \eta) \, g_\varepsilon(\eta) \, g_\delta(\theta) \, ds(\eta) \, ds(\theta) \right| \ge \frac{c}{d(z, \partial D)} \,. \qquad (7.34)$$

This result assures that for z sufficiently close to boundary ∂D (and regions G_z chosen appropriately) the quantity

$$\lim_{\delta \to 0} \lim_{\varepsilon \to 0} \left| \int_{S^2} \int_{S^2} u^\infty(-\theta, \eta) \, g_\varepsilon(\eta) \, g_\delta(\theta) \, ds(\eta) \, ds(\theta) \right|$$

becomes large.

In [160] Roland Potthast used domains G_z of the special form

$$G_z = G_{z,p} = (z + \rho p) + \left\{ x \in \mathbb{R}^3 : |x| < R, \ \frac{x}{|x|} \cdot p > -\cos\beta \right\}$$

for some (large) radius $R > 0$, opening angle $\beta \in (-\pi/2, \pi/2)$, direction of opening $p \in S^2$, and $\rho > 0$. The dependence on β, ρ, and R is not indicated since they are kept fixed. This domain $G_{z,p}$ is a ball centered at $z + \rho p$ with radius R from which the cone of direction $-p$ and opening angle β has been removed. Obviously, it is chosen such that $z \notin \overline{G_{z,p}}$. These sets $G_{z,p}$ are translations and rotations of the reference set

$$\hat{G} = \rho\hat{p} + \left\{ x \in \mathbb{R}^3 : |x| < R, \ \frac{x}{|x|} \cdot \hat{p} > -\cos\beta \right\}$$

for some fixed $\hat{p} \in S^2$. Indeed, it is easily seen that for any $p = M\hat{p} \in S^2$ where $M \in \mathbb{R}^{3\times 3}$ is orthogonal, and for any $z \in \mathbb{R}^3$ it holds that $G_{z,p} = z + M\hat{G}$. Therefore, assume that for some $\varepsilon > 0$ a function $\hat{g} \in L^2(S^2)$ has been found such that

$$\left\| v_{\hat{g}}\big|_{\hat{G}} - \Phi(\cdot, 0) \right\|_{C(\hat{G})} \leq \varepsilon.$$

Setting $g(\theta) := \exp(-ikz \cdot \theta)\, \hat{g}(M^{\top}\theta)$ we have for $x = z + Mx' \in G_{z,p}$ where $x' \in \hat{G}$,

$$\Phi(x, z) - v_g(x) = \Phi(0, Mx') - \int_{S^2} g(\theta)\, e^{ik(z + Mx')\cdot\theta}\, ds(\theta)$$

$$= \Phi(0, x') - \int_{S^2} e^{ikz\cdot\theta}\, g(\theta)\, e^{ik(Mx')\cdot\theta}\, ds(\theta)$$

$$= \Phi(0, x') - \int_{S^2} e^{ikz\cdot(M\theta)}\, g(M\theta)\, e^{ikx'\cdot\theta}\, ds(\theta)$$

$$= \Phi(0, x') - \int_{S^2} \hat{g}(\theta)\, e^{ikx'\cdot\theta}\, ds(\theta)$$

$$= \Phi(0, x') - v_{\hat{g}}(x')$$

and thus

$$\left\| v_g\big|_{G_{z,p}} - \Phi(\cdot, z) \right\|_{C(G_{z,p})} = \left\| v_{\hat{g}}\big|_{\hat{G}} - \Phi(\cdot, 0) \right\|_{C(\hat{G})} \leq \varepsilon.$$

Therefore, for given small values of $\varepsilon > 0$ and $\delta > 0$ the computations of the approximations $g_\varepsilon, g_\delta \in L^2(S^2)$ for arbitrary z and $G_{z,p}$ are very simple once the approximations for $z = 0$ and $\hat{G}$ have been computed.

With these transformations, we can consider the singular sources method as a sampling method with sampling objects z and M.

We note that from the arguments used in the proof of Theorem 7.8 we are not able to show that the common limit $\lim_{\varepsilon,\delta\to 0}$ exists. Using the factorization (1.55), the following stronger result than (7.34) is obtained as a corollary.[4]

Theorem 7.9 *Let again F be the far field operator (7.30). Fix $z \notin \overline{D}$ and a bounded domain $G_z \subset \mathbb{R}^3$ such that its exterior is connected and $z \notin \overline{G_z}$ and $\overline{D} \subset G_z$. For any $\varepsilon > 0$ choose $g_\varepsilon \in L^2(S^2)$ with (7.33) with respect to the H^1-norm, i.e.,*

$$\left\| v_{g_\varepsilon} \big|_{G_z} - \Phi(\cdot, z) \right\|_{H^1(G_z)} \leq \varepsilon .$$

Assume furthermore that k^2 is not a Dirichlet eigenvalue of $-\Delta$ in D. Then there exists a constant $c > 0$ depending only on D and k such that

$$\left| \lim_{\varepsilon\to 0} \int\limits_{S^2} \int\limits_{S^2} u^\infty(\theta, \eta)\, g_\varepsilon(\eta)\, \overline{g_\varepsilon(\theta)}\, ds(\eta)\, ds(\theta) \right| = \lim_{\varepsilon\to 0} \left| (Fg_\varepsilon, g_\varepsilon)_{L^2(S^2)} \right| \geq \frac{c}{d(z, \partial D)} .$$

$$(7.35)$$

Proof: From (1.55) and Corollary 1.18 and the fact that $Hg_\varepsilon = v_{g_\varepsilon}\big|_{\partial D}$ we conclude that

$$\lim_{\varepsilon\to 0} \left| (Fg_\varepsilon, g_\varepsilon)_{L^2(S^2)} \right| = \lim_{\varepsilon\to 0} \left| \langle S^{-1} Hg_\varepsilon, Hg_\varepsilon \rangle \right| = \left| \langle S^{-1} \Phi(\cdot, z), \Phi(\cdot, z) \rangle \right|$$

$$\geq c_1 \left\| S^{-1} \Phi(\cdot, z) \right\|^2_{H^{-1/2}(\partial D)} \geq c_2 \left\| \Phi(\cdot, z) \right\|^2_{H^{1/2}(\partial D)} \geq \frac{c}{d(z, \partial D)} .$$

For a proof of the last estimate we refer to [11]. This proves the theorem. $\square$

7.4 The probe method

In this section we describe the Probe Method. Originally, it has been proposed by Masaru Ikehata for the inverse problem of impedance tomography of Chapter 6. In the first subsection we will treat this inverse problem and follow Ikehata's paper [96] but generalize it to more general admittivity functions γ. In the second subsection we will modify it according to the approach presented in [77] and treat the inverse scattering problem for mixed boundary conditions as we have discussed it in Chapter 3.

7.4.1 The probe method in impedance tomography

First, we recall the formulation of the problem from Chapter 6.

Let $B \subset \mathbb{R}^3$ be a bounded and connected domain with smooth boundary ∂B. For the *direct problem* the (in general complex and matrix valued) admittivity function

---

[4] At least if k^2 is not an eigenvalue.

$\gamma \in L^{\infty}(B, \mathbb{C}^{3\times 3})$ and the boundary data $f \in H_{\diamond}^{-1/2}(\partial B)$ are given and $u \in H_{\diamond}^{1}(B)$ has to be determined such that

$$\operatorname{div}\left[\gamma(x)\nabla u(x)\right] = 0 \quad \text{in } B \tag{7.36}$$

and

$$\partial_{\gamma} u = \nu \cdot \gamma \nabla u = f \quad \text{on } \partial B. \tag{7.37}$$

The spaces $H_{\diamond}^{-1/2}(\partial B)$ and $H_{\diamond}^{1}(B)$ are defined in (6.6) and (6.8), respectively[5]. Again, the solution of this problem has to be understood in the variational sense:

$$\iint_{B} \nabla \psi^{*} \gamma \nabla u \, dx = \langle f, \psi \rangle \quad \text{for all } \psi \in H_{\diamond}^{1}(B). \tag{7.38}$$

We assume that $\gamma \in L^{\infty}(B, \mathbb{C}^{3\times 3})$ satisfies the condition (6.9) of Chapter 6, i.e., there exists $c_0 > 0$ such that

$$\operatorname{Im}\left[z^{*}\gamma(x)z\right] \leq 0 \quad \text{and} \quad \operatorname{Re}\left[z^{*}\gamma(x)z\right] \geq c_0 \, |z|^2 \tag{7.39}$$

for all $z \in \mathbb{C}^3$ and almost all $x \in B$. The unique solvability of this boundary value problem (7.36) and (7.37) assures the existence of the *Neumann-to-Dirichlet* operator $\Lambda : H_{\diamond}^{-1/2}(\partial B) \rightarrow H_{\diamond}^{1/2}(\partial B)$ which assigns to each $f \in H_{\diamond}^{-1/2}(\partial B)$ the trace $u|_{\partial B} \in H_{\diamond}^{1/2}(\partial B)$ of the solution $u \in H_{\diamond}^{1}(B)$ of (7.38). The most important properties of Λ are listed in Theorem 6.1 of Chapter 6.

As in Chapter 6 we assume that γ is the perturbation of a known background admittivity function which we take again to be the identity $\gamma_0 = I$. This $\gamma_0 = I$ corresponds to the Neumann-to-Dirichlet operator Λ_0. We make Assumption 6.4 on γ which we recall for the convenience of the reader.

Assumption 7.10 *Let* $\gamma \in L^{\infty}(B, \mathbb{C}^{3\times 3})$ *has the form*

$$\gamma(x) = \begin{cases} I + q(x), & x \in D, \\ I, & x \in B \setminus D, \end{cases}$$

for some domain D with $\overline{D} \subset B$ and some matrix-valued function $q \in L^{\infty}(D, \mathbb{C}^{3\times 3})$ with support $\overline{D}$. Furthermore, we assume that

(a) there exists $c_0 \in (0, 1)$ with

$$\operatorname{Im}[z^{*}\gamma(x)z] \leq 0 \quad and \quad \operatorname{Re}[z^{*}\gamma(x)z] \geq c_0 \, |z|^2 \tag{7.40}$$

for all $z \in \mathbb{C}^3$ and almost all $x \in B$, and

---

5 Recall that the diamond indicates that only functions with vanishing means on ∂B are considered.

(b) there exists $c_1 > 0$ such that either

$$z^*\big[I - \gamma(x)\big(\operatorname{Re}\gamma(x)\big)^{-1}\gamma(x)^*\big]z \geq c_1|z|^2 \tag{7.41}$$

for all $z \in \mathbb{C}^3$ and almost all $x \in D$, or

$$z^*\big[\operatorname{Re}\gamma(x) - I\big]z \geq c_1|z|^2 \tag{7.42}$$

for all $z \in \mathbb{C}^3$ and almost all $x \in D$.

We note that the matrices $\operatorname{Re}\gamma(x) = \frac{1}{2}\big(\gamma(x)+\gamma(x)^*\big)$ are self-adjoint and uniformly positive-definite by assumption (7.40). Therefore, their inverses $\big(\operatorname{Re}\gamma(x)\big)^{-1}$ exist and are essentially bounded on B.

In the *inverse problem* one tries to determine the support D of $\gamma - I$ from the knowledge of $\Lambda - \Lambda_0$.

In the probe method the sampling objects are curves in B starting at the boundary ∂B of B. They are called "needles" in the original work by Ikehata.

Definition 7.11 *A needle is a continuously differentiable function $\eta : [0,1] \to \overline{B}$ such that $\eta(0) \in \partial B$ and $\eta(t) \in B$ for all $t \in (0,1]$ and $\eta'(t) \neq 0$ for all $t \in [0,1]$ and $\eta(t) \neq \eta(s)$ for $t \neq s$.*

The following estimates are the basic ingredients for the Probe Method. They correspond to the estimates (6.31) and (6.32) for the operators T and T_0 of Chapter 6.

Theorem 7.12 *We assume that $\gamma \in L^\infty(D, \mathbb{C}^{3\times3})$ satisfies (7.40) of Assumption 7.10.*

(a) For any $f \in H_\diamond^{-1/2}(\partial B)$ it holds that

$$\operatorname{Re}\langle f, (\Lambda - \Lambda_0)f\rangle \leq \iint_D \nabla u_0^*\big[\gamma(\operatorname{Re}\gamma)^{-1}\gamma^* - I\big]\nabla u_0\,dx\,, \tag{7.43}$$

$$\operatorname{Re}\langle f, (\Lambda - \Lambda_0)f\rangle \geq \iint_D \nabla u_0^*\big[I - (\operatorname{Re}\gamma)^{-1}\big]\nabla u_0\,dx\,, \tag{7.44}$$

where $u_0 \in H_\diamond^1(B)$ denotes the unique solution of (7.36), (7.37) for the background case, i.e., for $\gamma_0 = I$.

(b) Under the additional assumption (7.41) we have that

$$-\left(\frac{1}{c_0} - 1\right)\iint_D |\nabla u_0|^2\,dx \leq \operatorname{Re}\langle f, (\Lambda - \Lambda_0)f\rangle \leq -c_1\iint_D |\nabla u_0|^2\,dx\,. \tag{7.45}$$

(c) Under the additional assumption (7.42) we have that

$$\frac{c_1}{\|\gamma\|_{L^\infty(B)}} \iint_D |\nabla u_0|^2 \, dx \le \mathrm{Re}\langle f, (\Lambda - \Lambda_0)f\rangle \le \left(\frac{\|\gamma\|_{L^\infty(B)}}{c_0} - 1\right) \iint_D |\nabla u_0|^2 \, dx.$$

$$(7.46)$$

Proof: (a) From (7.38) for $\psi = u$ we note that $\iint_B \nabla u^* \gamma \nabla u \, dx = \langle f, \Lambda f\rangle$ and, analogously, for Λ_0. We write γ_0 for the identity I to emphasize the analogous forms of $\langle f, \Lambda f\rangle$ and $\langle f, \Lambda_0 f\rangle$. Therefore,

$$\langle f, (\Lambda - \Lambda_0)f\rangle = \iint_B \left[\nabla u^* \gamma \nabla u - \nabla u_0^* \gamma_0 \nabla u_0\right] dx$$

$$= 2 \iint_B (\nabla u - \nabla u_0)^* \gamma \nabla u \, dx$$

$$\qquad - \iint_B \left[\nabla u_0^* \gamma_0 \nabla u_0 - 2\nabla u_0^* \gamma \nabla u + \nabla u^* \gamma \nabla u\right] dx$$

$$= - \iint_B \left[\nabla u_0^* \gamma_0 \nabla u_0 - 2\nabla u_0^* \gamma \nabla u + \nabla u^* \gamma \nabla u\right] dx$$

since the first integral vanishes by (7.38) (set $\psi = u - u_0 \in H^1_\diamond(B)$ which vanishes on ∂B). Since the matrices $\mathrm{Re}\,\gamma$ are self-adjoint and uniformly positive-definite there exist self-adjoint and positive-definite square roots $(\mathrm{Re}\,\gamma)^{1/2} \in L^\infty(B, \mathbb{R}^{3\times 3})$. Taking the real part of the previous formula yields

$$\mathrm{Re}\langle f, (\Lambda - \Lambda_0)f\rangle = - \iint_B \left|(\mathrm{Re}\,\gamma)^{1/2}\nabla u - (\mathrm{Re}\,\gamma)^{-1/2}\gamma^*\nabla u_0\right|^2 dx$$

$$+ \iint_B \nabla u_0^* \left[\gamma(\mathrm{Re}\,\gamma)^{-1}\gamma^* - \gamma_0\right]\nabla u_0 \, dx$$

$$\le \iint_B \nabla u_0^* \left[\gamma(\mathrm{Re}\,\gamma)^{-1}\gamma^* - \gamma_0\right]\nabla u_0 \, dx$$

This proves (7.43). Similarly, by interchanging the orders of γ and γ_0,

$$\langle f, (\Lambda_0 - \Lambda)f\rangle = \iint_B \left[\nabla u_0^* \gamma_0 \nabla u_0 - \nabla u^* \gamma \nabla u\right] dx$$

$$= - \iint_B \left[\nabla u^* \gamma \nabla u - 2\nabla u^* \gamma_0 \nabla u_0 + \nabla u_0^* \gamma_0 \nabla u_0\right] dx$$

and thus, taking the real part,

$$\mathrm{Re}\langle f,(\Lambda_0-\Lambda)f\rangle = -\iint_B \left|(\mathrm{Re}\,\gamma)^{1/2}\nabla u - (\mathrm{Re}\,\gamma)^{-1/2}\gamma_0^*\nabla u_0\right|^2 dx$$

$$-\iint_B \nabla u_0^*\left[\gamma_0 - \gamma_0(\mathrm{Re}\,\gamma)^{-1}\gamma_0\right]\nabla u_0\,dx$$

$$\leq -\iint_B \nabla u_0^*\left[\gamma_0 - \gamma_0(\mathrm{Re}\,\gamma)^{-1}\gamma_0\right]\nabla u_0\,dx\,.$$

The observation that the integrands vanish on $B\setminus D$ yields the desired estimates.

(b) The upper estimate follows directly from assumption (7.41). For the lower estimate we note that assumption (7.40) yields $z^*\left(\mathrm{Re}\,\gamma(x)\right)^{-1}z \leq (1/c_0)\,|z|^2$ for all $z\in\mathbb{C}^3$ and almost all $x\in D$ and thus $z^*\left[I-\left(\mathrm{Re}\,\gamma(x)\right)^{-1}\right]z \geq (1-1/c_0)\,|z|^2$ which yields the lower estimate of (7.45).

(c) The upper estimate (7.46) follows directly from

$$z^*\gamma(x)\left(\mathrm{Re}\,\gamma(x)\right)^{-1}\gamma(x)^*z - |z|^2 \leq \frac{1}{c_0}|\gamma(x)^*z|^2 - |z|^2 \leq \left(\frac{|\gamma(x)|_F^2}{c_0}-1\right)|z|^2\,.$$

For the lower estimate we write

$$z^*\left[I-\left(\mathrm{Re}\,\gamma(x)\right)^{-1}\right]z = \left[\left(\mathrm{Re}\,\gamma(x)\right)^{-1/2}z\right]^*\left[\mathrm{Re}\,\gamma(x)-I\right]\left[\left(\mathrm{Re}\,\gamma(x)\right)^{-1/2}z\right]$$

$$\geq c_1\left|\left(\mathrm{Re}\,\gamma(x)\right)^{-1/2}z\right|^2 \geq \frac{c_1}{|\gamma(x)|_F}\,|z|^2$$

for all $z\in\mathbb{C}^3$ and almost all $x\in D$. Here, $|\gamma(x)|_F = \sqrt{\sum_{i,j=1}^3 \gamma(x)_{ij}^2}$ denotes the Frobenius norm of the matrix $\gamma(x)$. Observing that $\mathrm{essup}_{x\in B}\,|\gamma(x)|_F = \|\gamma\|_{L^\infty(B)}$ yields the assertion. $\square$

The following corollary of Theorem 7.2 is needed in the Probe Method.

Corollary 7.13 *Let $\eta:[0,1]\to\overline{B}$ be a needle. For any $t\in(0,1]$ there exists a sequence $w_n\in H^1(B)$ of harmonic functions in B such that*

$$\left\|w_n - \Phi_0(\cdot,\eta(t))\right\|_{H^1(U)} \longrightarrow 0,\quad n\to\infty, \tag{7.47}$$

for every open subset U with $\overline{U}\subset B\setminus C_t$. Here, and in the following,

$$C_t = \left\{\eta(s):0\leq s\leq t\right\}$$

is the part of the needle from $s=0$ to $s=t$ and $\Phi_0(x,y)$ denotes the fundamental solution of the Laplace equation in $\mathbb{R}^3$, i.e.,

$$\Phi_0(x,y) = \frac{1}{4\pi|x-y|},\quad x\neq y\,.$$

Proof: Since C_t is compact and $C_t \cap \partial B = \{\eta(0)\}$ the set $B \setminus C_t$ is open and connected. For $n \in \mathbb{N}$ we define the open set $G_n := \{x \in B \setminus C_t : d(x, \partial B) > 1/n, \; d(x, C_t) > 1/n\}$. For sufficiently large n the complement $B \setminus \overline{G_n}$ of G_n is connected. Therefore, by Theorem 7.2 applied to $u = \Phi_0(\cdot, \eta(t))$ in G_n there exists a harmonic function $w_n \in H^1(B)$ in B with

$$\left\| w_n - \Phi_0(\cdot, \eta(t)) \right\|_{H^1(G_n)} \leq \frac{1}{n} .$$

The observation that every compact set $\overline{U} \subset B \setminus C_t$ is contained in G_n for some sufficiently large n yields the assertion. $\square$

We define $f_n \in H_\diamond^{-1/2}(\partial B)$ by $f_n = \partial w_n / \partial \nu$ on ∂B where w_n has the approximation property (7.47) from the previous Corollary 7.13. We note that f_n depends on C_t and indicate this in the following by writing $f_n(C_t)$. We note that it does not depend on the unknown domain D. Therefore, it can be – at least in principle – computed beforehand.

Substituting f_n into (7.45) and (7.46) yields

$$\frac{1}{c} \iint\limits_{D} |\nabla w_n|^2 \, dx \leq \left| \mathrm{Re}\langle f_n, (\Lambda - \Lambda_0) f_n \rangle \right| \leq c \iint\limits_{D} |\nabla w_n|^2 \, dx \qquad (7.48)$$

for some $c > 1$ provided Assumption 7.10 holds. Now we can formulate and prove the main theorem of this section.

Theorem 7.14 *Let Assumption 7.10 hold and fix a needle* $\eta : [0, 1] \to \overline{B}$. *Define the set* $T \subset [0, 1]$ *by*

$$T = \left\{ t \in [0, 1] : \sup_{n \in \mathbb{N}} \left\{ \left| \mathrm{Re}\langle f_n(C_t), (\Lambda - \Lambda_0) f_n(C_t) \rangle \right| < \infty \right\} . \right. \qquad (7.49)$$

Here, $f_n(C_t) = \partial w_n / \partial \nu \in H^{-1/2}(\partial B)$ *and* $w_n \in H^1(B)$ *are determined as in Corollary 7.13.*

Then there exists $\varepsilon > 0$ *with* $[0, \varepsilon] \subset T$. *In particular,* $T \neq \emptyset$.

Furthermore, define $t^* = \sup\{t \in [0, 1] : [0, t] \in T\}$. *Then*

$$t^* = \begin{cases} \min\{t \in [0, 1] : \eta(t) \in \partial D\}, & \text{if } C_1 \cap \overline{D} \neq \emptyset, \\ 1, & \text{if } C_1 \cap \overline{D} = \emptyset. \end{cases} \qquad (7.50)$$

We recall that $C_1 = \{\eta(t) : t \in [0, 1]\}$ *denotes the trace of the needle* η.

Proof: The proof is divided into three parts.

First we show that $C_t \cap \overline{D} = \emptyset$ implies $t \in T$. This would prove the first assertion because $\eta(0) \notin \overline{D}$ and thus also $C_\varepsilon \cap \overline{D} = \emptyset$ for sufficiently small ε. Therefore, let $C_t \cap \overline{D} = \emptyset$. Choose an open set $U \subset B$ which contains $\overline{D}$ such that $\overline{U} \cap C_t = \emptyset$. Since

w_n converges to $\Phi_0(\cdot, \eta(t))$ in $H^1(U)$ it also converges in $H^1(D)$. Therefore, the upper bound in (7.48) converges to

$$c \iint_D |\nabla \Phi_0(x, \eta(t))|^2 dx$$

which yields an upper bound on $|\mathrm{Re}\langle f_n(C_t), (\Lambda - \Lambda_0) f_n(C_t)\rangle|$. Therefore, $t \in T$.

Second, let now $\eta(t) \in \overline{D}$. We show that $t \notin T$ by contradiction. Assume on the contrary that $t \in T$. By the lower estimate of (7.48) there exists $M > 0$ such that

$$\iint_D |\nabla w_n|^2 \, dx \leq M \quad \text{for all } n \in \mathbb{N}. \tag{7.51}$$

Define the (truncated) cone K_0 by

$$K_0 = \left\{ \eta(t) + r\hat{x} : 0 < r \leq \rho \, , \, \hat{x} \in V \right\}$$

where $\rho > 0$ and the (relatively) open subset $V \subset S^2$ are chosen such that $K_0 \subset D \setminus C_t$. It is here that we need differentiability of η and smoothness of ∂D. Then the subset $K_\varepsilon = \left\{ \eta(t) + r\hat{x} \in K_0 : \varepsilon \leq r \leq \rho \right\}$ is compact and contained in $D \setminus C_t$ for every $\varepsilon \in (0, \rho)$. Furthermore, we compute by using polar coordinates with respect to $\eta(t)$,

$$\iint_{K_\varepsilon} |\nabla_x \Phi_0(x, \eta(t))|^2 dx = |V| \int_\varepsilon^\rho \frac{1}{(4\pi)^2 r^4} r^2 dr = \frac{|V|}{(4\pi)^2} \left(\frac{1}{\varepsilon} - \frac{1}{\rho} \right).$$

Therefore, we can choose ε such that

$$\iint_{K_\varepsilon} |\nabla_x \Phi_0(x, \eta(t))|^2 dx \geq 2M \, .$$

This is a contradiction since

$$\iint_{K_\varepsilon} |\nabla w_n|^2 dx \leq \iint_D |\nabla w_n|^2 dx \leq M$$

and $w_n \to \Phi_0(\cdot, \eta(t))$ in $H^1(K_\varepsilon)$ by the approximation property of w_n.

In the third part we prove (7.50). It suffices to consider the case where $C_1 \cap \overline{D} \neq \emptyset$ since otherwise $t^* = 1$ by the first part. Set $\hat{t} = \min\left\{ t \in [0, 1] : \eta(t) \in \partial D \right\}$. From the second part we conclude that $\hat{t} \notin T$ since $\eta(\hat{t}) \in \partial D \subset \overline{D}$. Therefore, $t^* \leq \hat{t}$. Finally, assume on the contrary that $t^* < \hat{t}$. Then $C_{\hat{t}} \cap \overline{D} = \emptyset$ and thus, by the first part, $\hat{t} \in T$. This contradicts the definition of T. $\square$

Formula (7.50) allows it – at least in principle – to compute ∂D from $\Lambda - \Lambda_0$. One has to choose a family of needles which covers the domain B, and for each needle η one computes t^* from the formula of the preceding theorem. If $t^* < 1$ then $\eta(t^*) \in \partial D$.

This procedure is very expensive from the computational point of view. However, if we restrict ourselves to "linear" needles, i.e., rays of the form

$$C = \{z + ta : t \geq 0\} \cap B, \tag{7.52}$$

for $z \in B$ and unit vectors $a \in \mathbb{R}^3$ then we can reduce the computational effort considerable as we indicate in the following. However, by using only rays one can not expect to detect the boundary of D completely. Only the "visible points" on ∂D can be detected. We refer to the remarks at the end of this section.

Let B be contained in a ball of radius R with center at the origin. Define $\hat{B} = K(0, 2R)$ to be the ball of radius $2R$ and $\hat{C} = \{te : t \geq 0\} \cap \hat{B}$ where $e = (1, 0, 0)^\top \in \mathbb{R}^3$ to be the reference needle. Construct a sequence $\hat{w}_n \in H^1(\hat{B})$ of functions with

$$\left\| \hat{w}_n - \Phi_0(\cdot, 0) \right\|_{H^1(V)} \longrightarrow 0$$

for every open set V such that $\overline{V} \subset \hat{B} \setminus \hat{C}$. Let C be a needle of the form (7.52). Choose an orthogonal matrix $Q \in \mathbb{R}^{3 \times 3}$ such that $Qa = e$. Define $w_n \in H^1(B)$ by

$$w_n(x) = \hat{w}_n(Qx - Qz) \quad \text{for } x \in B. \tag{7.53}$$

Then $|Qx - Qz| \leq |Qx| + |Qz| = |x| + |z| \leq 2R$, i.e., $Qx - Qz \in \hat{B}$ and, furthermore, for any open set U with $\overline{U} \in B \setminus C$ we conclude that $\overline{V} \subset \hat{B} \setminus \hat{C}$ where $V = \{Qx - Qz : x \in U\}$. Therefore, since $\Phi_0(x, z) = \Phi_0(Qx - Qz, 0)$,

$$\left\| w_n - \Phi_0(\cdot, z) \right\|_{H^1(U)} = \left\| \hat{w}_n - \Phi_0(\cdot, 0) \right\|_{H^1(V)} \longrightarrow 0$$

as n tends to infinity.

Therefore, the computation of the approximating sequence has to be done only once for the reference needle $\hat{C}$ in the reference ball $\hat{B}$. Then $f_n \in H^{1/2}(\partial B)$ is given by

$$f_n(x) = w_n(x) = \hat{w}_n(Qx - Qz) \quad \text{for } x \in \partial B$$

and $n \in \mathbb{N}$.

For the computation of $t^* = \sup\{t \in [0, 1] : [0, t] \in T\}$ with T from (7.49) one has to decide whether a supremum is finite or infinite. Numerically, this is certainly not an easy task. In [96] Ikehata proposed (formulated in our setting) to replace the set T of (7.49) by

$$T_M = \left\{ t \in [0, 1] : \sup_{n \in \mathbb{N}} \left\{ | \operatorname{Re} \langle f_n(C_t), (\Lambda - \Lambda_0) f_n(C_t) \rangle | \leq M \right\} \right.$$

for some $M > 0$ and proved a result analogously to the one in Theorem 7.14. We refer to [96] for more details.

7.4.2 The probe method for the inverse scattering problem with mixed boundary conditions

In this subsection we study the situation of Chapter 3, i.e., the inverse scattering problem where D consists of several components, and on the boundary of some of these (which we denote by $\Gamma_1 = \partial D_1$) we impose Dirichlet boundary conditions while on the others (which we denote by $\Gamma_2 = \partial D_2$) impedance boundary conditions. For the application of the Factorization Method we had to assume the knowledge of separators Ω_1 and/or Ω_2. The Probe Method does not need this a priori information. In the main ideas we follow the presentation in [77] but adopt it to the treatment of the previous subsection for clarity.

In the present situation of the inverse scattering problem a needle is defined as follows (compare Definition 7.11).

Definition 7.15 *Let $R > 0$ such that $\overline{D} \subset K(0,R)$ where $K(0,R)$ denotes the ball of radius centered at the origin. A needle for $K(0,R)$ is a continuously differentiable function $\eta : [0,1] \to \mathbb{R}^3$ such that $|\eta(0)| = R$ and $|\eta(t)| < R$ for all $t \in (0,1]$ and $\eta'(t) \neq 0$ for all $t \in [0,1]$ and $\eta(t) \neq \eta(s)$ for $t \neq s$.*

In contrast to the previous subsection we take Herglotz waves v_ψ for the approximation of the fundamental solution. We recall that a Herglotz wave function is defined as

$$v_\psi(x) = \int_{S^2} e^{ik\, x \cdot \theta}\, \psi(\theta)\, ds(\theta)\,, \quad x \in \mathbb{R}^3\,,$$

Corollary 7.16 *Let $\eta : [0,1] \to \mathbb{R}^3$ be a needle for $K(0,R)$. For any $t \in (0,1]$ there exists a sequence $\psi_n \in L^2(S^2)$ such that the corresponding Herglotz wave functions $v_n = v_{\psi_n}$ satisfy*

$$\left\| v_n - \Phi\big(\cdot, \eta(t)\big) \right\|_{H^1(U)} \longrightarrow 0\,, \quad n \to \infty\,, \tag{7.54}$$

for every open subset U with $\overline{U} \subset K(0,R) \setminus C_t$. Again, $\Phi = \Phi(x,z)$ denotes the fundamental solution of the Helmholtz equation and

$$C_t = \big\{\eta(s) : 0 \leq s \leq t\big\}$$

is the part of the needle from $s = 0$ to $s = t$.

Proof: The proof follows exactly the same arguments of the proof of Corollary 7.13. On the region G_n one applies the approximation Theorem 7.3 instead of Theorem 7.2. We omit the details. $\square$

Now we are able to prove a characterization which is quite analogous to Theorem 7.14.

Theorem 7.17 *Let $\eta : [0,1] \to \mathbb{R}^3$ be a needle for $K(0,R)$. For any $t \in (0,1]$ choose a sequence $\psi_n = \psi_n(C_t) \in L^2(S^2)$ according to Corollary 7.16 and define the set $T \subset [0,1]$ by*

$$T = \Big\{ t \in [0,1] : \sup_n \big| \big(F_{mix}\psi_n(C_t), \psi_n(C_t)\big)_{L^2(S^2)} \big| < \infty \Big\}\,.$$

Then there exists $\varepsilon > 0$ with $[0, \varepsilon] \subset T$. In particular, $T \neq \emptyset$.
Furthermore, define $t^* = \sup\{t \in [0, 1] : [0, t] \in T\}$. Then

$$t^* = \begin{cases} \min\{t \in [0, 1] : \eta(t) \in \partial D\}, & \text{if } C_1 \cap \overline{D} \neq \emptyset, \\ 1, & \text{if } C_1 \cap \overline{D} = \emptyset. \end{cases} \tag{7.55}$$

Proof: We follow the proof of Theorem 7.14. The fundamental estimates (7.48) are now replaced by estimates which follow from the factorization

$$F_{mix} = -\begin{pmatrix} H_1 \\ (\partial H)_2 + \lambda H_2 \end{pmatrix}^* T_{mix}^{-1} \begin{pmatrix} H_1 \\ (\partial H)_2 + \lambda H_2 \end{pmatrix}. \tag{7.56}$$

of F_{mix} which has been derived in Chapter 3 (compare (3.26)). We recall the definitions of $H_j : L^2(S^2) \to H^{1/2}(\Gamma_j)$ and $(\partial H)_j : L^2(S^2) \to H^{-1/2}(\Gamma_j)$ for $j = 1, 2$, i.e.,

$$(H_j \psi)(x) = \int_{S^2} \psi(\theta)\, e^{ik\,\theta \cdot x}\, ds(\theta), \quad x \in \Gamma_j,$$

$$(\partial H)_j \psi(x) = \frac{\partial}{\partial \nu} \int_{S^2} \psi(\theta)\, e^{ik\,\theta \cdot x}\, ds(\theta), \quad x \in \Gamma_j.$$

From (7.56) we have that

$$(F_{mix}\psi, \psi)_{L^2(S^2)} = -\left\langle T_{mix}^{-1} \begin{pmatrix} H_1 \\ (\partial H)_2 + \lambda H_2 \end{pmatrix} \psi, \ \begin{pmatrix} H_1 \\ (\partial H)_2 + \lambda H_2 \end{pmatrix} \psi \right\rangle \tag{7.57}$$

for all $\psi \in L^2(S^2)$.

First we show again that $C_t \cap \overline{D} = \emptyset$ implies $t \in T$. This would prove the first assertion because $\eta(0) \notin \overline{D}$ and thus also $C_\varepsilon \cap \overline{D} = \emptyset$ for sufficiently small ε. Therefore, let $C_t \cap \overline{D} = \emptyset$. Choose an open set $U \subset K(0, R)$ which contains $\overline{D}$ such that $\overline{U} \cap C_t = \emptyset$. From the construction of ψ_n we note that v_n converges to $\Phi(\cdot, \eta(t))$ in $H^1(U)$. Since D is contained in U we conclude that $H_1 \psi_n = v_n|_{\Gamma_1}$ converges to $\Phi(\cdot, \eta(t))$ in $H^{1/2}(\Gamma_1)$ and $(\partial H)_2 \psi_n + \lambda H_2 \psi_n = \partial v_n/\partial \nu + \lambda v_n$ converges to $\partial \Phi(\cdot, \eta(t))/\partial \nu + \lambda \Phi(\cdot, \eta(t))$ in $H^{-1/2}(\Gamma_2)$. From (7.57) and the boundedness of the dual form $\langle \cdot, \cdot \rangle$ we conclude that $(F_{mix}\psi_n, \psi_n)_{L^2(S^2)}$ remains bounded. This means $t \in T$ by the definition of T.

Second, let now $\eta(t) \in \overline{D}$ but $\eta(s) \notin \overline{D}$ for all $0 \leq s < t$. We assume on the contrary that $t \in T$. We restrict ourselves to the case where $\eta(t) \in \overline{D_1}$. The case $\eta(t) \in \overline{D_2}$ is treated analogously. Since $C_t \cap \overline{D_2} = \emptyset$ we conclude as before that $(\partial H)_2 \psi_n + \lambda H_2 \psi_n$ converges to $\partial \Phi(\cdot, \eta(t))/\partial \nu + \lambda \Phi(\cdot, \eta(t))$ in $H^{-1/2}(\Gamma_2)$. Now we recall the form of the operator T_{mix}.

We write T_{mix}^{-1} in matrix form as $T_{mix}^{-1} = \begin{pmatrix} \widetilde{T}_{11} & \widetilde{T}_{12} \\ \widetilde{T}_{21} & \widetilde{T}_{22} \end{pmatrix}$. From (7.57) we estimate

$$\left|(F_{mix}\psi_n, \psi_n)_{L^2(S^2)}\right| \geq \left|\langle \widetilde{T}_{11} H_1 \psi_n, H_1 \psi_n \rangle\right|$$

$$- \left[\|\widetilde{T}_{21}\| + \|\widetilde{T}_{12}\|\right] \|H_1 \psi_n\|_{H^{1/2}(\Gamma_1)} \|(\partial H)_2 \psi_n + \lambda H_2 \psi_n\|_{H^{-1/2}(\Gamma_2)}$$

$$- \|\widetilde{T}_{22}\| \|(\partial H)_2 \psi_n + \lambda H_2 \psi_n\|_{H^{-1/2}(\Gamma_2)}^2.$$

By (3.32) the operator $\tilde{T}_{11}$ has the form $\tilde{T}_{11} = S_1^{-1} + K$ where K is compact. The operator S_1^{-1} is a sum of a coercive and a compact operator. Furthermore, in part (c) of Theorem 3.4 we have shown that $\mathrm{Im}\langle T_{mix}\varphi, \varphi\rangle > 0$ for all $\varphi \in H^{-1/2}(\Gamma_1) \times H^{1/2}(\Gamma_2)$ with $\varphi \neq 0$. Therefore, also $\mathrm{Im}\langle \varphi, T_{mix}^{-1}\varphi\rangle > 0$ for all $\varphi \in H^{1/2}(\Gamma_1) \times H^{-1/2}(\Gamma_2)$ with $\varphi \neq 0$ and thus $\mathrm{Im}\langle \tilde{T}_{11}\varphi_1, \varphi_1\rangle < 0$ for all $\varphi_1 \in H^{1/2}(\Gamma_1)$ with $\varphi_1 \neq 0$. Therefore, we can apply Lemma 1.17 from Chapter 1 which yields existence of a constant $c > 0$ such that

$$\left|\langle \tilde{T}_{11}\varphi_1, \varphi_1\rangle\right| \geq c\,\|\varphi_1\|_{H^{1/2}(\Gamma_1)}^2, \quad \varphi_1 \in H^{1/2}(\Gamma_1)\,.$$

Combining this with the above estimate of $\left|(F_{mix}\psi, \psi)_{L^2(S^2)}\right|$ yields

$$
\begin{aligned}
\left|(F_{mix}\psi_n, \psi_n)_{L^2(S^2)}\right| \geq\;& c\|H_1\psi_n\|_{H^{1/2}(\Gamma_1)}^2 \\
&- \left[\|\tilde{T}_{21}\| + \|\tilde{T}_{12}\|\right]\|H_1\psi_n\|_{H^{1/2}(\Gamma_1)}\|(\partial H)_2\psi_n + \lambda H_2\psi_n\|_{H^{-1/2}(\Gamma_2)} \\
&- \|\tilde{T}_{22}\|\|(\partial H)_2\psi_n + \lambda H_2\psi_n\|_{H^{-1/2}(\Gamma_2)}^2 \\
=\;& \|H_1\psi_n\|_{H^{1/2}(\Gamma_1)}\Big\{c\|H_1\psi_n\|_{H^{1/2}(\Gamma_1)} \\
&- \left[\|\tilde{T}_{21}\| + \|\tilde{T}_{12}\|\right]\|(\partial H)_2\psi_n + \lambda H_2\psi_n\|_{H^{-1/2}(\Gamma_2)}\Big\} \\
&- \|\tilde{T}_{22}\|\|(\partial H)_2\psi_n + \lambda H_2\psi_n\|_{H^{-1/2}(\Gamma_2)}^2
\end{aligned}
$$

Since we assumed that $t \in T$ the left hand side of this inequality remains bounded as n tends to infinity. We have also shown that $\|(\partial H)_2\psi_n + \lambda H_2\psi_n\|_{H^{-1/2}(\Gamma_2)}$ is bounded. A simple application of the binomial formula implies boundedness of $\|H_1\psi_n\|_{H^{1/2}(\Gamma_1)}$. We show that this is not the case just as in the proof of Theorem 7.14. Indeed, choose again a set K_0 of the form

$$K_0 = \left\{\eta(t) + r\hat{x} : 0 < r \leq \rho\,, \ \hat{x} \in V\right\}$$

where $\rho > 0$ and the (relatively) open subset $V \subset S^2$ are chosen such that $K_0 \subset D_1 \setminus C_t$. The subset $K_\varepsilon = \left\{\eta(t) + r\hat{x} \in K_0 : \varepsilon \leq r \leq \rho\right\}$ is compact and contained in $D_1 \setminus C_t$ for every $\varepsilon \in (0, \rho)$. Using polar coordinates with respect to $\eta(t)$ we compute again

$$\iint\limits_{K_\varepsilon} |\nabla_x \Phi(x, \eta(t))|^2 dx \geq |V| \int\limits_\varepsilon^\rho \frac{1}{(4\pi)^2 r^4} r^2 dr = \frac{|V|}{(4\pi)^2}\left(\frac{1}{\varepsilon} - \frac{1}{\rho}\right).$$

Let now $M > 0$ arbitrary large but fixed. Choose $\varepsilon > 0$ such that

$$\iint\limits_{K_\varepsilon} |\nabla_x \Phi(x, \eta(t))|^2 dx \geq 2M\,.$$

Since $K_\varepsilon \cap C_t = \emptyset$ we conclude that v_n converges to $\Phi(\cdot, \eta(t))$ in $H^1(K_\varepsilon)$. Therefore, for large enough n we have that $\iint_{K_\varepsilon} |\nabla v_n|^2 dx \geq M$, i.e., also $\|v_n\|_{H^1(D_1)} \geq \|v_n\|_{H^1(K_\varepsilon)} \geq M$. Therefore, $\|v_n\|_{H^1(D_1)}$ tends to infinity and also the traces $\|H_1\psi_n\|_{H^{1/2}(\Gamma_1)}$.

In the third part we prove (7.55). It suffices to consider the case where $C_1 \cap \overline{D} \neq \emptyset$ since otherwise $t^* = 1$ by the first part. Set $\hat{t} = \min\{t \in [0,1] : \eta(t) \in \partial D\}$. From the second part we conclude that $\hat{t} \notin T$ since $\eta(\hat{t}) \in \partial D \subset \overline{D}$. Therefore, $t^* \leq \hat{t}$. Finally, assume on the contrary that $t^* < \hat{t}$. Then $C_{\hat{t}} \cap \overline{D} = \emptyset$ and thus, by the first part, $\hat{t} \in T$. This contradicts the definition of T. $\square$

As already indicated in the previous subsection, it is not clear how this form of the Probe Method can be used to construct a numerical algorithm. It is much easier, and faster, to use linear needles, i.e., rays, only. The approximation of the fundamental solution with respect to an arbitrary ray can be transformed to the case of a reference ray by a simple rotation and translation. Indeed, let

$$C = \{z + ta : t \geq 0\} \cap K(0,R), \tag{7.58}$$

be a ray where $|z| < R$ and $a \in \mathbb{R}^3$ denotes some unit vector. Define again $\hat{B} = K(0,2R)$ to be the ball of radius $2R$ and $\hat{C} = \{te : t \geq 0\} \cap \hat{B}$ where $e = (1,0,0)^\top \in \mathbb{R}^3$ to be the reference needle. Construct a sequence $\hat{\psi}_n \in L^2(S^2)$ such that the corresponding Herglotz wave functions $\hat{v}_n \in H^1(\hat{B})$ satisfy

$$\left\| \hat{v}_n - \Phi_0(\cdot,0) \right\|_{H^1(V)} \longrightarrow 0$$

for every open set V such that $\overline{V} \subset \hat{B} \setminus \hat{C}$. Let C be a needle of the form (7.58). Choose an orthogonal matrix $Q \in \mathbb{R}^{3\times3}$ such that $Qa = e$. Define $\psi_n \in L^2(S^2)$ by

$$\psi_n(\theta) = e^{ik\,\theta\cdot z}\,\hat{\psi}_n(Q\theta) \quad \text{for } \theta \in S^2. \tag{7.59}$$

For the corresponding Herglotz functions we compute

$$v_n(x) = \int_{S^2} \psi_n(\theta)\,e^{ik\,x\cdot\theta}\,ds(\theta) = \int_{S^2} \hat{\psi}_n(Q\theta)\,e^{ik\,\theta\cdot(x+z)}\,ds(\theta) = \hat{v}_n\big(Q(x+z)\big).$$

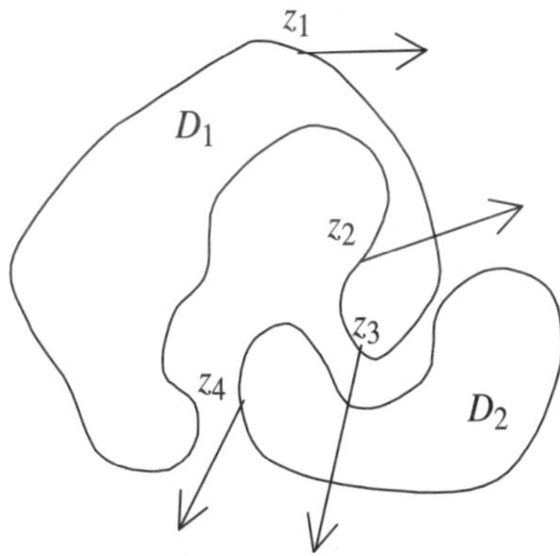

Figure 7.1 Visible and invisible boundary points

Furthermore, $|Qx - Qz| \leq |Qx| + |Qz| = |x| + |z| \leq 2R$, i.e., $Qx - Qz \in \hat{B}$, and for any open set U with $\overline{U} \in K(0, R) \setminus C$ we conclude that $\overline{V} \subset \hat{B} \setminus \hat{C}$ where $V = \{Qx - Qz : x \in U\}$. Therefore, since $\Phi(x, z) = \Phi(Qx - Qz, 0)$,

$$\left\| v_n - \Phi(\cdot, z) \right\|_{H^1(U)} = \left\| \hat{v}_n - \Phi(\cdot, 0) \right\|_{H^1(V)} \longrightarrow 0$$

as n tends to infinity.

If we allow only rays as needles we can only detect boundary points z which are *visible* from the sphere $\{x \in \mathbb{R}^3 : |x| = R\}$, i.e., for which a ray of the form $C = \{z + ta : t \geq 0\}$ exists for some $a \in S^2$ such that $C \cap \overline{D} = \{z\}$.

For example, in Figure 7.1 the points z_1 and z_4 are visible while z_2 and z_3 are invisible. If every boundary point is visible then we can reconstruct the obstacle completely by the Probe Method with rays only.

Bibliography

[1]   G. Alessandrini. Stable determination of conductivity by boundary measurements. *Appl. Anal.*, 27:153–172, 1988.

[2]   H. Ammari and H. Kang. *Reconstruction of Small Inhomogeneities from Boundary Measurements*, volume 1846 of *Lecture Notes in Mathematics*. Springer, 2004.

[3]   T.S. Angell, X. Jiang, and R.E. Kleinman. A distributed source method for inverse acoustic scattering. *Inverse Probl.*, 13:531–545, 1997.

[4]   T.S. Angell and A. Kirsch. *Optimization Methods in Electromagnetic Radiation*. Springer, 2004.

[5]   T.S. Angell and R.E. Kleinman. The Helmholtz equation with L_2—boundary values. *SIAM J. Math. Anal.*, 16:259–278, 1985.

[6]   T.S. Angell, R.E. Kleinman, B. Kok, and G.F. Roach. A constructive method for identification of an impenetrable scatterer. *Wave Motion*, 11:185–200, 1989.

[7]   T. Arens. Linear sampling methods for 2d inverse elastic wave scattering. *Inverse Probl.*, 17:1445–1464, 2001.

[8]   T. Arens. Why linear sampling works. *Inverse Probl.*, 20:163–173, 2004.

[9]   T. Arens and N. Grinberg. A complete factorization method for scattering by periodic surfaces. *Computing*, 75:111–132, 2005.

[10] T. Arens and A. Kirsch. The factorization method in inverse scattering from periodic structures. *Inverse Probl.*, 19:1195–1211, 2003.

[11] T. Arens and A. Lechleiter. The linear sampling method revisited. *J. Int. Eqn. Appl.* to appear.

[12] K. Astala, M. Lassas, and L. Päivärinta. Calderón's inverse problem for anisotropic conductivities in the plane. *Comm. Partial Differential Equations*, 30:207–224, 2005.

[13] K. Astala and L. Päivärinta. Calderón's inverse conductivity problem in the plane. *Ann. Math.*, 163:265–299, 2006.

[14] G. Bal. Reconstructions in impedance and optical tomography with singular interfaces. *Inverse Probl.*, 21:113–131, 2005.

[15] M. Bonnet and B. Guzina. Sounding of finite solid bodies by way of topological derivative. *Int. J. Num. Meth. Eng.*, 61:2344–2373, 2004.

[16] L. Borcea. Electrical impedance tomography. *Inverse Probl.*, 18:R99–R136, 2002.

[17] D. Braess. *Finite Elemente*. Springer, New York etc., 1992.

[18] M. Brühl. Gebietserkennung in der elektrischen Impedanztomographie. Dissertion, University of Karlsruhe, Karlsruhe, 1999.

[19] M. Brühl. Explicit characterization of inclusions in electrical impedance tomography. *SIAM J. Math. Anal.*, 32:1327–1341, 2001.

[20] M. Brühl and M. Hanke. Numerical implementation of two noniterative methods for locating inclusions by impedance tomography. *Inverse Probl.*, 16:1029–1042, 2000.

[21] M. Brühl, M. Hanke, and M. Pidcock. Crack detection using electrostatic measurements. *Math. Model. Numer. Anal.*, 35:595–605, 2001.

[22] M. Burger and S. Osher. A survey on level set methods for inverse problems and optimal design. *Euro. J. Appl. Math.*, 16:263–301, 2005.

[23] F. Cakoni and D. Colton. The linear sampling method for cracks. *Inverse Probl.*, 19:279–295, 2003.

[24] F. Cakoni and D. Colton. On the mathematical basis of the linear sampling method. *Georgian Math. J.*, 10:411–425, 2003.

✳ [25] F. Cakoni and D. Colton. *Qualitative Methods in Inverse Scattering Theory.* Springer, 2006.

[26] F. Cakoni, D. Colton, and H. Haddar. The linear sampling method for anisotropic media. *J. Comp. Appl. Math.*, 146:285–299, 2002.

[27] A. Charalambopoulos, D. Gintides, and K. Kiriaki. The linear sampling method for the transmission problem in three-dimensional linear elasticity. *Inverse Probl.*, 18:547–558, 2002.

[28] A. Charalambopoulos, D. Gintides, and K. Kiriaki. The linear sampling method for non-absorbing penetrable elastic bodies. *Inverse Probl.*, 19:549–561, 2003.

[29] A. Charalambopoulos, A. Kirsch, K. Anagnostopoulos, D. Gintides, and K. Kiriaki. The factorization method in inverse elastic scattering from penetrable bodies. *Inverse Probl.*, 23:27–51, 2007.

[30] M. Cheney. The linear sampling method and the MUSIC algorithm. *Inverse Probl.*, 17:591–596, 2001.

[31] M. Cheney and D. Isaacson. Inverse problems for a perturbed dissipative half-plane. *Inverse Probl.*, 11:865–888, 1995.

[32] M. Cheney, D. Isaacson, and J. Newell. Electrical impedance tomography. *SIAM Rev.*, 41:85–101, 1999.

[33] J. Cheng, J. J. Liu, and G. Nakamura. The numerical realization of the probe method for the inverse scattering problems from the near-field data. *Inverse Probl.*, 21:839–855, 2005.

[34] K.-S. Cheng, D. Isaacson, J. Newell, and D. Gisser. Electrode models for electric current computed tomography. *IEEE Trans. Eng.*, 36:918–924, 1989.

[35] F. Collino, M. Fares, and H. Haddar. Numerical and analytical studies of the linear sampling method in electromagnetic inverse scattering problems. *Inverse Probl.*, 19:1279–1298, 2003.

[36] D. Colton, J. Coyle, and P. Monk. Recent developments in inverse acoustic scattering theory. *SIAM Rev.*, 42:396–414, 2000.

[37] D. Colton, H. Haddar, and P. Monk. The linear sampling method for solving the electromagnetiv inverse scattering problem. *SIAM J. Sci. Comput.*, 24:719–731, 2002.

[38] D. Colton, H. Haddar, and M. Piana. The linear sampling method in inverse electromagnetic scattering theory. *Inverse Probl.*, 19:S105–S137, 2003.

[39] D. Colton and A. Kirsch. A simple method for solving inverse scattering problems in the resonance region. *Inverse Probl.*, 12:383–393, 1996.

[40] D. Colton, A. Kirsch, and P. Monk. The linear sampling method in inverse scattering theory. In D. Colton, H.W. Engl, A. Louis, J.R. McLaughlin, and W. Rundell, editors, *Surveys on Solution Methods for Inverse Probl.*, pages 107–118, Vienna, New York, 2000. Springer.

[41] D. Colton, A. Kirsch, and L. Päivärinta. Far field patterns for acoustic waves in an inhomogeneous medium. *SIAM J. Math. Anal.*, 20:1472–1483, 1989.

[42] D. Colton and R. Kress. *Integral Equation Methods in Scattering Theory*. Wiley, 1983.

[43] D. Colton and R. Kress. *Inverse Acoustic and Electromagnetic Scattering Theory*. Springer, second edition, 1998.

[44] D. Colton and R. Kress. Using fundamental solutions in inverse scattering. *Inverse Probl.*, 22:R49–R66, 2006.

[45] D. Colton and P. Monk. A novel method for solving the inverse scattering problem for time—harmonic acoustic waves in the resonance region. *SIAM J. Appl. Math.*, 45:1039–1053, 1985.

[46] D. Colton and P. Monk. A novel method for solving the inverse scattering problem for time—harmonic acoustic waves in the resonance region II. *SIAM J. Appl. Math.*, 46:506–523, 1986.

[47] D. Colton and P. Monk. The inverse scattering problem for time-harmonic acoustic waves in a penetrable medium. *Q. J. Mech. Appl. Math.*, 40:189–212, 1987.

[48] D. Colton and P. Monk. The inverse scattering problem for time-harmonic acoustic waves in an inhomogeneous medium. *Q. J. Mech. Appl. Math.*, 41:97–125, 1988.

[49] D. Colton and P. Monk. A new method for solving the inverse scattering problem for acoustic waves in an inhomogeneous medium. *Inverse Probl.*, 5:1013–1026, 1989.

[50] D. Colton and P. Monk. A new algorithm in electromagnetic inverse scattering theory with an application to medical imaging. *Math. Methods Appl. Sci.*, 20:385–401, 1997.

[51] D. Colton and P. Monk. A linear sampling method for the detection of leukemia using microwaves. *SIAM J. Appl. Math.*, 58:926–941, 1998.

[52] D. Colton and P. Monk. A linear sampling method for the detection of leukemia using microwaves II. *SIAM J. Appl. Math.*, 60:241–255, 2000.

[53] D. Colton and L. Päivärinta. The uniqueness of a solution to an inverse scattering problem for electromagnetic waves. *Arch. Rational Mech. Anal.*, 119:59–70, 1992.

[54] D. Colton and L. Päivärinta. Transmission eigenvalues and a problem of Hans Levy. *J. Comp. Appl. Math.*, 117:91–104, 2000.

[55] D. Colton, L. Päivärinta, and J. Sylvester. The interior transmission problem. *Inverse Probl. Imaging*, 1:13–28, 2007.

[56] D. Colton, M. Piana, and R. Potthast. A simple method using Morozov's discrepancy principle for solving inverse scattering problems. *Inverse Probl.*, 13:1447–1493, 1997.

[57] D. Colton and B.D. Sleeman. Uniqueness theorems for the inverse problem of acoustic scattering. *IMA J. Appl. Math*, 31:253–259, 1983.

[58] D. Colton and B.D. Sleeman. An approximation property of importance in inverse scattering theory. *Proc. Edinb. Math. Soc.*, 44:449–454, 2001.

[59] P.-M. Cutzach and C. Hazard. Existence, uniqueness and analyticity properties for electromagnetic scattering in a two-layered medium. *Math. Methods Appl. Sci.*, 21:433–461, 1998.

[60] A.J. Devaney. Super-resolution processing of multi-static data using time reversal and MUSIC. Unpublished, 2000.

[61] J. Elschner and M. Yamamoto. Uniqueness in determining polygonal sound-hard obstacles with a single incoming wave. *Inverse Probl.*, 22:355–364, 2006.

[62] H.W. Engl, M. Hanke, and A. Neubauer. *Regularization of Inverse Problems*. Kluwer Academic Press, Dordrecht, 1996.

[63] K. Erhard and R. Potthast. A numerical study of the probe method. *SIAM J. Sci. Comput.*, 28:1597–1612, 2006.

[64] G. Eskin and J. Ralston. Inverse coefficient problems in perturbed half spaces. *Inverse Probl.*, 15:683–699, 1999.

[65] S.N. Fata and B.B. Guzina. A linear sampling method for near field inverse problems in elastodynamics. *Inverse Probl.*, 20:713–736, 2004.

[66] F. Frühauf, B. Gebauer, and O. Scherzer. Detecting interfaces in a parabolic-elliptic problem from surface measurements. *SIAM J. Numer. Anal.* 45: 810–836, 2007.

[67] S. Garreau, P. Guillaume, and M. Masmoudi. The topological derivative for PDE systems: the elastic case. *SIAM J. Control Optimization*, 39:1756–1778, 2001.

[68] B. Gebauer. The factorization method for real elliptic systems. *J. Anal. Anwend.*, 25:81–102, 2006.

[69] B. Gebauer, M. Hanke, A. Kirsch, W. Muniz, and C. Schneider. A sampling method for detecting buried objects using electromagnetic scattering. *Inverse Probl.*, 21:2035–2050, 2005.

[70] D. Gilbarg and N.S. Trudinger. *Elliptic Partial Differential Equations of Second Order*. Springer, 1983.

[71] N.I. Grinberg. Obstacle localization in an homogeneous half-space. *Inverse Probl.*, 17:1113–1125, 2001.

[72] N.I. Grinberg. Obstacle visualization via the factorization method for the mixed boundary value problem. *Inverse Probl.*, 18:1687–1704, 2002.

[73] N.I. Grinberg. On the inverse obstacle scattering problem with Robin or mixed boundary condition: Application of the modified Kirsch factorization method. Preprint 02/4, University of Karlsruhe, Department of Mathematics, Karlsruhe, 2002.

[74] N.I. Grinberg. The factorization method in inverse obstacle scattering. Habilitation thesis, University of Karlsruhe, Karlsruhe, 2004.

[75] N.I. Grinberg. The operator factorization method in inverse obstacle scattering. *Integral Equations Oper. Theory*, 54:333–348, 2006.

[76] N.I. Grinberg and A. Kirsch. The linear sampling method in inverse obstacle scattering for impedance boundary conditions. *J. Inverse Ill-Posed Probl.*, 10:171–185, 2002.

[77] N.I. Grinberg and A. Kirsch. The factorization method for obstacles with a-priori separated sound-soft and sound-hard parts. *Math. Comput. in Simul.*, 66:267–279, 2004.

[78] B. Guzina and M. Bonnet. Topological derivative for the inverse scattering of elastic waves. *Q. J. Mech. Appl. Math.*, 57:161–179, 2004.

[79] H. Haddar. The interior transmission problem for anisotropic Maxwell's equations and its applications to the inverse problem. *Math. Methods Appl. Sci.*, 27:2111–2129, 2004.

[80] H. Haddar and P. Monk. The linear sampling method for solving the electromagnetic inverse medium problem. *Inverse Probl.*, 18:891–906, 2002.

[81] P. Hähner. An inverse problem in electrostatics. *Inverse Probl.*, 15:961–975, 1999.

[82] M. Hanke and M. Brühl. Recent progress in electrical impedance tomography. *Inverse Probl.*, 19:565–590, 2003.

[83] M. Hanke, F. Hettlich, and O. Scherzer. The Landweber iteration for the inverse scattering problem. In K.-W. Wang et al., editor, *Proc. of the 1995 Design Engineering Technical Conference*, volume 3, Part C, pages 909–915, New York, 1995. The American Society of Mechanical Engineers.

[84] F. Hettlich. Frechet derivatives in inverse obstacle scattering. *Inverse Probl.*, 11:371–382, 1995.

[85] F. Hettlich. The Landweber iteration applied to inverse conductive scattering problems. *Inverse Probl.*, 14:931–947, 1998.

[86] F. Hettlich and W. Rundell. A second degree method for nonlinear inverse problems. *SIAM J. Numer. Anal.*, 37:587–620, 2000.

[87] T. Hohage. Regularization of exponentially ill-posed problems. *Numer. Funct. Anal. Optimization*, 21:439–464, 2000.

[88] N. Honda, R. Nakamura, R. Potthast, and M. Sini. The no-response approach and its relation to non-iterative methods in inverse scattering. *Ann. Mat. Pura Appl.* at press, published online, 2006.

[89] N. Hyvönen. Characterization of inclusions in optical tomography. *Inverse Probl.*, 20:737–751, 2004.

[90] N. Hyvönen. Complete electrode models of electrical impedance tomography: Approximation properties and characterization of inclusions. *SIAM J. Appl. Math.*, 64:902–931, 2004.

[91] N. Hyvönen. Application of a weaker formulation of the factorization method to the characterization of absorbing inclusions in optical tomography. *Inverse Probl.*, 21:1331–1343, 2005.

[92] N. Hyvönen. Application of the factorization method to the characterization of weak inclusions in electrical impedance tomography. *Adv. in Appl. Math.* 39:197–221, 2007.

[93] N. Hyvönen. Locating transparent regions in optical absorption and scattering tomography. *SIAM J. Appl. Math.* 67:1101–1123, 2007.

[94] N. Hyvönen, H. Hakula, and S. Pursiainen. Numerical implementation of the factorization method within the complete electrode model of electrical impedance tomography. *Inverse Probl. Imaging*, 1:299–317, 2007.

[95] M. Ikehata. Reconstruction of an obstacle from the scattering amplitude at a fixed frequency. *Inverse Probl.*, 14:949–954, 1998.

[96] M. Ikehata. Reconstruction of the shape of the inclusion by boundary measurements. *Commun. Partial Differ. Equations*, 23:1459–1474, 1998.

[97] M. Ikehata. Size estimation of inclusions. *J. Inverse Ill-Posed Probl.*, 6:127–140, 1998.

[98] M. Ikehata. How to draw a picture of an unknown inclusion from boundary measurements. two mathematical inversion algorithms. *J. Inverse Ill-Posed Probl.*, 7:255–271, 1999.

[99] M. Ikehata. Reconstruction of obstacles from boundary measurements. *Wave Motion*, 30:205–223, 1999.

[100] M. Ikehata. The probe method and its applications. In G. Nakamura, S. Saitoh, J. Seo, and M. Yamamoto, editors, *Inverse Problems and Related Topics*, volume 419 of *Research Notes in Mathematics*, London, 2000. CRC Press.

[101] M. Ikehata. Reconstruction of the support function for inclusion from the boundary measurements. *J. Inverse Ill-Posed Probl.*, 8:367–378, 2000.

[102] M. Ikehata. A new formulation of the probe method and related problems. *Inverse Probl.*, 21:413–426, 2005.

[103] M. Ikehata. Identification of the shape of the inclusion having essentially bounded conductivity. *J. Inverse Ill-Posed Probl.* 7:533–540, 1999.

[104] M. Ikehata. Inverse crack problem and probe method. *Cubo* 8:29–40, 2006.

[105] V. Isakov. On uniqueness in the inverse transmission scattering problem. *Commun. Partial Differ Equations*, 15:1565–1587, 1990.

[106] V. Isakov. *Inverse Problems for Partial Differential Equations*. Springer-Verlag, Berlin, Heidelberg, New York, 1998.

[107] S. Järvenpää and E. Somersalo. Impedance imaging and electrode models. In *Inverse Problems in Medical Imaging and Nondestructive Testing*, pages 65–74, Vienna, 1996. Springer. Proceedings of the conference in Oberwolfach.

[108] S.N. Karp. Far field amplitudes and inverse diffraction theory. In Langer, editor, *Electromagnetic Waves*, pages 291–300, Madison, 1962. Univ. of Wisconsin Press.

[109] J.B. Keller. *Survey of Wave Propagation and Underwater Acoustics*, volume 70 of *Lecture Notes in Physics*. Springer, Berlin, 1977.

[110] H. Kersten. Grenz- und Sprungrelationen für Potentiale mit quadratsummierbarer Dichte. *Resultate d. Math.*, 3:17–24, 1980.

[111] A. Kirsch. Remarks on some notions of weak solutions for the Helmholtz equation. *Appl. Anal.*, 47:7–24, 1992.

[112] A. Kirsch. The domain derivative and two applications in inverse scattering theory. *Inverse Probl.*, 9:81–96, 1993.

[113] A. Kirsch. *An Introduction to the Mathematical Theory of Inverse Probl.*, volume 120 of *Applied Mathematical Sciences*. Springer, 1996.

[114] A. Kirsch. Characterization of the shape of a scattering obstacle using the spectral data of the far field operator. *Inverse Probl.*, 14:1489–1512, 1998.

[115] A. Kirsch. Factorization of the far field operator for the inhomogeneous medium case and an application in inverse scattering theory. *Inverse Probl.*, 15:413–429, 1999.

[116] A. Kirsch. New characterizations of solutions in inverse scattering theory. *Appl. Anal.*, 76:319–350, 2000.

[117] A. Kirsch. A new class of methods for solving inverse scattering problems. In Alfredo et al. Bermúdez, editor, *Mathematical and Numerical Aspects of Wave Propagation*, pages 499–503, Santiago de Compostela, Spain, 2000. Society for Industrial and Applied Mathematics. 5th international conference.

[118] A. Kirsch. The MUSIC-algorithm and the factorization method in inverse scattering theory for inhomogeneous media. *Inverse Probl.*, 18:1025–1040, 2002.

[119] A. Kirsch. The factorization method for Maxwell's equations. *Inverse Probl.*, 20:S117–S134, 2004.

[120] A. Kirsch. The factorization method for a class of inverse elliptic problems. *Math. Nachr.*, 278:258–277, 2005.

[121] A. Kirsch. An integral equation approach and the interior transmission problem for Maxwell's equations. *Inverse Probl. Imaging*, 1:159–180, 2007.

[122] A. Kirsch. An integral equation for Maxwell's equations in a layered medium with an application to the factorization method. to appear in: *J. Int. Equ. Appl.*, 2007.

[123] A. Kirsch and R. Kress. An optimization method in inverse acoustic scattering. In Brebbia et al., editor, *Boundary Elements IX, Vol 3, Fluid Flow and Potential Applications*, pages 3–18, Springer, 1987.

[124] A. Kirsch and R. Kress. Uniqueness in inverse scattering. *Inverse Probl.*, 9:285–299, 1993.

[125] A. Kirsch, R. Kress, P. Monk, and A. Zinn. Two methods for solving the inverse acoustic scattering problem. *Inverse Probl.*, 4:749–769, 1988.

[126] A. Kirsch and P. Monk. An analysis of the coupling of finite element and Nyström methods in acoustic scattering. *IMA J. Num. Math.*, 14:523–544, 1994.

[127] A. Kirsch and P. Monk. A finite element/spectral method for approximating the time harmonic Maxwell system in $\mathbb{R}^n$. *SIAM J. Appl. Math.*, 55:1324–1344, 1995.

[128] A. Kirsch and L. Päivärinta. On recovering obstacles inside inhomogeneities. *Math. Methods Appl. Sci.*, 21:619–651, 1998.

[129] A. Kirsch and S. Ritter. A linear sampling method for inverse scattering from an open arc. *Inverse Probl.*, 16:89–105, 2000.

[130] R. Kress. Scattering by obstacles. In R. Pike and P. Sabatier, editors, *Scattering. Scattering and Inverse Scattering in Pure and Applied Science*, chapter 1.2.2. Academic Press, New York, London, 2002.

[131] R. Kress. A factorization method for an inverse Neumann problem for harmonic vector fields. *Georgian Math. J.*, 10:549–560, 2003.

[132] R. Kress. Newton's method for inverse obstacle scattering meets the method of least squares. *Inverse Probl.*, 19:91–104, 2003.

[133] R. Kress and L. Kühn. Linear sampling methods for inverse boundary value problems in potential theory. *Appl. Numer. Math.*, 43:161–173, 2002.

[134] R. Kress and W. Rundell. A quasi-Newton method in inverse obstacle scattering. *Inverse Probl.*, 10:1145–1157, 1994.

[135] R. Kress and W. Rundell. Inverse obstacle scattering using reduced data. *SIAM J. Appl. Math*, 59:442–454, 1998.

[136] M. Lassas, M. Cheney, and G. Uhlmann. Uniqueness for a wave propagation inverse problem in a half-plane. *Inverse Probl.*, 14:679–684, 1998.

[137] P. Lax. A stability theorem for solutions of abstract differential equations, and its application to the study of the local behavior of solutions of elliptic equations. *Comm. Pure and Appl. Math.*, 9:747–766, 1956.

∗ [138] P.D. Lax and R.S. Phillips. *Scattering Theory*. Academic Press, New York, London, 1967.

[139] A. Lechleiter. A regularization technique for the factorization method. *Inverse Probl.*, 22:1605–1625, 2006.

[140] H.Y. Liu and J. Zou. Uniqueness in an inverse acoustic obstacle scattering problem for both sound-hard and sound-soft polyhedral scatterers. *Inverse Probl.*, 22:515–524, 2006.

[141] R. Luke and R. Potthast. The no response test—a sampling method for inverse scattering problems. *SIAM J. Appl. Math.*, 63:1292–1312, 2003.

[142] B. Malgrange. Existence et approximation des solutions des équations aux dérivées partielles et des équations de convolution. *Ann. Inst. Fourier*, 6:271–355, 1956.

[143] R. McGahan and R. Kleinman. Special session on image reconstruction using real data. *IEEE Trans. Antennas Propag.*, 38:39–40, 1996.

[144] W. McLean. *Strongly Elliptic Systems and Boundary Integral Equations*. Cambridge University Press, Cambridge, 2000.

[145] C. Miranda. *Partial Differential Equations of Elliptic Type*. Springer, New York etc., second revised edition, 1970.

[146] P. Monk. *Finite Element Methods for Maxwell's Equations*. Oxford Science Publications, Oxford, 2003.

[147] C. Müller. *Foundations of the Mathematical Theory of Electromagnetic Waves*. Springer, New York etc., 1969.

[148] A.I. Nachman, L. Päivärinta, and A. Teirilä. On imaging obstacles inside inhomogeneous media. Submitted.

[149] G. Nakamura, G. Uhlmann, and J.-N. Wang. Oscillating-decaying solutions, Runge approximation property for the anisotropic elasticity system and their

applications to inverse problems. *Journal de Mathematiques Pures et Appliquees*, 85:21–54, 2005.

[150] J.-C. Nédélec. *Acoustic and Electromagnetic Equations*. Springer, New York etc., 2001.

[151] R.G. Novikov. Multidimensional inverse spectral problems. *Funct. Anal. Appl.*, 22:263–272, 1988.

[152] T. Okaji. Strong unique continuation property for the time harmonic Maxwell equations. *J. Math. Soc. Japan*, 54:89–122, 2002.

[153] M. Orispää. On point sources and near field measurements in inverse acoustic obstacle scattering. Doctorial thesis, University of Oulu, Oulu, Finland, 2002.

[154] S. Osher and J. Sethian. Fronts propagating with curvature dependent speed, algorithms based on Hamilton-Jacobi formulations. *J. Comp. Phys.*, 79:12–49, 1988.

[155] R. Pike and P. Sabatier. *Scattering. Scattering and Inverse Scattering in Pure and Applied Science*. Academic Press, New York, London, 2002.

[156] O. Pironneau. *Optimal Shape Design for Elliptic Systems*. Springer, New York etc., 1984.

[157] R. Potthast. A fast new method to solve inverse scattering problems. *Inverse Probl.*, 12:731–742, 1996.

[158] R. Potthast. A point-source method for inverse acoustic and electromagnetic scattering problems. *IMA J. Appl. Math.*, 61:119–140, 1998.

[159] R. Potthast. Stability estimates and reconstructions in inverse acoustic scattering using singular sources. *J. Comput. Appl. Math.*, 114:247–274, 2000.

[160] R. Potthast. *Point Sources and Multipoles in Inverse Scattering Theory*. Chapman & Hall/CRC, Boca Raton, 2001.

[161] R. Potthast. Sampling and probe methods – an algorithmical view. *Computing*, 75:215–235, 2005.

[162] R. Potthast. A survey on sampling and probe methods for inverse problems. *Inverse Probl.*, 22:R1–R47, 2006.

[163] R. Potthast, J. Sylvester, and S. Kusiak. A "range test" for determining scatterers with unknown physical properties. *Inverse Probl.*, 19:533–547, 2003.

[164] A.G. Ramm. Recovery of the potential from fixed energy scattering data. *Inverse Probl.*, 4:877–886, 1988.

[165] A.G. Ramm and E. Sommersalo. Electromagnetic inverse problems with surface measurements at low frequencies. *Inverse Probl.*, 5:1107–1116, 1989.

[166] F. Rellich. Über das asymptotische Verhalten von Lösungen von $\Delta u + \lambda u = 0$ in unendlichen Gebieten. *Jber. Deutsch. Math. Verein.*, 53:57–65, 1943.

[167] M. Renardy and R.C. Rogers. *An Introduction to Partial Differential Equations*. Springer, 1993.

[168] J.R. Ringrose. *Compact Non-Self-Adjoint Operators*. Van Nostrand Reinhold Co., London, 1971.

[169] B.P. Rynne and B.D. Sleeman. The interior transmission problem and inverse scattering from inhomogeneous media. *SIAM J. Math. Anal.*, 22:1755–1762, 1991.

[170] J. Saranen. On an inequality of Friedrichs'. *Math. Scand.*, 51:310–322, 1982.

[171] J. Sokolowski and J.P. Zochowski. On the topological derivatives in shape optimization. *SIAM J. Control Optim.*, 37:1251–1272, 1999.

[172] J. Sokolowski and J.P. Zochowski. Topological derivatives for elliptic problems. *Inverse Probl.*, 15:123–134, 1999.

[173] J. Sokolowski and J.P. Zolesio. *Introduction to Shape Optimization.* Springer, New York etc., 1992.

[174] E. Somersalo, M. Cheney, and D. Isaacson. Existence and uniqueness for electrode models for electric current computed tomography. *SIAM J. Appl. Math.*, 52:1023–1040, 1992.

[175] A. Sommerfeld. *Partial Differential Equations in Physics.* Academic Press, New York, 1957.

[176] J. Sylvester and G. Uhlmann. A global uniqueness theorem for an inverse boundary value problem. *Ann. Math.*, 125:153–169, 1987.

[177] J. Sylvester and G. Uhlmann. Inverse boundary value problems at the boundary—continuous dependence. *Commun. Pure Appl. Math.*, XLI:197–219, 1988.

[178] A. Tacchino, J. Coyle, and M. Piana. Numerical validation of the linear sampling method. *Inverse Probl.*, 18:511–527, 2002.

[179] H. Triebel. *Höhere Analysis.* VEB Deutscher Verlag der Wissenschaften, Berlin, 1972.

[180] G. Uhlmann. Developments in inverse problems since Calderón's foundational paper. In *Harmonic Analysis and Partial Differential Equations*, Chicago Lectures in Math., pages 295–345, 1999.

[181] V. Vogelsang. On the strong unique continuation principle for inequalities of Maxwell type. *Math. Ann.*, 289:285–295, 1991.

[182] C. Weber. A local compactness theorem for Maxwell's equations. *Math. Methods Appl. Sci.*, 2:12–25, 1980.

[183] C.H. Wilcox. *Scattering Theory for the d'Alembert Equation in Exterior Domains*, volume 442 of *Lecture Notes in Mathematics.* Springer, New York etc., 1975.

[184] A. Wirgin. Special section: Inverse problems in underwater acoustics. *Inverse Probl.*, 16, 2000.

[185] Y. Xu. Generalized dual space indicator method for inverse scattering problems in underwater acoustics. In A. Bermudes et al., editor, *Mathematical and Numerical Aspects of Wave Propagation*, pages 543–547, Philadelphia, 2000. SIAM. Proceedings of the 5th International Conference (WAVES 2000).

Index